Geochemistry

Edited by Miloš René,
Gemma Aiello and Gaafar El Bahariya

Published in London, United Kingdom

IntechOpen

Supporting open minds since 2005

Geochemistry
http://dx.doi.org/10.5772/intechopen.87688
Edited by Miloš René, Gemma Aiello and Gaafar El Bahariya

Contributors
Ali Ismail Al-Juboury, Rana Mahmood, Abulaziz M. Al-Hamdani, Victor Gdalievich Ganelin, Olufemi S. Bamigboye, Václav Procházka, Miroslav Žáček, Petr Sulovsky, Tomáš Vaculovič, Lenka Rukavičková, Dobroslav Matějka, Salih Muhammad Muhammad Awadh, Miloš René, Gaafar A. El Bahariya, Leonardo Gonçalves, Cristiane Gonçalves, Dmitry Vlasov, Elena Panova, Marina Zelenskaya, Oksana Rodina, Alexey Vlasov, Katerina Sazanova, Benjamin Ntieche, Margaret Brocx, Vic Semeniuk, Haofu Zheng, Bo Liu, Kaibo Shi, Gemma Aiello, Pauline Wokwenmendam Nguet, Zakari Nchouwet, Mahomed Aziz Mounjouohou, Daouda Mfepat, Weimin Jiang, Jinxing Yu, Yue Kong, Tong Li, Changhai Li

First published in London, United Kingdom, 2021 by IntechOpen
IntechOpen is the global imprint of INTECHOPEN LIMITED, registered in England and Wales, registration number: 11086078, 5 Princes Gate Court, London, SW7 2QJ, United Kingdom
Printed in Croatia

British Library Cataloguing-in-Publication Data
A catalogue record for this book is available from the British Library

Additional hard and PDF copies can be obtained from orders@intechopen.com

Geochemistry
Edited by Miloš René, Gemma Aiello and Gaafar El Bahariya
p. cm.
Print ISBN 978-1-83962-850-4
Online ISBN 978-1-83962-851-1
eBook (PDF) ISBN 978-1-83962-852-8

For EU product safety concerns:
IN TECH d.o.o., Prolaz Marije Krucifikse Kozulić 3, 51000 Rijeka, Croatia, info@intechopen.com or visit our website at intechopen.com.

Meet the editor

Dr. Miloš René, Ph.D. is a scientist at the Institute of Rock Structure and Mechanics, Czech Academy of Sciences. He obtained his MSc in Mineralogy in 1968 and his Ph.D. in Economic Geology from Charles University, Prague, the Czech Republic in 1981. He has published about eighty-five scientific papers in peer-reviewed journals in the fields of mineralogy, petrology, geochemistry, and economic geology. His recent scientific activity is concentrated on petrology and geochemistry of granitic rocks, mineralogy of accessory minerals, and mineralogy of uranium deposits, especially from the Central European Variscan Belt.

Dr. Gemma Aiello was born in Aversa (CE), Italy, on 24 October 1964. In 1989, she graduated in Geological Sciences at the University of Naples "Federico II." In 1993, she earned a Ph.D. in Sedimentary Geology at the University of Naples "Federico II," Department of Earth Sciences, Faculty of Geological Sciences. She completed a two-year postdoctoral fellowship at the University of Naples "Federico II," a CNR-CEE fellowship and several contracts at the Research Institute "Geomare Sud," CNR, Naples, Italy. Since 1998, she has been a full-time researcher at the Italian CNR. Dr. Aiello has twenty-five years of experience in the field of sedimentary geology, marine geology and geophysics, participating in different research projects for the Italian National Research Council (CARG, Vector, Centri Regionali di Competenza). She was a contract professor of sedimentology and stratigraphy at the Parthenope University of Naples, Italy, and a teacher in formation courses of technicians in marine science and engineering in Naples, Italy.

Dr. Gaafar El Bahariya is a Professor of Petrology and former head of the Department of Geology, Tanta University, Tanta, Egypt. He holds a Ph.D. in Petrology and Geochemistry from Tanta University through a co-supervision channel with Oklahoma University, Norman, USA. He has published many papers and chapters in books, as well as delivered invited talks. He is the principal investigator on a project funded by the University of Tanta on gold and iron mineralization and hosting rocks of the Eastern Desert of Egypt. Dr. El Bahariya has attended many international conferences in Japan, Turkey, Morocco, Libya, Saudi Arabia and France, and is a reviewer for many international journals of geosciences. He is a member of the Geological Society of America, Goldschmidt, and Egyptian Geological society. He is a judge member of the Egyptian Promotion Committee. He has good field geology experience of Precambrian geology and his research interests focus on petrology, geochemistry and tectonics.

Contents

Preface

Geochemistry is a branch of geological sciences that deals with chemistry rocks and minerals in various igneous, metamorphic and sedimentary processes, from high to low temperatures and in various environments.

Geochemistry is also an important background for a detailed description of rocks' origin and evolution. Geochemistry is also significant for other geological sciences, especially for dating various rock formations, geological processes and exploration of various mineral deposits.

Geochemistry includes thirteen chapters that discuss the chemical composition of various rock complexes, their structural evolution, the methodology of specific geochemical methods (isotopic geology) and use of geochemical methods for the exploration of mineral deposits, structural geology, stratigraphy and lithology.

Chapter 1, "Geochemistry of Granitic Rocks of the Moldanubian Batholith (Central European Variscides)," by Miloš René focuses on detailed geochemistry of four magmatic suites evolved in one of the largest plutonic complexes within the European Variscan Belt.

Chapter 2, "An Overview on the Classification and Tectonic Setting of Neoproterozoic Granites of the Nubian Shield, Eastern Desert, Egypt," by Gaafar A. El Bahariya presents a comprehensive description of granitoid rocks exposed in the Eastern desert of Egypt and their geotectonic classification.

Chapter 3, "Microstructure Features in Paleo and Neoproterozoic Granitic Rocks, Southeastern Region of Brazil," by Leonardo Gonçalves and Christiane Castro Gonçalves provides a mineralogical and microstructural description of granitic rocks exposed in the south-eastern region of Brazil, including detailed information on the geochemical evolution of feldspars during magmatic differentiation and high-temperature deformation.

Chapter 4, "Deformation Features and Structures in Some Igneous and Metamorphic Rocks: A Case Study of Central African Fold Belt in Cameroon," by Benjamin Ntieche, Wokwenmendam Nguet, Zakari Nchouwet, Mahomed Azis Mounjouohou and Daouha Mfepat, is a detailed tectonic study of selected magmatic and metamorphic complexes in Central Africa.

Chapter 5, "The Importance of Mechanical Transport, Rock Texture, and Mineral Chemistry in Chemical Weathering of Granites: The Melechov Massif, Czech Republic," by Václav Procházka, Miroslav Žáček, Petr Sulovský, Tomáš Vaculovič, Lenka Rukavičková and Dobroslav Matějka describes the geochemical evolution of selected granitic rocks from the northern part of the Moldanubian batholith, Central European Variscan Belt, during their chemical weathering. The study also partly focuses on the fractionation of the rare-earth elements in these granites.

Chapter 6, "Changes of Granite Rapakivi under the Biofouling Influence," by Dmitry Yu. Vlasov, Elena G. Panova, Marina S. Zelenskaya, Oksana A. Rodina, Alexey D. Vlasov and Katerina V. Sazanova is a detailed study of the influence of biochemical weathering on the destruction of rapakivi granite stones in several Russian and Finish cities. A significant part of this study presents the chemical composition of fresh rocks in comparison with the composition of crusts evolved during the biodegradation of these granitic rocks.

Chapter 7, "Geochemistry of Radioactive Isotopes," by Salih Muhammad Awadh examines the geochemistry of selected isotopes of argon, carbon, kalium, neodymium, rubidium, samarium, strontium, uranium, and thorium, which are predominantly used to date magmatic, metamorphic and sedimentary rock complexes. The chapter also contains the basic characteristics of radioactivity, radioactive decay, and the methodology of rock dating.

Chapter 8, "Exploration for Fe-Mn Oxides Using Geochemical Signatures in Soil: A Case Study of Part of Northwestern Nigeria," by Olufemi Sijuade Bamigboye focuses on detailed geochemical analyses of soil samples for geochemical exploration of Fe-Mn mineralization evolved in north-western Nigeria. It also presents results of detailed multi-componential geochemical analyses using principal component analysis.

Chapter 9, "Main Features of Sedimentogenesis and Ecogenesis in Late Paleozoic Sea Pools of Northeast Asia," by Victor Gdal'evich Ganelin presents sedimentological analyses of Late Paleozoic sediments of the Taimyr-Kolyma paleogeographic region. The evolution of limestone bodies is characterized using the composition of carbon and oxygen isotope ratios.

Chapter 10, "Structural Differentiation and Sedimentary System of the Permian Sichuan Cratonic Basin," by Haofu Zheng and Bo Liu focuses on detailed sedimentological and structural analyses of the Sichuan basin evolved in the western region of the Yangtze tectonic block. A significant part of this chapter details the tectono-sedimentological study of the Permian carbonate gas reservoirs evolved in this basin.

Chapter 11, "Late Neo-Proterozoic Tectono-Sedimentary Evolution of the Tarim Block, NW China," by Kaibo Shi, Bo Liu, Weimin Jiang, Jinxing Yu, Yue Kong, Tong Li and Changhai Li presents a sedimentology study of the Late Neo-Proterozoic and Early Cambrian sediments. It also characterizes the geochemical evolution of carbonate rocks by fractionation of carbon isotopes. For some sediments enriched inorganic matter, the content of total carbon is presented.

Chapter 12, "The Onshore Southern Carnarvon Basin in Coastal Western Australia during the Quaternary: Tectonic Setting and Facies-Complicated Heterogeneous Stratigraphic Patterns," by Vic Semeniuk and Margaret Broex present very detailed stratigraphy, lithology, and sedimentology of the Quaternary sediments.

Chapter 13, "Middle Miocene Evaporites from Northern Iraq: Petrography, Geochemistry, and Cap Rock Efficiency," by Ali I. Al-Juboury, Rana A. Mahmood, and Abulaziz M. Al-Hamdani examines petrography, geochemistry, and petroleum cap rock potential of middle-Miocene evaporites (gypsum and anhydrite). The gypsum enriched evaporites are characterized by their chemical composition.

Chapter 14, "Bioclastic Deposits in the NW Gulf of Naples (Southern Tyrrhenian Sea, Italy): A Focus on New Sedimentological and Stratigraphic Data around the Island of Ischia," by Gemma Aiello discusses sedimentology and stratigraphic studies of bioclastic deposits evolved in some parts of the Gulf of Naples.

This work was made possible with the support and fruitful suggestions from expert reviewers whose comments and contributions play a recognizable role in finalizing the high quality of this book.

Dr. Miloš René
Institute of Rock Structure and Mechanics,
Czech Academy of Sciences,
Prague, Czech Republic

Gemma Aiello
Institute of Marine Sciences - National Research Council ISMAR-CNR,
Italy

Gaafar El Bahariya
Tanta University,
Egypt

Section 1

Geochemistry of Granitic Rocks

Chapter 1

Geochemistry of Granitic Rocks of the Moldanubian Batholith (Central European Variscides)

Miloš René

Abstract

Granitic rocks of the Moldanubian batholith are represented by four magmatic suites: I- to S-type Weinsberg biotite granites-granodiorites, S-type Eisgarn two-mica granites, S-type Melechov/Zvůle two mica, highly fractionated granites and I/S-type Freistadt biotite granites-granodiorites. The biotite granites of the Weinsberg suite are subaluminous to weakly peraluminous granites, enriched in Ba, Sr, and Zr. The two-mica granites of the Eisgarn suite are represented by a peraluminous, the low-Th Deštná granites, intermediate-Th Mrákotín, Číměř/Aalfang granites, and the high-Th Lipnice/Steinberg granites. The alkali feldspar and two-mica granites of the Melechov/Zvůle suite are highly fractionated, peraluminous granitic rocks, depleted especially in Ba, Zr, and Th. The granites to granodiorites of the Freistadt suite are subaluminous on Ba- and Sr-enriched granitic rocks.

Keywords: granitic rocks, petrology, geochemistry, Moldanubian zone, Bohemian Massif

1. Introduction

Generation of granitic rocks batholiths and plutons is a characteristic late-stage feature in the evolution of collisional orogens [1–4]. One of the classic examples is the European Variscan Belt, produced as a result of the late Palaeozoic convergence of Gondwana and Laurussia. The Moldanubian batholith that forms one of the largest plutonic complexes within the European Variscan Belt, covering 10,000 km^2 in the central part of the Moldanubian Zone of the Bohemian Massif, provides an excellent insight into origin and evolution of such crustally derived magmas [5–10]. The most significant previous geochemistry studies of the Moldanubian batholith are concentrated in papers Liew et al., Vellmer and Wedepohl, and Breiter [5, 6, 9]. However, these papers are concentrated only on selected parts (the Austrian and South Bohemian) of this batholith.

The aim of the presented paper is detailed classification of individual magmatic suites of the whole Moldanubian batholith and description of petrology, geochemistry, and origin of its magmatic suites.

2. Geological setting

In map view, the Moldanubian batholith resembles a "V" shape having two nearly perpendicular segments oriented ~NNE-SSW and ~NW-SE (**Figure 1**). The ~NNE-SSW trending segment as the eastern part of the batholith is formed by large continuous exposures of granitic rock pluton, whereas the ~NW-SE as the western part of the batholith is formed by a number of isolated smaller plutons and bodies, some of which seem to be roughly parallel to the ~NW-SE trending regional shear zones (the Pfahl and Danube shear zones [13–15]). Both batholith branches are closely associated with the host migmatites [10, 16, 17]. Estimates of the pluton emplacement depths range from a number of separate smaller plutons, some of which seem to be from 18 to 20 km in the NW-SE segment and/or 7 to 9 km in the NNE-SSW segment of the Moldanubian batholith [10, 12].

The most significant S-type granites of the Eisgarn suite occurred in the NNE-SSW branch of the Moldanubian batholith are hosted in a complex of cordierite-bearing migmatites and migmatitized paragneisses of the Pelhřimov complex [16].

The batholith consists of multiple intrusive units (plutons and stocks), predominantly composed of felsic to intermediate, granitic to granodioritic rocks with either S- or transitional I/S-type character [5, 6, 8, 9, 18]. All these granitoids can be classified into four main suites [6, 8, 9]:

a. Coarse-grained, mainly porphyritic I- to I/S-type biotite granites to granodiorites of the Weinsberg suite including four subtypes (coarse-grained, porphyritic Weinsberg I, very coarse-grained porphyritic biotite to muscovite-biotite Weinsberg II, coarse-grained biotite-amphibole granite to granodiorite "Schlieren granite" and slightly porphyritic Srní granite)

b. S-type two-mica granites of the Eisgarn suite including the equigranular Mrákotín variety, porphyritic Čímĕř/Aalfang variety, highly porphyritic Lipnice/Steinberg variety and equigranular Deštná variety, occurred only in the Klenov pluton

c. Younger group of highly fractionated, coarse-grained S-type two-mica granites of the Melechov/Zvůle suite formed ring-shaped bodies or stocks, often with zoned internal structure. From the NE to the SW, there are following bodies in the ~NNE-SSW segment of the Moldanubian batholith (the Melechov, Čeřínek, and Zvůle bodies). According to its geologic position, mineralogical and geochemical composition could be to these bodies attached also the coarse-grained, weakly porphyritic biotite-muscovite Plechý granite in central part of the Plechý pluton (Verner et al. [12]) and the fine-grained two-mica Sulzberg granite in the Bärenstein pluton of the ~NW-SE segment of the Moldanubian batholith [19] (**Figure 2**).

d. Fine- to coarse-grained I/S-type-biotite granites and granodiorites of the Freistadt/Mauthausen suite including the coarse-grained "marginal variety," medium-grained "central variety" of the Freistadt pluton, and the muscovite bearing biotite granite of the Graben granite variety occurred only on the eastern margin of the Freistadt town [21, 22].

Klötzli et al. [23] proposed that the origin of the Moldanubian batholith begun with the partial melting of mid—to lower crustal pre-Variscan rocks, polyphase

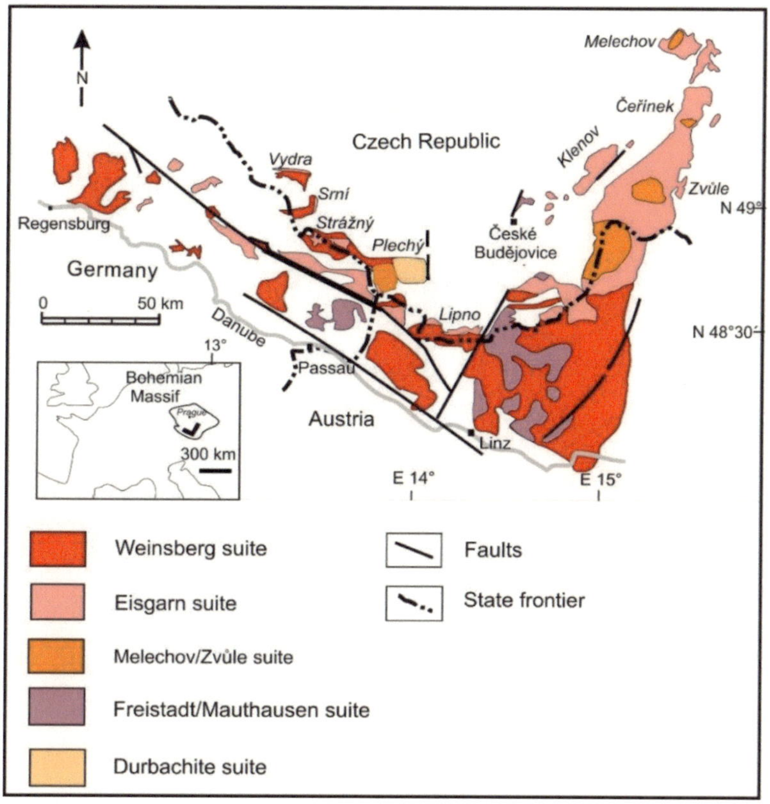

Figure 1.
Geological map of the Moldanubian batholith (after [11, 12], modified by author).

exhumation, anatexis, and deformation, closely followed by emplacement of large volumes S/I-type and S-type granodiorites and granites between 331 and 326 Ma.

According to the high-precision ID-TIMS U-Pb zircon and monazite dating the oldest Weinsberg granite (330.7 ± 0.4 Ma) intruded in southern part of the ~NNE-SSW segment of the Moldanubian batholith. Neighbouring two-mica granites of the Eisgarn suite were dated at 327.0 ± 0.3 Ma (Mrákotín variety), 327.14 ± 0.21 Ma (Deštná variety), and 328.5 ± 2.1 Ma (Číměř/Aalfang variety). However, the Weinsberg granite in the ~NW-SE segment of the Moldanubian batholith is comparably younger (327.7 ± 0.4 to 325.75 ± 0.39 Ma). The two-mica granite of the Eisgarn suite from this magmatic segment in the Lipno/Sternstein pluton was dated at 326.4 ± 0.6 Ma. Thus, the NW-SE segment of the Moldanubian batholith was formed ~3 Ma later than the NNE-SSW segment. However, the Weisberg granite from southern part of NW-SE segment, south of the Pfahl fault, was dated at 322.7 ± 0.7 Ma, implying that the processes of crustal melting migrated in the area of the Moldanubian batholith further toward south with time [24]. These U-Pb ages of granitoid rocks occurred in the NW-SE segment of the Moldanubian batholith and their structural relations with the Pfahl shear zone indicate the initial stages of dextral shearing at c. 342–327 Ma and mylonitic deformation coeval with granite emplacement during c. 326–327 Ma [10, 24]. The I/S-type granodiorites of the Freistadt/Mauthausen suite are distinctly younger. Age of monazite from the Freistadt granodiorite is 302 ± 2 Ma [25], and the age of monazite from the fine-grained, I-type Mauthausen granite is 316 ± 1 Ma [26].

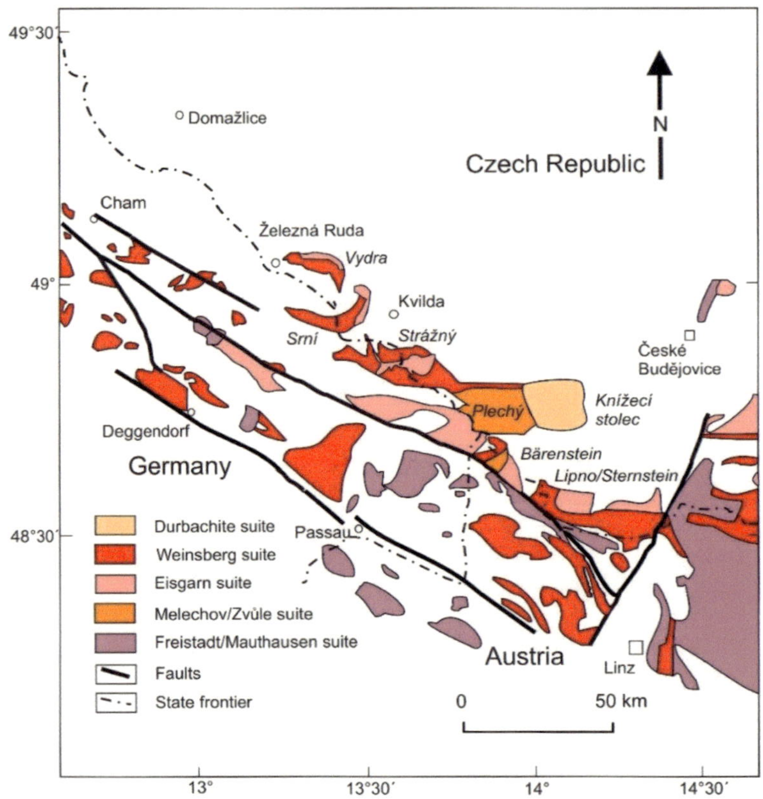

Figure 2.
Schematic geological map of the NW-SE segment of the Moldanubian batholith (after [20] modified by author).

3. Analytical methods

Rock samples of 2–5 kg in weight were crushed in a jaw crusher and a representative split of these materials was ground in an agate ball mill. For geochemical study of analysed granitoids were used 75 representative rock samples (**Table 1**, **Figures 3–7**). Major elements and some trace elements (Ba, Rb, Sr, Zr, Nb, Y, U, and Th) were determined using a Bruker AXS S4 Pioneer X-ray fluorescence spectrometer at the University of Salzburg, Austria, on fused glass discs and pressed rock powder pellets, respectively. The FeO was determined by titration, H_2O^+ and H_2O^- were analyzed gravimetrically in the chemical laboratory of the Institute of Rock Structure and Mechanics, Academy of Sciences of the Czech Republic, Prague. Rare earth elements were quantified by inductive coupled plasma mass spectrometry (ICP-MS) techniques at Activation Laboratories Ltd., Ancaster, Canada, using a Perkin Elmer Sciex ELAN 6100 ICP mass spectrometer, following standard lithium metaborate/tetraborate fusion and acid decomposition sample preparative procedures.

Microprobe analyses of selected minerals (feldspars and biotite) were performed in polished thin sections using a CAMECA SX-100 microprobe operated in wavelength-dispersive mode at the Institute of Geology, Academy of Sciences of the Czech Republic. The raw analytical data were corrected using the PAP procedure [28]. Operation conditions were accelerating potential of 15–20 kV, beam current of 10–70 nA (measured on

DOI: http://dx.doi.org/10.5772/intechopen.93189

Suite	Variety	CaO w. %	Zr ppm	Th ppm	ΣREE	La_N/Yb_N	Eu/Eu*
Weinsberg	Weinsberg I	1.5–3.0	214–441	19–45	269–443	10.0–30.5	0.30–0.60
	Weinsberg II	1.1–2.6	58–491	4–31	58–491	4.9–30.3	0.18–0.53
	Schlierengranite	1.9–4.6	133–652	5–42	n.a.	n.a.	n.a.
	Srní	1.5–4.3	190–331	25–31	266–430	12.6–33.4	0.38–0.58
Eisgarn	Deštná	0.4–1.1	39–81	2–7	33–69	3.2–19.0	0.35–1.18
	Mrákotín	0.4–1.1	72–157	9–41	91–222	11.2–67.7	0.16–0.48
	Číměř/Aalfang	0.5–1.0	96–175	17–54	117–242	21.8–44.2	0.15–0.31
	Lipnice/ Steinberg	0.6–1.1	143–291	39–110	207–423	19.9–64.3	0.09–0.31
Melechov/ Zvůle	Melechov	0.6–0.8	52–60	3	25–54	6.1–8.9	0.11–0.69
	Čeřínek	0.3–0.7	18–114	6–14	93–137	19.9–26.1	0.28–0.32
	Zvůle	0.5–0.9	31–88	7–13	88	29.9	0.40
	Plechý	0.5–0.7	55–113	9–23	66–118	12.5–20.6	0.13–0.46
	Sulzberg	0.5–0.6	130–136	26–35	149–177	27.2–31.8	0.26–0.29
Freistadt	Central	1.9–3.4	102–188	4–19	141–180	12.2–20.9	0.68–0.81
	Marginal	2.0–4.1	101–214	7–18	129–199	9.9–16.6	0.56–0.84
	Graben granite	1.8–2.3	98–171	10–15	155–238	19.7–30.7	0.58

n. a.– not available.

Table 1.
Chemical composition of granitic rock of the Moldanubian batholith.

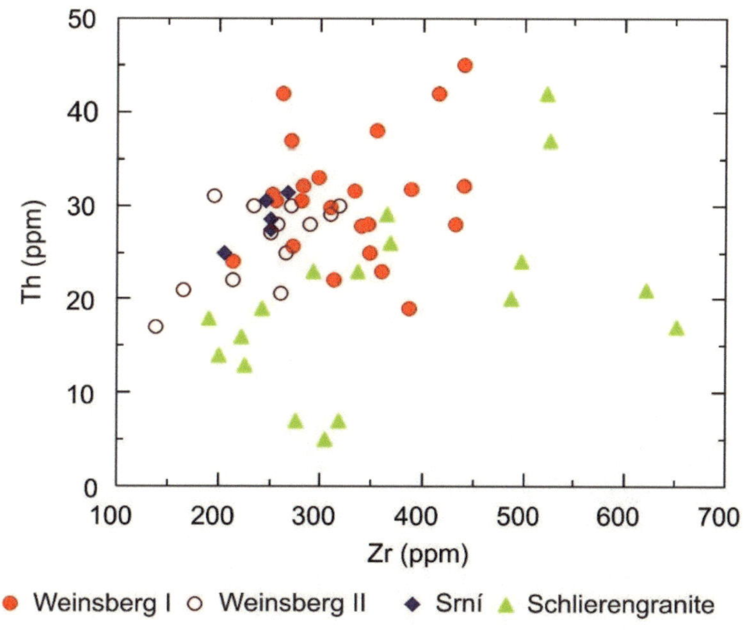

Figure 3.
Distribution Th and Zr in granitic rocks of the Weinsberg suite.

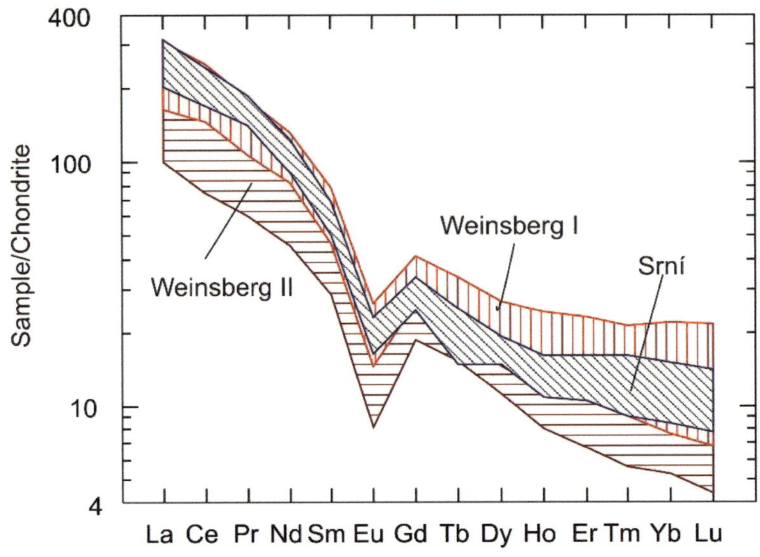

Figure 4.
Chondrite-normalized REE pattern for the Weinsberg suite. Normalizing values are from Boyton [27].

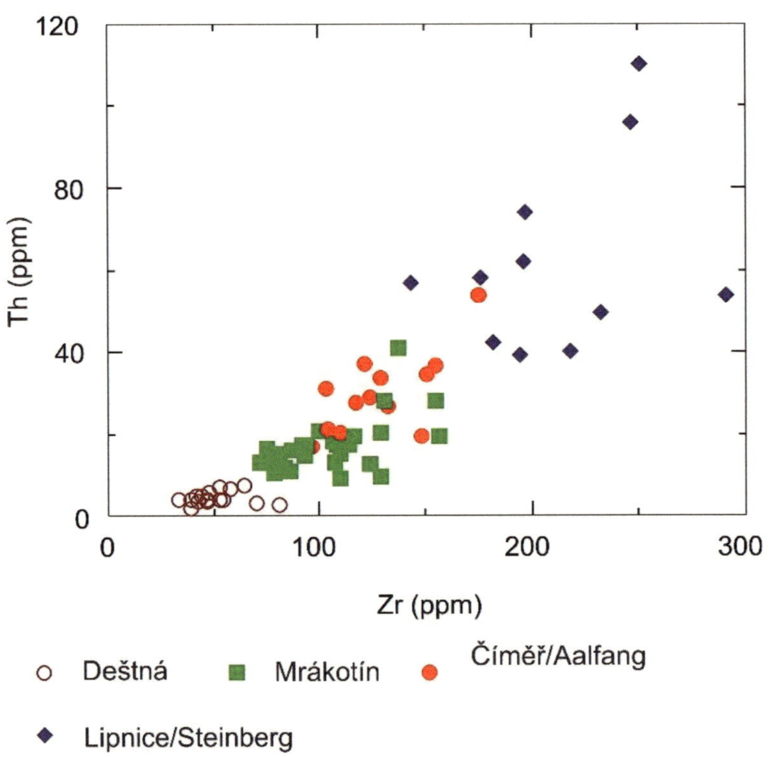

Figure 5.
Distribution of Th and Zr in granitic rocks of the Eisgarn suite.

a Faraday cup), and a beam diameter of 2 μm. Both synthetic and natural minerals were used as a reference material. Mineral formulae were recalculated using the MinPet 2.02 software [29].

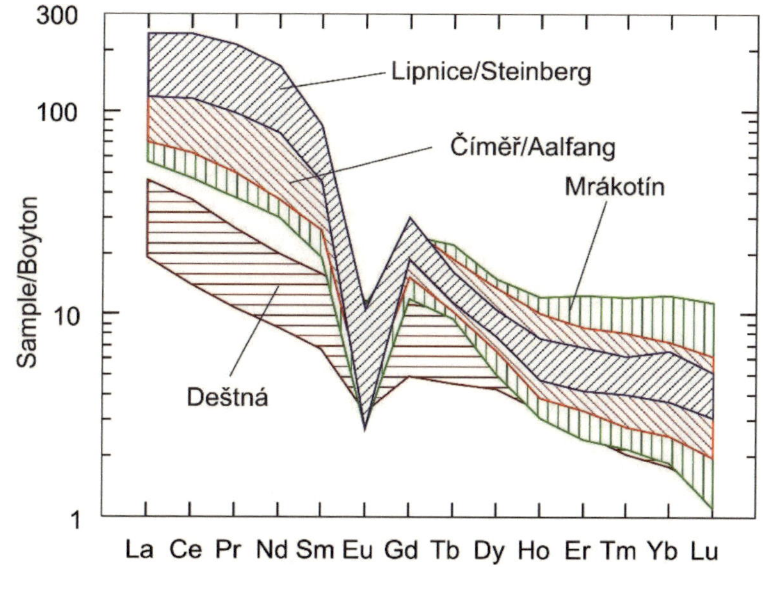

Figure 6.
Chondrite-normalized REE pattern for the Eisgarn suite. Normalizing values are from Boyton [27].

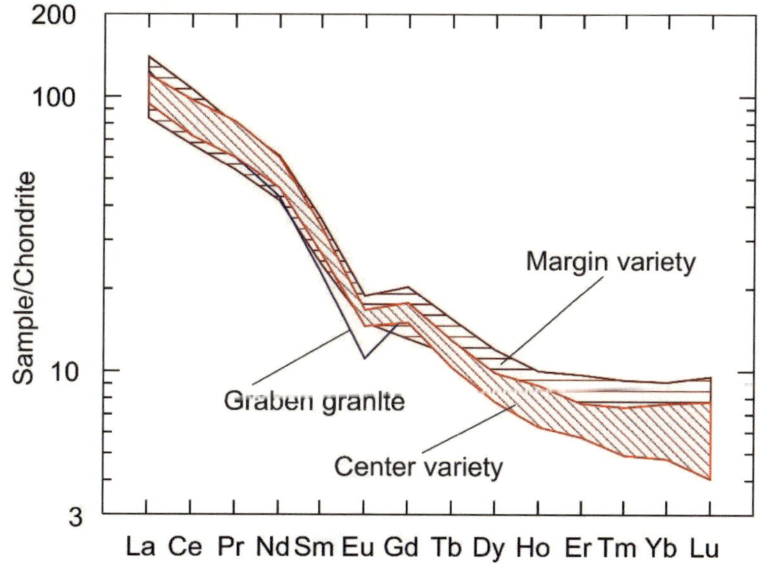

Figure 7.
Chondrite-normalized REE pattern for the Freistadt suite. Normalizing values are from Boyton [27].

4. Results

4.1 Petrography

The coarse-grained I/S-type granites and granodiorites of the Weinsberg suite are in the Moldanubian batholith represented by four varieties—Weinsberg I, Weisberg II, Schlieren granite, and Srni variety. The subtypes Weinsberg I and II, represented by coarse-grained porphyritic biotite granite to quartz monzodiorite

(Weisberg I) and very coarse-grained porphyritic biotite to muscovite-biotite granite (Weinsberg II), were distinguished based on zircon morphology and minor differences in composition [30, 31]. The Schlieren granite represent predominantly in situ evolved intrusion occurred in the Mühlviertel area (Upper Austria) and attached area of the Bavaria (Germany). The Schlieren granite, however, was originally described as "coarse-grained" gneisses [32]. The Srní variety occurred only in the Vydra and Prášily plutons of the NW-SE segment of the Moldanubian batholith (**Figure 2**) is medium-grained, weakly porphyritic granite to granodiorite. Major components of the Weinsberg granites to quartz monzodiorites are quartz, K-feldspar, plagioclase, biotite and rare amphibole (Schlieren granite), muscovite, or garnet (Srní variety). The accessories include apatite, zircon, ilmenite, titanite, allanite, monazite, and rare xenotime (**Table 2**).

The two-mica granites of the Eisgarn suite are the most abundant granitic rocks in the Moldanubian batholith. Several petrographic varieties in the Eisgarn suite were identified, containing variable texture and biotite to muscovite proportions. In the Czech part of the Moldanubian batholith, the equigranular Mrákotín variety predominates. Similar equigranular two-mica granites occur in the Bavarian part of the Moldanubian batholith (Haidmühle, Theresienreut variety). In the Austrian part of the Moldanubian batholith, the Číměř/Aalfang porphyritic variety with biotite over muscovite predominated. In the Klenov pluton occur the equigranular two-mica leucogranites, described as the Deštná variety [33, 34]. In the Melechov and Plechý plutons occur highly porphyritic granites of the Lipnice/Steinberg variety. Major components of the Eisgarn monzogranites are quartz, K-feldspar, plagioclase, biotite, and muscovite. Monazite accessory minerals are represented by apatite, andalusite, ilmenite, zircon, and rare cordierite, sillimanite and xenotime (**Table 3**).

The younger group of highly fractionated S-type two-mica granites of the Melechov/Zvůle suite are composed by coarse-grained two-mica alkali feldspar granites to monzogranites that forms ring-shaped stocks, often with zoned internal structure. There are in the Moldanubian batholith represented by the Melechov, Čeřínek, Zvůle, Plechý plutons, and the Sulzberg granite in the Bärenstein pluton. Major rock-forming minerals of these granites are plagioclase (28–36 vol. %) (An_{1-23}), quartz (27–34 vol. %), K-feldspar (26–31 vol. %),

Variety	Modal composition (vol. %)	Accessories	Plagioclase	Biotite
Weinsberg I	Qtz 16–43, Kfs 12–54, Pl 11–44, Bt 6–27	Apatite, zircon, ilmenite, titanite, monazite, allanite	An_{15-42}	Annite, Fe/(Fe+ Mg) 0.60–0.67, Al^{4+}2.05–2.39, Ti 0.23–0.45 apfu
Weinsberg II	Qtz 24–30, Kfs 22–33, Pl 23–37, Bt 7–15	Apatite, zircon, ilmenite, monazite	An_{24-41}	Annite, Fe/(Fe+ Mg) 0.79–0.81, Al^{4+} 2.43–2.47, Ti 0.34–0.42 apfu
Schlieren granite	Pl 32–50, Kfs 7–37, Qtz 18–34, Bt 6–32, Amp 0–3	Apatite, zircon, titanite, ilmenite, magnetite, allanite	An_{20-40}	Annite, Fe/(Fe+ Mg) 0.0.53–0.55, Al^{4+} 2.10–2.13, Ti 0.23–0.42 apfu
Srní	Pl 19–53, Pl 19–53, Kfs 7–41, Bt 4–27, Ms 0–6, Gt 0–3	Apatite, zircon, ilmenite, monazite, xenotime	An_{9-31}	Annite-siderophyllite, Fe/(Fe+ Mg) 0.64–0.81, Al^{4+} 2.24–2.57, Ti 0.26–0.42 apfu

Table 2.
Modal composition and mineralogy of granitic rocks of the Weinsberg suite.

Variety	Modal composition (vol. %)	Accessories	Plagioclase	Biotite
Mrákotín	Qtz 23–46, Kfs 14–42, Pl 10–37, Ms 3–17, Bt 3–11	Apatite, andalusite, ilmenite, zircon, monazite	$An_{9–25}$	siderophyllite, Fe/(Fe + Mg) 0.63–0.72, Al^{4+} 2.45–2.73, Ti 0.25–0.38 apfu
Číměř/Aalfang	Kfs 14–50, Qtz 22–41, Pl 8–38, Bt 3–16, Ms. 1–8	Andalusite, apatite, ilmenite, zircon, monazite, cordierite	$An_{9–23}$	annite–siderophyllite, Fe/(Fe + Mg) 0.64–0.72, Al^{4+}2.44–2.61, Ti 0.24–0.39 apfu
Lipnice/Steinberg	Kfs 23–40, Qtz 25–37, Pl 19–37, Bt 6–10, Ms. 2–7	Apatite, ilmenite, monazite, zircon, rutile, sillimanite	$An_{13–20}$	Annite Fe/(Fe + Mg) 0.59–0.75, Al^{4+} 2.10–2.42, Ti 0.22–0.35 apfu
Deštná	Qtz 28–42, Kfs 21–41, Pl 15–50, Ms. 1–8, Bt 1–5	Apatite, ilmenite, monazite, zircon, xenotime	$An_{12–25}$	Annite-siderophyllite, Fe/(Fe + Mg) 0.64–0.74, Al^{4+} 2.27–2.67, Ti 0.17–0.42 apfu

Table 3.
Modal composition and mineralogy of granitic rocks of the Eisgarn suite.

muscovite (3–9 vol. %) and biotite (3–4 vol. %) [annite-siderophyllite, Fe/(Fe + Mg) 0.67–0.77, Al^{4+} 0.67–0.0.74, Ti 0.16–0.41]. Accessory minerals are represented by apatite, ilmenite, rutile, zircon, monazite, tourmaline, and garnet.

The I/S-type Freistadt biotite granodiorites are medium- to coarse-grained granitic rocks. Major rock-forming minerals of the both main varieties ("central" and "marginal") of the Freistadt suite are quartz, plagioclase, K-feldspar, and biotite. The Graben granite contains also 3 vol. % of muscovite. The accessory minerals in granitic rocks of the Freistadt suite are represented by apatite, ilmenite, zircon, titanite, allanite, and rare xenotime (**Table 4**). The Graben granite is more fractionated granitic rock with relatively high LREE/HREE ratio (20–31).

4.2 Geochemistry

The granites and granodiorites of the Weinsberg suite are subaluminous and weakly peraluminous rocks with A/CNK [mol. $Al_2O_3/(CaO + Na_2O + K_2O)$] of 0.9–1.2. For all varieties of the Weinsberg granite suite are significant high Ba (291–2016 ppm), Sr. (75–680 ppm), Zr (89–652 ppm), and ΣREE (133–491 ppm) concentrations (**Table 1**). The highest concentrations of Ba, Sr, and Zr are significant especially for the Schlieren granite (354–2016 ppm Ba, 196–680 ppm Sr, 133–652 ppm Zr) (**Figure 3**, **Table 1**). For individual varieties of the Weinsberg suite, the partly different fractionation of REE is significant (**Figure 4**).

The granites of the Eisgarn suite are subaluminous to strongly peraluminous rocks (A/CNK 1.0–1.3) with low CaO concentrations (0.4–1.2 wt. %) (**Table 1**). In the strongly peraluminous varieties, magmatic andalusite, sillimanite, and rarely occurred cordierite are present. In the Mrákotín variety, the peraluminosity is also expressed by widespread dominance of muscovite over biotite. Three main geochemical varieties of two-mica granites could be distinguished by their concentrations of Th on the low-Th (2–7 ppm) Deštná granites, the intermediate Th Mrákotín/Číměř/Aalfang granites (8–54 ppm), and the high-Th Lipnice/Steinberg granites

Variety	Modal composition (vol. %)	Accessories	Plagioclase	Biotite
Freistadt central	Qtz 23–29, Pl 34–49, Kfs 13–25, Bt 9–14, Ms 0–2	Apatite, zircon, ilmenite, monazite, titanite, xenotime	An_{20-22}	Annite Fe/(Fe + Mg) 0.58–0.59, Al^{4+} 2.20–2.24, Ti 0.34–0.47 apfu
Freistadt marginal	Qtz 12–32, Pl 32–68, Kfs 3–27, Bt 6–17, Ms 0–1	Apatite, zircon, ilmenite, titanite, monazite, allanite	An_{25-37}	Annite Fe/(Fe + Mg) 0.44–0.62, Al^{4+}2.09–2.28, Ti 0.30–0.50 apfu
Graben granite	Qtz 29–31, Pl 33–40, Kfs 19–25, Bt 5–9, Ms 2–6	Apatite, zircon, ilmenite, monazite	An_{23-26}	Annite Fe/(Fe + Mg) 0.66–0.69, Al^{4+} 2.33–2.41, Ti 0.31–0.41 apfu

Table 4.
Modal composition and mineralogy of granodiorites and granites of the Freistadt suite.

(39–110 ppm) (**Table 1**, **Figure 5**). The highest ΣREE was found in the Lipnice/Steinberg granite variety (207–242 ppm), whereas the lowest ΣREE is for the Deštná granites significant (33–69 ppm). The highest LREE/HREE ratio is for the Lipnice/Steinberg variety significant (**Figure 6**).

The granites of the Melechov/Zvůle granite suite are peraluminous rocks (A/CNK 1.1–1.3) with low CaO concentrations (0.3–0.9 wt. %) (**Table 1**). The individual varieties of these granites could be distinguished especially by their concentrations of Zr and Th. The highest concentrations of both elements are significant for the Plechý (55–113 ppm Zr, 9–23 ppm Th) and Sulzberg (130–136 ppm Zr, 26–35 ppm Th) granites. The lowest concentrations of the both elements occur in the Stvořidla granites, which form the central stock in the Melechov pluton (36–67 ppm Zr, 2.8–2.9 ppm Th). These granites also have the lowest concentrations of ΣREE (27–35 ppm) and the lowest LREE/HREE ratio (3.2–8.6). In the opposite, the highest content of ΣREE (149–177 ppm) and the highest LREE/HREE ratio (27–32) are for the Sulzberg granites significant.

The granites and granodiorites of the Freistadt suite are subaluminous rocks (A/CNK = 1.0–1.1) with partly elevated CaO concentrations (1.8–4.1 wt. %). The central and marginal granodiorites of this suite are enriched in Ba (616–1017 ppm) and Sr (253–471 ppm). Their total concentrations of REE in granites and granodiorites of the Freistadt suite are 129–238 ppm and the central and marginal granodiorites display relatively low LREE/HREE ratio (9–21) with moderate negative Eu anomaly (Eu/Eu* = 0.56–0.84) (**Table 1**, **Figure 7**).

5. Discussion

In the past, the orig
in and fractionation of granitic rocks of the Moldanubian batholith was discussed by geochemical modelling based on trace-element fractionation [6, 18, 31, 35, 36] and experimental study of biotite stability [34]. According to these studies, granitic rocks of the Moldanubian batholith could originated by LP-HT partial melting of various metasediments and/or by melting of a mixture of metasediments and amphibolites [6, 18, 31, 36]. According to the majority of these studies, the granitic rocks of the individual magmatic suites occurred in the Moldanubian batholith were also differently fractionated [35, 36]. The fractionation of these magmatic suites

could be well documented by distribution some trace elements (e.g., Ba, Sr, Th, Zr, REE) (**Figures 3–7**).

For distinguishing, source rock series (greywackes vs. pelites) could be used CaO/Na_2O and Al_2O_3/TiO_2 ratios [37]. According to this study, discussed in detail by René et al. [34] and René [18], the granitic rocks of the Eisgarn suite originated by partial melting of metapelites, whereas granites and granodiorites of the Weinsberg and Freistadt suites originated by partial melting of a metagreywackes-metabasalt mixture.

The estimation of melting temperatures of granitic melts is usually based on saturation thermometers based on melting of zircon and monazite [38–40]. According to zircon saturation, thermometry granitic rocks are usually divided on the hot and cold granites [41]. The most detailed study of all problems connected with the use of zircon thermometry was published by Siégel et al. [42] and Clemens et al. [43]. For all granitic rocks from the Moldanubian batholith, the T_{Zrnsat} was calculated according to the revisited formula published by Boehnke et al. [39] and T_{Mnzsat} according to the model of Montel [40]. The highest saturation temperatures from both models were found for granitic rocks of the Weinsberg suite (681–900°C) and for the Zr- and Th-enriched Lipnice/Steinberg granites of the Eisgarn suite °C (760–910°C). The lowest saturation temperatures were found for some two-mica granites of the Eisgarn suite, especially for the Zr- and Th-depleted Deštná granites (628–740°C). The saturation temperatures for the Mrákotín granites varied between 682 and 859°C. However, the crystallisation temperature derived from experimental melting of biotite for the Mrákotín granite is 830–850°C [34]. It is also interested that in all these cases that the T_{Mnzsat} temperatures are usually partly higher that the T_{Zrnsat} temperatures. These differences could be explained by restitic (inherited) monazite crystals from original metasediments. On the other hand, during detailed study of zircon crystals from granites of the Eisgarn suite in rare cases, inherited zircon cores overgrowth by younger magmatic zircon were found [44].

Partly controversial are sources of heating, which were used for the generation of magmatic rocks of the Moldanubian batholith. According to Velmer and Wedepohl [6], the main heating source of the Moldanubian crust was intrusion of tholeiitic rocks during underplating of an upper crust. However, Gerdes et al. [8] proposed as a main heat source the contribution of internal radiogenic heat production within the thickened upper crust. Some other source of heating needed for generation of granitic rocks in the NW-SE segment of the Moldanubian batholith, together with LP-HT metamorphism rock series southern of the Pfahl zone, was a late Variscan delamination of upper mantle [10, 45]. Recent structural studies that granite plutons in the NW-SE segment of the Moldanubian batholith were emplaced at a shallow level into previously exhumed upper crust during dextral strike-slip shear zones that are later multiply reactivated [13, 15].

6. Conclusion

Granitic rocks of the Moldanubian batholith are represented by four magmatic suites. The biotite granites of the Weinsberg suite are subaluminous to weakly peraluminous granites enriched especially in Ba, Sr, and Zr. The two-mica granites of the Eisgarn suite are represented by the low-Th Deštná granites, the intermediate-Th Mrákotín, Číměř/Aalfang granites, and the high-Th Lipnice/Steinberg granites. The highly fractionated alkali feldspar and two-mica granites of the Melechov/Zvůle suite are depleted in Ba, Zr, and Th. The granites to granodiorites of the Freistadt suite are of Ba- and Sr-enriched granitic rocks. The granitic rocks of the individual magmatic suites occurred in the Moldanubian batholith were variable fractionated. The fractionation of these magmatic suites is documented by distribution some trace elements, especially by the Ba, Sr, Th, Zr, and REE. The granitic rocks of the Eisgarn suite originated by partial melting of metapelites, whereas granites and granodiorites of the Weinsberg and Freistadt suites originated by partial melting of a metagreywackes-metabasalt mixture. The estimation of melting temperatures of granitic melts for all magmatic suites of the Moldanubian batholith, based on zircon and monazite saturation thermometers, show that the highest temperatures from both models were found for granitic rocks of the Weinsberg suite and for Zr- and Th-enriched Lipnice/Steinberg granites from the Eisgarn suite (681–910°C). The partial differences between both saturation thermometers could be explained especially by restitic (inherited) zircon and monazite occurred in investigated granitic rocks.

Acknowledgements

Thanks are extended to the support of the long-term conceptual development research organization RVO 67985891 and the project of Ministry of Education, Youth and Sports (ME10083). The author is also grateful to Z. Korbelová from the Geological Institute of the Academy of Sciences of the Czech Republic for her technical assistance by using electron microprobe analyses of selected minerals (plagioclase, biotite). I am grateful to anonymous reviewer for numerous comments and recommendations that helped to improve this paper.

Conflict of interest

The author declares no conflict of interests.

Author details

Miloš René
Institute of Rock Structure and Mechanics, v.v.i., Academy of Sciences of the Czech Republic, Prague, Czech Republic

*Address all correspondence to: rene@irsm.cas.cz

References

[1] Williamson BJ, Shaw A, Downes H, Thirlwall MF. Geochemical constraints on the genesis of Hercynian two-mica leucogranites from the Massif Central, France. Chemical Geology. 1996;**127**:25-42. DOI: 10.1016/0009-2541(95)00105-0

[2] Clemens JD. S-type granitic magmas—Petrogenetic issues, models and evidence. Earth-Science Reviews. 2003;**61**:1-18. DOI: 10.1016/S0012-8252(02)00107-1

[3] Chappell BW, Hine T. The Cornubian batholith: An example of magmatic fractionation on a crustal scale. Resource Geology. 2006;**56**:203-244. DOI: 10.1111/j.1751-3928.2006.tb00281.x

[4] Janoušek V, Moyen JF, Martin H, Erban V, Farrow C. Geochemical modelling of igneous processes—Principles and recipes in R language. In: Bringing the Power of R to a Geochemical Community. 1st ed. Berlin, Heidelberg: Springer Geochemistry. p. 346. DOI: 10.1007/978-3-662-46792-3

[5] Liew TC, Finger F, Höck V. The Moldanubian granitoid plutons of Austria: Chemical and isotopic studies bearing on their environmental setting. Chemical Geology. 1989;**76**:41-55. DOI: 10.1016/0009-2541(89)90126-5

[6] Vellmer C, Wedepohl KH. Geochemical characterization and origin of granitoids from the south bohemian batholith in Lower Austria. Contributions to Mineralogy and Petrology. 1994;**118**:13-32. DOI: 10.1007/BF00310608

[7] Finger F, Roberts MP, Haunschmid B, Schermaier A, Steyer HP. Variscan granitoids of Central Europe: Their typology, potential sources and tectonothermal relations. Mineralogy and Petrology. 1997;**61**:67-96. DOI: 10.1007/BF01172478

[8] Gerdes A, Wörner G, Henk A. Postcollisional granite generation and HT/LP metamorphism by radiogenic heating: The Variscan south bohemian batholith. Journal of the Geological Society of London. 2000;**157**:577-587. DOI: 10.1144/jgs.157.3.577

[9] Breiter K. Geochemical classification of Variscan granitoids in the Moldanubicum (Czech Republic, Austria). Abhandlungen der Geologischen Bundesanstalt. 2010;**65**:19-25

[10] Žák J, Verner K, Janoušek V, Holub FV, Kachlík V, Finger F, et al. A plate-kinematic model for the assembly of the bohemian massif constrained by relationships around granitoid plutons. Geological Society of London, Special Publication. 2014;**405**:169-196. DOI: 10.1144/SP405.9

[11] Siebel W, Shang CK, Reiter E, Rohrmüller J, Breiter K. Two distinctive granite suites in the SW bohemian massif and their record of emplacement: Constraints from geochemistry and zircon $^{207}Pb/^{208}Pb$ chronology. Journal of Petrology. 2008;**49**:1853-1872. DOI: 10.1093/petrology/egn049

[12] Verner K, Žák J, Pertoldová J, Šrámek J, Trubač J, Týcová P. Magmatic history and geophysical signature of a post-collisional intrusive center emplaced near a crustal-scale shear zone: The Plechý granite pluton (Moldanubian batholith, bohemian massif). International Journal of Earth Sciences. 2009;**98**:517-532. DOI: 10.1007/s00531-007-0285-9

[13] Brandmayr M, Dallmeyer RD, Handler R, Wallbrecher E. Conjugate shear zones in the southern bohemian massif (Austria):

Implications for Variscan and alpine tectonothermal activity. Tectonophysics. 1995;**248**:97-116. DOI: 10.1016/0040-1951(95)00003-6

[14] Büttner SH. The geometric evolution of structures in granite during continuous deformation from magmatic to solid-state conditions: An example from the central European Variscan belt. American Mineralogist. 1999;**84**:1781-1792

[15] Büttner SH. Late Variscan stress-field rotation initiating escape tectonics in the south-western bohemian massif: A far field response to late-orogenic extension. Journal of Geosciences. 2007;**52**:29-43. DOI: 10.3190/j.geosci.004

[16] Žák J, Verner K, Faryad W, Chlupáčová M, Veselovský F. The generation of voluminous S-type granites in the Moldanubian unit, bohemian massif, by rapid isothermal exhumation of the metapelitic middle crust. Lithos. 2011;**121**:25-40. DOI: 10.1016/j.lithos.2010-10.002

[17] Žák J, Verner K, Ježek J, Trubač J. Complex mid-crustal flow within a growing granite-migmatite dome: An example from the Variscan belt illustrated by the anisotropy of magnetic susceptibility and fabric modelling. Geological Journal. 2019;**54**:3681-3699. DOI: 10.1002/gj.3335

[18] René M. Source compositions and melting temperatures of the main granitic suites from the Moldanubian batholith. Journal of Geosciences. 2016;**61**:355-370. DOI: 10.3190/jgeosci.223

[19] Fuchs G. Exkursion III/4: Kristallin Mühlviertel und Sauwald, südliche Böhmische Masse. Mitteilungen der Geologische Gesellschaft. 1964;**57**:281-289

[20] Klomínský J, Jarchovský T, Rajpoot GS. Atlas of Plutonic Rocks and Orthogneisses in the Bohemian Massif. 2. Moldanubicum. 1st ed. Prague: Czech Geological Survey; 2010. p. 200

[21] Klob H. Freistädter Granodiorit im österreichischen Moldanubikum. Verhandlungen der Geologische Bundesanstalt. 1971;**1**:98-142

[22] Friedl G. Geologisch-petrographische Untersuchungen in der Gegend nordöstlich von Freistadt (Oberösterreich) mit besonderer Berücksichtigung des "Grabengranits" [thesis]. Salzburg: University of Salzburg; 1990

[23] Klötzli US, Koller F, Scharbert S, Höck V. Cadomian lower-crustal contributions to Variscan Granit Petrogenesis (south bohemian pluton, Austria): Constraints from zircon typology and geochemistry, whole-rock, and feldspar Pb-Sr isotope systematics. Journal of Petrology. 2001;**42**:1621-1642. DOI: 10.1093/petrology/42.9.1621

[24] Finger F, Gerdes A, Verner K, Žák J. Resolving the growth history of the South Bohemian Batholith by means of high-precision ID-TIMS U-Pb zircon and monazite dating. In: Proceedings of the 3rd Workshop on Orogenic Processes in the Bohemian Massif; 15-17 June 2018; Presseck; 2018. pp. 13-14

[25] Gerdes A, Friedl G, Parrish RR, von Quadt A, Finger F. U-Pb-Datierungen (TIMS, SHRIMP und LA-MC-ICP-MS) an Zirkonen und Monaziten Südböhmischer Granitoide. Erlangen Geologische Abhandlungen. 2002;**Sonderband 4**:30-32

[26] Gerdes A, Friedl G, Parrish RR, Finger F. High-resolution geochronology of Variscan granite emplacement—The south bohemian batholith. Journal of the Czech Geological Society. 2003;**48**:53-54

[27] Boyton WV. Geochemistry of the rare earth elements: Meteorite

studies. In: Henderson P, editor. Rare Earth Element Geochemistry. 1st ed. Amsterdam: Elsevier; 1984. pp. 63-114

[28] Pouchou JL, Pichoir F. PAP (φ-ρ-Z) procedure for improved quantitative microanalysis. In: Armstrong JT, editor. Microbeam Analysis. San Francisco: San Francisco Press; 1985. pp. 104-106

[29] Richard R. MinPet: mineralogical and petrological data processing system, version 2.02. Québec, Canada: MinPet Geological Software; 1995

[30] Stöbich DM. Trachstudien an den akzessorischen Zirkonen des Weinsberger Granits im östlichen Mühlviertel und westlichen Waldviertel [thesis]. Salzburg: University of Salzburg; 1992

[31] Gerdes A. Magma homogenisation during anataxis, ascent and/or emplacement? Constraints from the Variscan Weinsberg granites. Terra Nova. 2001;**13**:305-312. DOI: 10.1046/j.1365-3121.2001.00365.x

[32] Fuchs G. Zur Altersgliederung des Moldanubikums Oberösterreichs. Vienna: Verhandlungen der Geologische Bundesanstalt; 1962. pp. 96-117

[33] René M, Matějka D, Nosek T. Geochemical constraints on the origin of a distinct type of two-mica granites (Deštná–Lásenice type) in the Moldanubian batholith (Czech Republic). Acta Montana, Geodynamics. Series A. 2003;**23**:59-76

[34] René M, Holtz F, Luo C, Beermann O, Stelling J. Biotite stability in peraluminous granitic melts: Compositional dependence and application to the generation of two-mica granites in the South Bohemian batholith (Bohemian Massif, Czech Republic). Lithos. 2008;**102**:538-553. DOI: 10.1016/j.lithos.2007.07.022

[35] Finger F, Clemens J. Migmatization and "secondary" granitic magmas: Effects of emplacement and crystallization of "primary" granitoids in southern Bohemian Massif, Austria. Contributions to Mineralogy and Petrology. 1995;**120**:311-326. DOI: 10.1007/BF00306510

[36] Matějka D, Janoušek V. Whole-rock geochemistry and petrogenesis of granites from the northern part of the Moldanubian Batholith (Czech Republic). Acta Universitatis Carolineae, Geology. 1998;**42**:73-79

[37] Sylvester PJ. Post-collisional strongly peraluminous granites. Lithos. 1998;**45**:29-44. DOI: 10.1016/S0024-4937(98)00024-3

[38] Watson EB, Harrison TM. Zircon saturation revisited: Temperature and composition effects in a variety of crustal magma types. Earth and Planetary Science Letters. 1983;**64**:295-304. DOI: 10.1016/0012-821X(83)90211-X

[39] Boehnke P, Watson EB, Trail D, Harrison TM, Schmitt AK. Zircon saturation re-revisited. Chemical Geology. 2013;**351**:324-334. DOI: 10.1016/j.chemgeo.2013.05.028

[40] Montel JM. A model for monazite/melt equilibrium and applications to the generation of granitic magmas. Chemical geology. 1993;**110**:127-146

[41] Miller CF. Hot and cold granites? Implications of zircon saturation temperatures and preservation of inheritance. Geology. 2003;**31**:529-532. DOI: 10.1130/0091-7613(2003)031

[42] Siégel C, Bryan SE, Allen CM, Gust DA. Use and abuse of zircon-based thermometers: A critical review and a recommended approach to identify atecrystic zircon. Earth-Science Reviews. 2018;**176**:87-116. DOI: 10.1016/j.earscirev.2017.08.011

[43] Clemens JD, Stevens G, Bryan SE. Conditions during the formation of granitic magmas by crustal melting—Hot or cold; drenched, damp or dry? Earth-Science Reviews. 2020;**200**:1-21. DOI: 10.1016/j.earscirev.2019.102982

[44] René M. Occurrence of Th, U, Y, Zr and REE-bearing accessory minerals on granites and their petrogenetic significance. In: Blasik M, Bogdaska H, editors. Granite, Occurrence, Mineralogy an Origin. 1st ed. New York: Nova Science Publishers, Inc.; 2012. pp. 27-56

[45] Finger F, Gerdes A, Janoušek V, René M, Riegler G. Resolving the Variscan evolution of the Moldanubian sector of the Bohemian Massif: The significance of the Bavarian and the Moravo-Moldanubian tectonometamorphic phases. Journal of Geosciences. 2007;**52**:9-28. DOI: 10.3190/j.geosci.005

Chapter 2

An Overview on the Classification and Tectonic Setting of Neoproterozoic Granites of the Nubian Shield, Eastern Desert, Egypt

Gaafar A. El Bahariya

Abstract

Granites constitute the main rock components of the Earth's continental crust, which suggested to be formed in variable geodynamics environments. The different types of granitic rocks, their compositional characteristics, tectonic settings and magma sources are outlined. Mineralogical classification of granites includes four rock types: tonalites, granodiorites, granite (monzogranite and syenogranites) and alkali-feldspar granites. Alphabetical classification subdivided granites into: I-type, S-type, A-type and M-type granites. Moreover, formation of granitic magmas requires distinctive geodynamic settings such as: volcanic arc granite (Cordilleran); collision-related granites (leucogranites); intra-plate and ocean ridge granites. The Eastern Desert of Egypt (ED) forms the northern part of Nubian Shield. Both older and younger granites are widely exposed in the ED. Old granites (OG) comprise tonalites and granodiorites of syn- to late-orogenic granitoid assemblages. They are calcalkaline, I-type, metaluminous and display island arc tectonic setting. Younger granites (YG) on the other hand, include granites, alkali-feldspar granites and minor granodiorites. They are of I- and A-type granites and of post-orogenic to anorogenic tectonic settings. The majority of the YG are alkaline, A-type granite and of within-plate tectonic setting (WPG). The A-type granites are subdivided into: A2-type postorogenic granites and A1-type anorogenic granites. Granite magma genesis involves: (a) fractional crystallization of mafic mantle-derived magmas; (b) anatexis or assimilation of old, upper crustal rocks (c) re - melting of juvenile mafic mantle – derived rocks underplating the continental crust. Generally, older I-type granitoids were interpreted to result from melting of mafic crust and dated at approximately 760–650 Ma, whereas younger granites suggested to be formed as a result of partial melting of a juvenile Neoproterozoic mantle source. Moreover, they formed from anatectic melts of various crustal sources that emplaced between 600 and 475 Ma.

Keywords: granitoid rocks, neoproterozoic, granites, Nubian shield, Eastern Desert, Egypt, older and younger granites, tectonic setting

1. Introduction

Granitoid rocks are the most abundant plutonic rocks in the continental crust, which are diverse and their magma sources have long been debated. There is a tectonomagmatic connection, where each granite type is related to definite tectonic settings, and all magmatism is tectonically motivated, so that such a division is simply a matter of emphasis [1]. Using the QAP diagram (**Figure 1**) granites are classified into four granite domains: tonalite, granodiorite, granite (monzogranite, and syenogranite) and alkali feldspar granites according to IUGS [2]. Genetic alphabetical classification includes: I-type granites (I = igneous); S-type granites (S = sedimentary); M-type granites products of mantle melts (M = mantle) and A-type (A = anorogenic) [3]: A = type granites were referred as alkaline or anorogenic granites [4]. The granitic rocks show an obvious alumina saturation from metaluminous, through peraluminous to peralkaline [5, 6]. Peraluminous granites have $Al_2O_3 > CaO + Na_2O + K_2O$; Metaluminous: $A_2O_3 < CaO + Na_2O + K_2O$ and $Al_2O_3 > Na_2O + K_2O$ and peralkaline $Na_2O + K_2O > Al_2O_3$. Chemically, S-type granites are equivalent to peraluminous granites, and I-type granites are equivalent to metaluminous granites. Whalen et al. [7] used the contents of Zr, Ce, Y and Nb (normalized to Ga/Al) to discriminate between A-type granites and both I- and S- type granites. A more comprehensive descriptive basis for chemically subdividing granitoids was given by Frost et al. (2001), in which they subdivided granitic rocks based on the so - called *Fe* - number (whole– rock Σ FeO/[Σ FeO + MgO] into ferroan and magnesian rocks.

Granites formed in a variety of tectonic settings around the world, either at plate margins or intraplate. They are subdivided according to their tectonic environments into four main groups—ocean ridge granites (ORG), volcanic arc granites (VAG), within plate granites (WPG) and collision granites (COLG) using trace element Y-Nb, Yb-Ta, Rb-(Y + Nb) and Rb— (Yb + Ta) diagrams [8]. Harris et al. [9] recognized four groups of collisional zone granites: (1) Pre-collision calc-alkaline (volcanic-arc) intrusions which are mostly derived from mantle modified by a subduction component; (2) Syn-collision peraluminous intrusions (leucogranites) which may be derived from the hydrated bases of continental thrust sheets; and

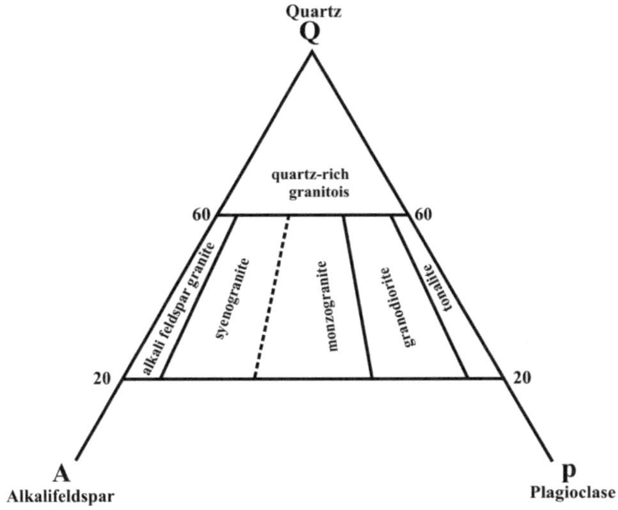

Figure 1.
A ternary QAP plot showing the relative modal proportions of quartz (Q), alkali feldspar (A) and plagioclase (P) defining the IUGS fields for granitic rocks (after Streckeisen, 1976).

(3) Late or post-collision calc-alkaline intrusions which may be derived from a mantle source but undergo extensive crustal contamination and (4) A fourth category, distinguished by Sylvester [10], consists of Alkaline granites in post - orogenic continental settings and fall within the A - type or *anorogenic* category of granitoids. **Figure 2** shows the different types of granitoid rocks related to various tectonic settings [11–14]. Sources of magma include two proposed processes: (i) fractional crystallization of mantle - derived basic magma; and (ii) partial melting of old sialic continental crust, leading either to the formation of anatectic plutons or to the contamination of mantle - derived magmas. The Cordilleran-type granites form in arc environments are characterized by magnesian compositions dictated by early crystallization of magnetite [15]. In contrast, ferroan granites are characteristic of extensional environments, where they form by partial melting or extreme differentiation of basaltic magma. *Island arc granitoids* (I-type granites) are produced above subduction zones, which characterized by large masses of batholiths made of diorites, quartz diorites, tonalites, granodiorites and minor granites. They are calc alkaline with small to high K contents and their Sr. initial ratio (Sr_i) are in the range 0.704–0.705 [16]. *S-type granites in continental collision* span compositions from granodiorites to granite, peraluminous granites without magnetite. *Collision - related leucogranites* of the High Himalayan have high (87 Sr./ 86 Sr) ratios in the range 0.743–0.762) and enriched in Rb and K but are depleted in Sr., Zr and LREE compared to Cordilleran granitoids [17]. *Intraplate (within plate) granites* can be subdivided based on tectonic criteria into intraoceanic, intracontinental and attenuated continental lithosphere. A-type granites are commonly alkali feldspar granites or syenogranites and often associated with syenites; these granites were defined by Loiselle and Wones [4] as alkaline or *anorogenic granites*. They are rich in silica and having high contents of LILE, HFSE (Zr, Ce, Nb, Hf, Ta, etc.), REE, K and Zr, but low in trace elements compatible in mafic silicates (Co, Sc, Cr, Ni) and feldspars (Ba, Sr., Eu) relative to the I-type ones. Intraplate A-type granitoids have significantly higher Fe/Mg ratios *(Fe* – number) than typical Cordilleran granitoids and accordingly fall in the “ferroan” granitoid category. Eby [18] divided A-type granites into two categories according to their Y/ Nb ratios: a group have low Y/Nb ratios and generally low initial 87 Sr./ 86 Sr. ratios that formed by differentiation of basaltic magmas of OIB – like (plume) mantle sources; the second group of A-type granitoids characterized by higher Y/Nb ratios (1.2–7) and highly variable initial 87 Sr./86 Sr. ratios. This group shows a complex petrogenetic history as having a significant mantle component or may be totally of crustal origin. **M-type granites** (plagiogranites) can be subdivided tectonically into subduction-related and subduction-unrelated on the basis of the chemistry of their associated basalts. Plagiogranites are characterized by normalized and REE patterns with a notable depletion in the most highly incompatible large-ion

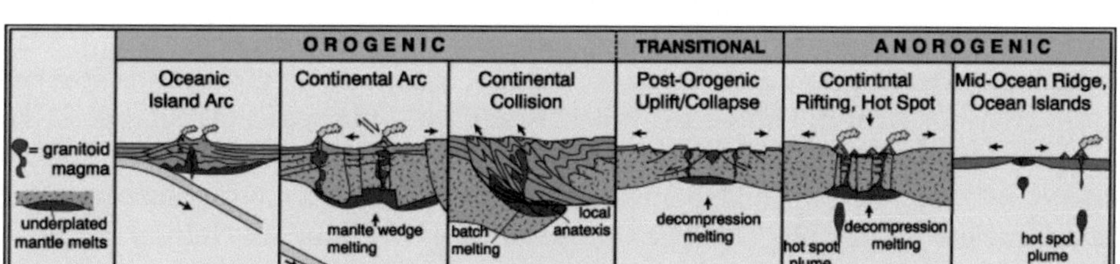

Figure 2.
A classification of Granitoid rocks based on tectonic setting (according to pitcher, 1993; Barbarin, 1990; winter, 2001).

lithophile (LIL) elements Rb, Ba and K, greater enrichment in Th, Nb and prominent negative anomalies in P and Ti.

The Nubian Shield (NS) consists mainly of juvenile Neoproterozoic crust, where the ED of Egypt constitutes its northern part. Four main rock assemblages characterize the NS, namely: a gneiss assemblage [19–24], an ophiolite and island arc assemblages [25–29], igneous intrusions and unmetamorphosed Dokhan volcanics and Hammamat molasse sediments. The Eastern Desert (ED) of Egypt include three domains: (a) the Northern Eastern Desert (NED), (b) Central Eastern Desert (CED) and (c) Southern Eastern Desert (SED), where the igneous activity began prior to 765 Ma and ended by 540 Ma [30–31]. Granitoid rocks, which are common in the Egyptian Eastern Desert (ED) and Sinai were emplaced between (*820 to 570 Ma) at various tectonic settings [32]. The granitoid rocks of the ED include both older granites, which constitute about 27% of the basement outcrop [33] and younger granites constituting approximately 30% of plutonic assemblages in Egypt [34]. The reconstruction of this chapter is based on a compiled data of published and previous geological, geochemical and geochronological studies. This to reviews and discuss the general geochemical characteristics, and classification of granitic rocks in general and in the ED of Egypt too. This could be helpful in understanding their compositional variation, tectonic environments and magma evolution.

2. Neoproterozoic granites of the Eastern Desert of Egypt

2.1 General outlines

Granites are of wide distributions between the different rock units of the Egyptian Neoproterozoic rocks, constituting approximately 60% of its plutonic assemblage [35]. The main exposures of granitic masses are concentrated in the ED of Egypt, where a huge masses of granite plutons intruded into the pre-existing country rocks (**Figure 3**; based on the Geological map of Egypt 1981; [36]). Granitoid rocks of the ED are classified into older *(750–610 Ma)* and younger *(620–540 Ma)* granites based on their composition, color, and relative age [37]. They further classified as: (1) Subduction-related older granites; (2) suture-related or Post-orogenic younger granites and (3) intraplate anorogenic younger granites [35]. *The Older granites (OG)* comprise mainly tonalites and granodiorites, and minor trondhjemite and quartz diorites. The *Younger granites* (YG) classified according to their geological setting and petrography [38] into: (i) phase I granodiorites with minor monzogranites, (ii) phase II (monzogranites and syenogranites), and phase III (alkali feldspar granites). Recently, part of the Younger granites (commonly phase III) is classified as A-type granites [39]. Stern and Hedge [31] proposed a major tectonic transition from a compressive to an extensional regime at 600 Ma. They concluded that the Egyptian granites are belonging to two main phases of the Pan-African Orogeny: (1) The older group (715–610 Ma) comprises syn- to late-tectonic granites forming batholithic masses that exhibit wide compositional variations (trondhjemites to granodiorites with minor granites), and (2) The younger group (600–540 Ma) comprises post-tectonic pluton to stock-sized granitic bodies, generally rich in K-feldspars and sometimes associated with rare metal mineralization. Bentor [32] classified the granites of the Arabian Nubian Shield into two groups: an older Syn- to late-orogenic granites (880–610 Ma), and younger post-orogenic to anorogenic granite (600–475 Ma. Loizenbauer et al. [40] identified three magmatic pulses in the Central Eastern Desert, dated as: 680 Ma; (2) 620 Ma; and (3) 585 Ma.

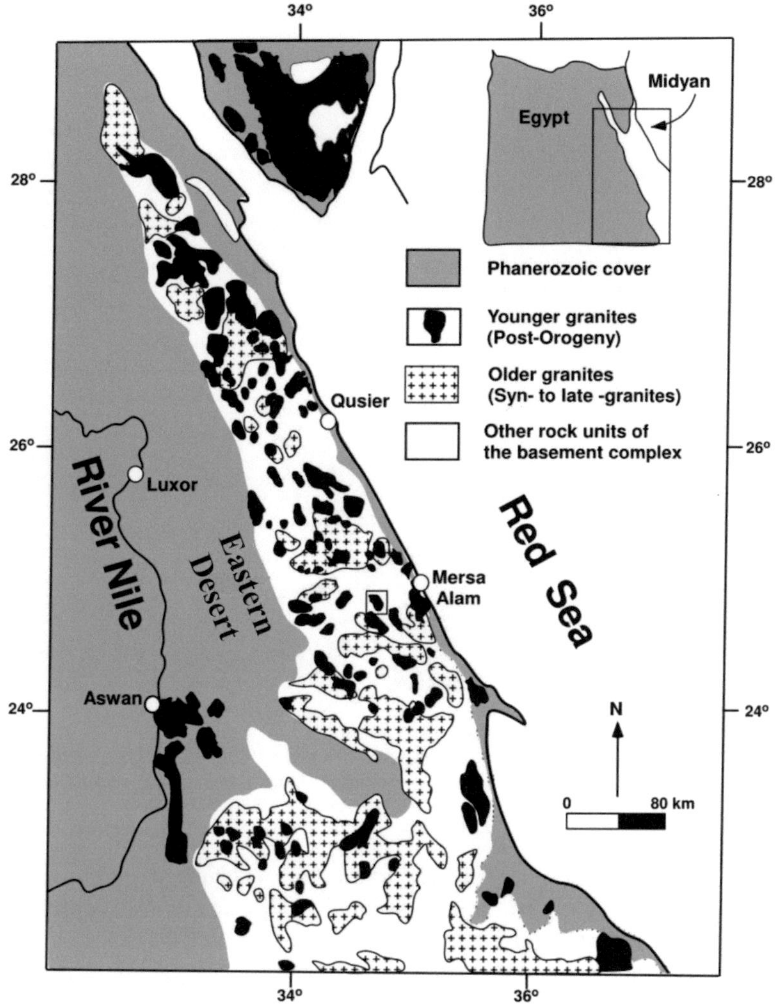

Figure 3.
Distribution of older and younger granites in the Eastern Desert of Egypt (based on the geological map of Egypt 1981; Asran, 2021).

2.2 Geological setting

The older granites of the ED constitute about 27% it's basement outcrops [33]. They occur as low relief igneous mountains (**Figure 4a**). They intrude the oldest rock types such as metavolcanics, and metasediments and have commonly have foliated margins concordant to wall rock structure. The granites include rounded to subrounded microgranular mafic enclaves of variable sizes (few cm up to meter (**Figure 4b**). The rocks are of gray to whitish gray colors, medium to coarse grained and composed mainly of tonalites and granodiorites. Examples of such granites are: Abu Ziran granites in the Central Eastern Desert and the Shaitian granite in the southern Eastern Desert.

The Younger granites are of wide distribution across the ED, where they form high relief bold mountains (**Figure 4c**). They intrude the earlier exposed rocks with sharp contacts and they commonly possess steep walls and oval or elongated outlines. They enclose mafic xenoliths, enclaves and roof pendants of country rocks with sharp contacts with the enclosing granitic rocks (**Figure 4d**). The rocks are of pink and red colors, medium to coarse grained and comprise monzogranites,

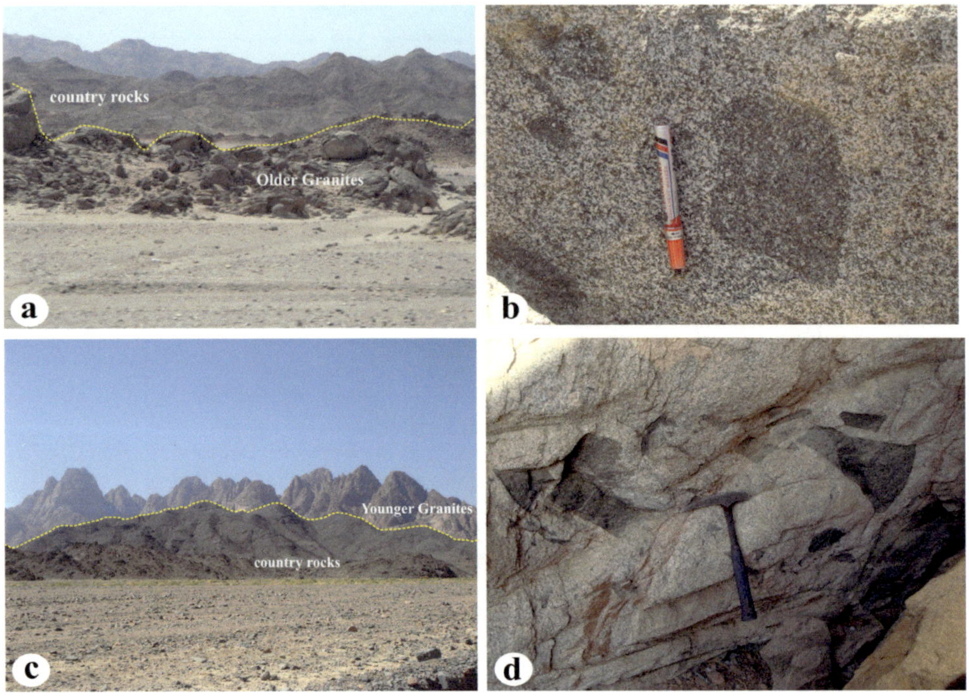

Figure 4.
Photographs showing the field observations of the Egyptian granites: (a) low relief older granite; (b) mafic enclaves with gradational contacts with the host older granite; (c) high relief younger granite (back); (d) angular xenoliths within host younger granites with sharp contacts.

syenogranite and alkali feldspar granites, even granodiorites are seldom reported. They are classified according to their geological setting and petrography into: (i) phase I granodiorites with minor monzogranites, (ii) phase II (monzogranites and syenogranites), and phase III (alkali feldspar granites) [38]. Examples of younger granites are El Sibai granite in the Central ED and Gattar granite in the northern ED.

2.3 Petrography

The Older granites are composed mainly of tonalite and granodiorites. They are consisting of quartz, plagioclase and K-feldspar as essential minerals as well as variable proportions of biotite and hornblende. Quartz occur as anhedral crystals, whereas plagioclase forms tabular, lamellar oligoclase and is occasionally zoned (**Figure 5a, b**). K-feldspar occurs as tabular microcline and/or Carlsbad orthoclase. Hornblende forms long prismatic crystals of green or yellowish green colors. Biotite occur as platy crystals of yellow or yellowish-brown colors.

The younger granites comprise monzogranite, syenogranite and alkali feldspar granite. They are composed of quartz, k-feldspar and plagioclase as essential minerals, together with subordinate biotite, muscovite, hornblende, riebeckite and arfvedsonite (**Figure 5c, d**). Quartz occurs as anhedral large crystals interstitial to other mineral constituents. Potash-feldspars include tabular orthoclase and microcline perthite crystals. Primary K-feldspar minerals are usually altered to sericite and clay minerals and corroded by quartz and plagioclase. Plagioclase is represented by subhedral tabular crystals with distinct albite-lamellar twining of albite to oligoclase composition and occasionally intergrown with the K-feldspar forming perthitic texture (**Figure 5c**). In alkali feldspar granite, quartz is actively

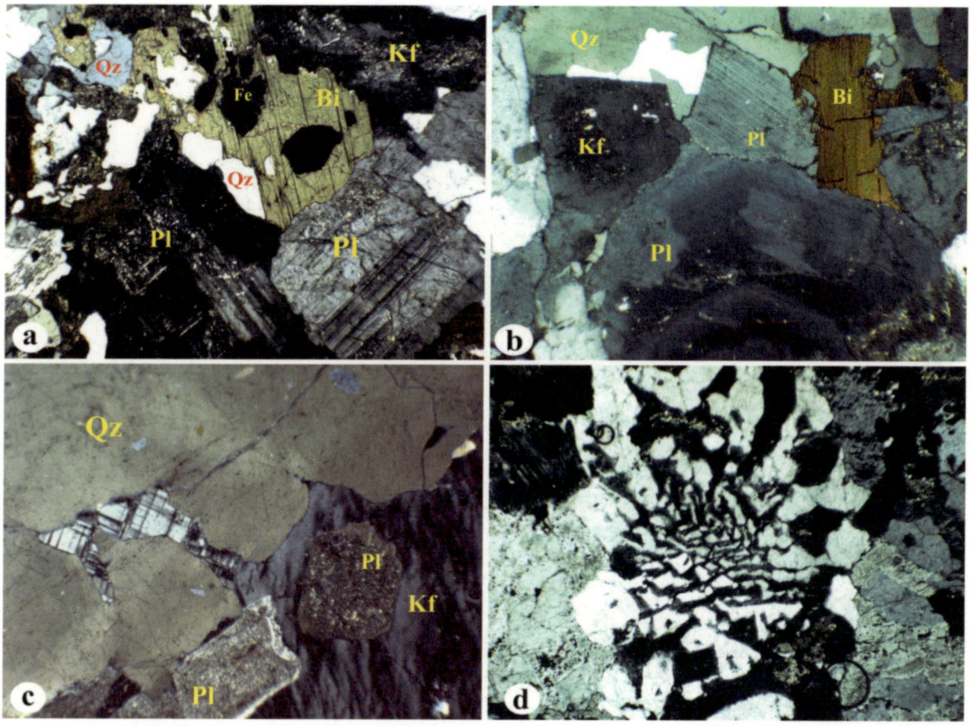

Figure 5.
Photomicrographs of the Egyptian granites: (a) tonalite consists of quartz, plagioclase, K-feldspar and biotite. Together with iron oxides; (b) granodiorite with the plagioclase showing zoning; (c) syenogranite with perthitic texture; (d) myrmekitic and micrographic textures in alkali feldspar granite (pl = plagioclase; k-f = feldspar; Q = quartz; Bi = biotite).

intergrowth by adjacent feldspar leaving blebs of quartz inside the replacing alkali feldspar forming micrographic and myrmekitic textures (**Figure 5c, d**). Biotite is subordinate and occurs as subhedral flaky or platy crystals with inclusions of zircon, apatite and sphene.

2.4 Geochemical characteristics

Compiled data of whole-rock major, trace and REE of representative granitic samples from different occurrences are presented [24, 41–44]. The OG are metaluminous to slightly peraluminous, and have calc-alkaline affinity, whereas most of YG have a peraluminous character and slightly metaluminous and peralkaline [24, 45]. According to Frost et al. [15], the majority of OG analyses fall within the magnesian field, while analyses of YG plot in ferroan field with few exceptions (**Figure 6**). The younger granites are either of calc-alkaline character, LILE-enriched, highly fractionated I-type granites, or alkaline rocks of A-type character. Commonly, the trace element characteristics for YG are marked by enriched contents of K, Rb, Ta, Th, Nb and Zr and depletion in Sr., Ba, P and Ti. They are enriched in the HFSE, Nb, Ta, Zr, Hf, Y, U, Th and total REEs and relatively depleted in Ba, Sr. with LREE-enriched to almost flat and prominent negative Eu anomaly. Noweir et al. [46] classified younger granites into four groups: (1) Group I of calc-alkaline to weakly alkaline I-type granites; (2) Group II alkaline A-type monzogranite to syenogranite; (3) Group III strongly alkaline alkali feldspar A-type granites; and (4) Group IV apogranites, enriched in Na_2O. The Phase I younger granites of Samadi and Um Rus are calc alkaline, I-type and formed in a compressional regime or continental arc setting [47].

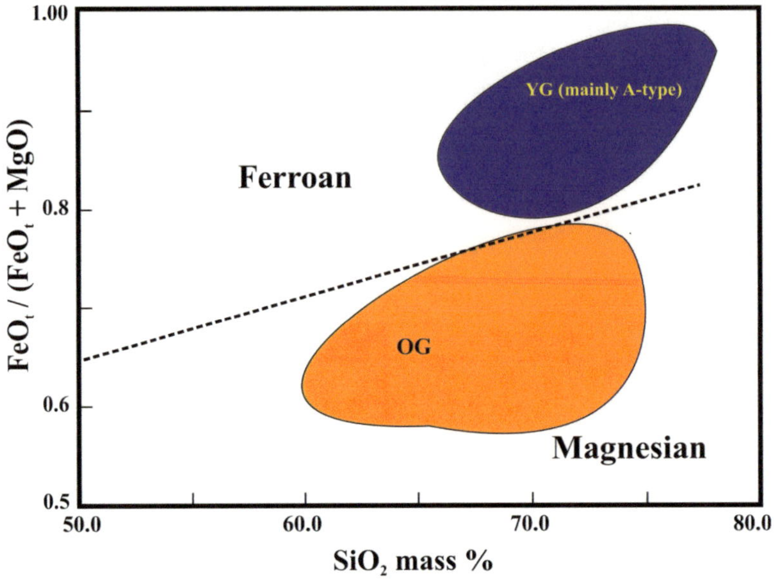

Figure 6.
Plot of Σ FeO/(Σ FeO + MgO) versus SiO_2 content (after frost et al., 2001; redrawn from Abd El Naby, 2021).

The phase-III younger granitoids (A-type) are characterized by higher SiO_2, Rb, Y, Nb and REE and lower CaO, MgO, Sr., and Ba contents than other phases of younger granites [45]. They are classified as alkaline, and peralkaline to mildly peraluminous A-type granites [48]. Commonly, the calc-alkaline rocks of YG are enriched in Sr. and Ba, but relatively depleted in Zr, Nb, Y, Zn and K in comparison with the alkaline suite granitoids (A-type granites). They are considered as fractionated I-type to A-type granites, magnesian and peraluminous to metaluminous, whereas the alkaline suite ones (A-type) are ferroan and commonly peralkaline.

The REE patterns of the granodiorites of the OG (**Figure 7a**) show enrichment in the LREE relative to HREE with small negative Eu anomaly (average Eu/Eu* = 0.674) [24]. The rocks display enrichment of LILE) HFSE, K, Sr., Rb, Ba, Th and Ta are, which are compatible with calc-alkaline trends found in island/continental arc settings. The alkali feldspar granites of the YG are characterized by LREE-enrichment or moderately fractionated LREE, flat heavy REE patterns (**Figure 7b**), and moderately to strongly negative Eu anomalies (Eu/Eu* = 0.14–0.63) [49]. REE patterns for YG show high contents of total REE and are enriched in LREEs, and depleted in HREEs, with negative Eu/Eu* [44]. Collision-related granites exhibited moderately negative Eu anomaly (Eu/Eu* = 0.093–0.436) whereas pattern of within plate granites showed moderately to strong negative Eu/Eu* values of approximately 0.026–0.211 for A1-type and 0.004–0.382 for A2-type (**Figure 6c, d**). Also, the calc-alkaline granites of the YG are characterized by higher Eu/Eu* values (0.5–1), giving rise to shallow negative anomalies. The alkaline granites appear to be differentiated from the calc-alkaline granitoids by higher RREE and much lower Eu/Eu* values [50].

2.5 Tectonic setting

Generally, the Egyptian granitoids can be classified into: (1) synorogenic calc-alkaline granitoids, (2) late- to post-orogenic calc-alkaline granitoids, and (3) post-orogenic alkaline granites [51], or moreover, granitic rocks subdivided into (1) subduction-related granites, (2) collision-related granites, (3) A2-type intraplate granites and (4) A1-type intraplate granites. Petro et al., [52] constructed

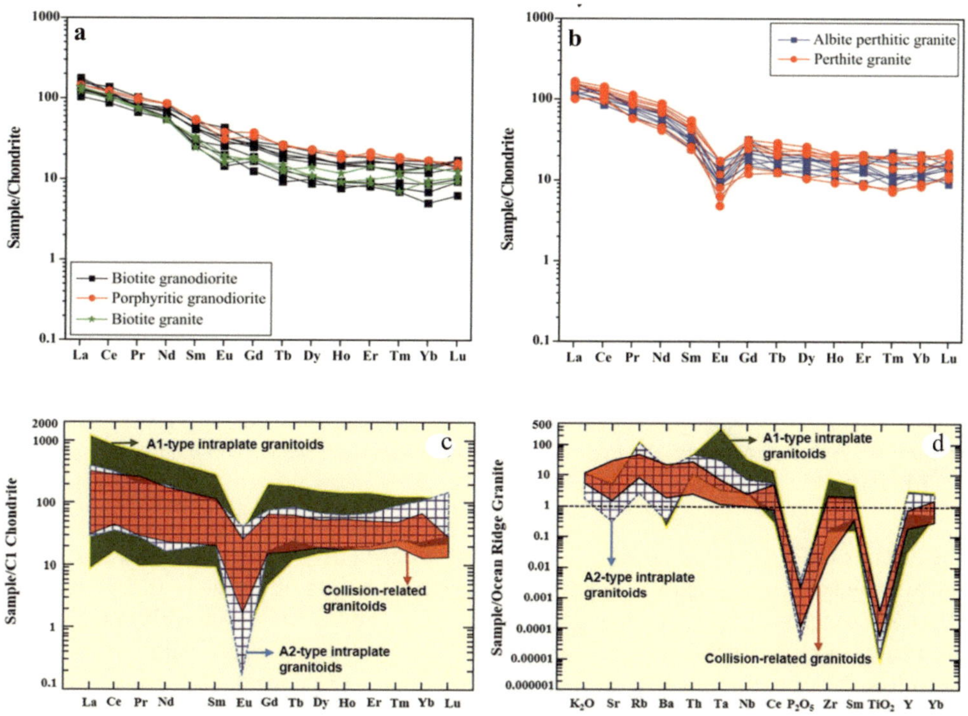

Figure 7.
(a) REE of older granites and (b) REE of younger and A-type granites (from Basta et al., 2017); (c) and (d) chondrite-normalized REE patterns (normalizing values from McDonough and sun 1995). b trace element patterns normalized to the oceanic ridge granite (from Abd El Naby, 2021).

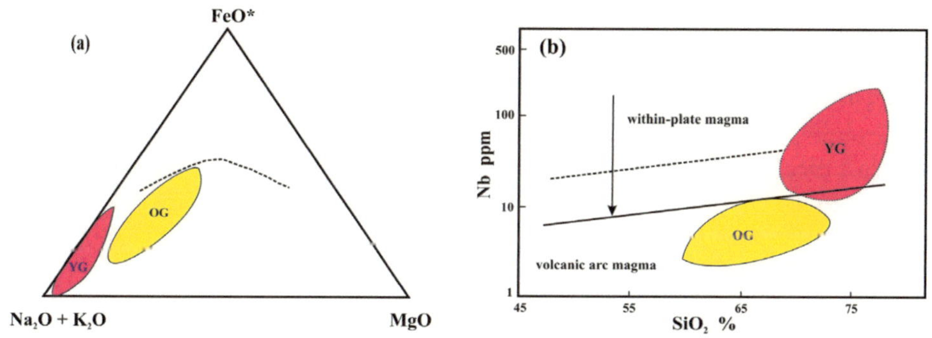

Figure 8.
(a) AFM diagram (after petro et al. (1979); (b) Nb-SiO_2 plot for the Egyptian granites (Pearce and Gale 1977; based on data from Abd El Naby, 2021).

the AFM triangular diagram to discriminate between granites formed in compressional (e.g., subduction related) and extensional regimes (e.g., intraplate environment). The data of the OG are plotted inclined to the AF side of the AFM diagram (**Figure 8a**) suggesting a compressional tectonic setting, while the data from the YG are plotted parallel to the AF line implying their extensional trend and that they formed in within plate environment. Pearce and Gale, [53] suggested Nb (ppm) versus SiO_2 (wt. %) diagram (**Figure 8b**) to discriminate tectonic environment of granites, in which the data of OG fall in the field of volcanic arc granites, whereas those of the YG plot in the field of the within–plate environment. The tectonic settings of the Egyptian granites can be deduced by using the discrimination diagrams of Pearce et al. [8]. In the Rb versus Y + Nb discrimination diagram (**Figure 9a**),

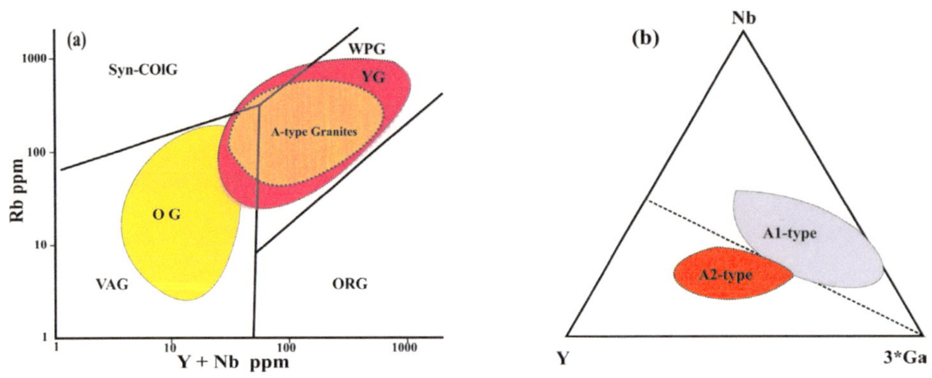

Figure 9.
(a) Yb + Nb versus Rb tectonic setting diagram of older and younger granitoid rocks of the ED of Egypt ((Pearce et al. 1984). Redrawn from El Bahariya, 2019); (b) Nb-Y-3 Ga ternary diagram (after Eby 1992). Redrawn from Abd El Naby, 2021.

the data of OG (tonalities, granodiorites) and phase I of younger granites plot within the volcanic-arc granite, while most obtained data of phase II and phase III of YG plot in the within-plate granite field. So, the YG include highly fractionated calc-alkaline I-type volcanic arc granites (postorogenic or collision-related) as well as A-type, intraplate granites. In other words, the younger granites, except phase I exhibit within plate tectonic setting. Most of the geochemical data of the A-type granites of the ED are generally consistent with a within-plate tectonic settings (**Figure 9a**). However, the majority of A-type granites such as El Atawi, Homrit are classified as A2 types with crustal sources, whose ratio Y/Nb is above 1.2 (**Figure 9b**) and appear to be formed mainly in a post-collisional setting., while A1-type granites such as Gattar, Abu Harba and Um Ara granites show Y/Nb ratio less than 1.2 and is characterized by continental intraplate environment [44]. *Alkaline granites* form mostly in intraoceanic system ocean islands or intracontinental rifts near the divergent boundaries of lithospheric plates in post-tectonic stage around 540 Ma [54]. The chemical characteristics of the A2-type intraplate granites indicate that they were derived from transitional stage between orogenic and anorogenic regimes, i.e., post-collision calc-alkaline granites. The change from compressional volcanic arc settings to extensional intraplate setting is likely to have occurred around 650–630 [55].

2.6 Origin and source of magma

Previous studies on granitoid rocks of Egypt suggested several models and scenarios for the sources of magmas in different tectonic sites. Some workers favor fractional crystallization of mantle-derived mafic magma [56]. Others suggested instead partial melting of various crustal sources [57] or combine mantle and crustal source components [58, 59]. Moreover, Lundmark et al. [54] also suggested three magmatic pulses with refined age brackets at 705–680, 660 and 635–630 Ma.

The older granites are classified as I-type granites and of volcanic-arc-granite tectonic setting [60]. Two contrasting petrogenetic models have been proposed for the origin of the OG: (1) magmas generation by partial melting of mantle sources previously metasomatized by slab-derived fluids through fractionation from mantle-derived, LILE-enriched basaltic melts in subduction settings with some crustal contamination [61–64]; (2) Fractional crystallization of mafic crustal melts generated by partial melting of amphibolite and mafic to intermediate igneous of lower crust [57]. However, the parental magmas may instead reflect mixing of mantle and crustal melts. The intermediate magma derived, by partial melting of the lower crust may promoted by heat from mantle-derived mafic melts producing a hybridized intermediate magma, that form tonalite and granodiorite by fractional crystallization during its ascent and cooling [65]. This metaluminous magnesian-rich magma eventually produce subduction-related granites (diorites, tonalities, granodiorites and minor monzogranites). Hassan and Hashad [34] proposed that emplacement of subduction-related granites was linked to three magmatic pulses at 850–800 Ma, 760–710 Ma and 630 Ma.

The magma of *the younger granites* appears to be derived by high degree of partial melting of crustal materials including: mafic lower crustal rocks [48], middle crust granodiorites-tonalites [66], metasedimentary protolith [67]. Alternatively, others-maintained mixing/mingling; of juvenile mantle-derived magma with felsic crustal melts [58]. *Post-collisional granites* of alkaline affinity are dominated at the final stages of the Pan-African orogeny [59]. The hot asthenosphere and crustal uplift causing extensive decompression melting and basaltic underplating and the heat promote partial melting of the lower crustal rocks to form the post-collisional calc-alkaline magma [68].

The alkaline A-type granites are considered as the product of either extensive fractional crystallization of mantle-derived mafic magmas [69] or partial melting crustal sources [70]. The A2-type intraplate granites appear to be derived from transitional stage between orogenic and anorogenic regimes. The A1-types are considered to be derived from differentiation of melts similar in composition to oceanic island basalts. Most of the A2 granites present in anorogenic environments and plot in the field of crust-derived A-type granites, which originate from a wide range of sources. However, the A2-type granites can also form at the sites of convergent (collision) plate boundaries [71]. Formation of A1-type intraplate granites is related to faulting [72] or magma is suggested to be produced by a melting process of lower crustal rocks in extension setting [73].

2.7 Age dating of granitoid rocks

Ages obtained for Older Granites from the Eastern Desert are younger than 750 Ma [74]. Stern and Hedge [31] gave Rb/Sr. whole-rock age of 674 ± 13 Ma for the time of intrusion of Wadi El-Miyah gray granites; and U/Pb zircon age 614 ± 8 Ma for the age of emplacement of Abu Ziran tonalite and granodiorite. Crystallization age for granodioritic batholiths from Humr Akarim and Humrat Mukbid, Eastern Desert of Egypt, is 630–620 Ma [75]. A comparable crystallization age (643 ± 9 Ma) of the Um Rus tonalite-granodiorite was determined by Zoheir et al. [76] although it is classified as phase I younger granite [38, 47].

Younger Granites (YG) include evolved island arc postorogenic and within plate anorogenic tectonic setting. The emplacement of the Egyptian late- to post-tectonic younger granites covers a time span between 600 and 550 Ma, [31] or 600 and 475 Ma [75]. The 635–580 Ma or 610 and 590 Ma period are characteristic for the postcollisional younger granite emplacement [77, 78]. They have been emplaced as two separates, but partially overlapping calc-alkaline and alkaline suites at 635–590 Ma and 608–580 Ma, respectively [79]. The Um Had granite has a U–Pb zircon age of 590–3.1 Ma [74]. Alkaline A-type granites from the ED (such as Al-Missikat, Abu Harba, and Gattar) dated ca. ~600 Ma [58].

3. Concluding remarks

Granites of different types have variable magma sources and tectonic environments. The Egyptian granites of the ED are classified into older and younger granites. The overall geological settings, geochemical characteristics and tectonic setting of the granitoid rocks of the ED of Egypt are presented and summarized in **Table 1**. *The older granites* occur as low relief igneous plutons intruded over a very long period of time from >850 to 615 Ma. Geochemistry of older granites reveals

	Older granites	Youger granites		
Geological setting	Low releif mountains, gray coulur, with gradational contacts with the country rocks and have ellipsoidal enclaves with gradational contacts with the host granite	High relief, pink or yeloowish white coulour, have sharp contacts with the country rocks and their xenoliths and inclaves are irregular in shape and have sharp contacts with the host granite.		
		Phase I	Phase II	Phase III
Granite rock types	Tonalite, granodiorite	Granodiotite, monzogranite	Syenogranite, monzogranites	alkali feldspar granites, syenogranites
Mineral composition	Quartz, plagioclase, k-feldspar, hornblende, biotite	Quartz, k-feldspar, plagioclase, biotite, hornblende	Quartz, k-eldspar, plagioclase, biotite, muscovite	Quartz, k-feldspar, plagioclase, arvedsonite, rebiekite, aegerine.
Geochemical characteristics	low-to medium-K calc-alkaline; I-type granites; subduction-related magnesian, granitoids display enrichment of LILE relative to HFSE and clear negative anomalies in Nb and Ta.	Highly fractionated calc-alkaline granites;.	alkaline to peralkaline granite, A-type; ferroan,	
			A2	A1
Tectonic setting and age dating	Compressional, Synorogenic; Volcanic arc granite or Magmatic arc/Active continental margin, *700–630 Ma (e.g., Lundmark et al. 2012; Ali et al. 2016)	Late-to post-orogenic, Collisional granite, or evolved volcani arc; Lundmark et al., 2012 590–610 Ma	Post-orogenic to anorogenic, Within plate 610 and 590 Ma (e.g., Ali et al. 2016; 590–540 Ma; Lundmark et al., 2012	Within plate, Anorogenic; 540 Ma Lundmark et al., 2012

Table 1.
Geological characteristics, geochemical features and tectonic setting of the granitoid rocks of the Eastern Desert of Egypt.

that they are metaluminous to slightly peraluminous, have calc-alkaline affinity, I-type granites and of volcanic-arc-granite tectonic setting. They are interpreted to result from melting of crustal rocks. Moreover, older I-type granites can form through fractionation from mantle-derived, LILE-enriched basaltic melts in subduction settings. *The younger granites* include highly fractionated calc-alkaline I-type granites as well as alkaline, A-type, granites. They are of peraluminous character and slightly metaluminous to peralkaline that have been emplaced as two suites: calc-alkaline (at 635–590 Ma) and alkaline (608–580 Ma) suites. Phase I younger granites (granodiorite and monzogranite) are suggested to be evolved island arc or post-orogenic subduction-related plutons, whereas Phase III and most of phase II exhibit within plate tectonic setting.

The diversity of granitoids rocks of the ED of Egypt and the variability of their chemical composition are controlled by the chemical composition of the source (crustal and mantle), P–T conditions, degree of partial melting, anatexis, as well as the extent of fractionation processes and crustal contamination. This implying various tectonic settings for the magma generation and emplacement. Granitic magmas formed during different tectonic regimes in compressional volcanic arc to extensional within-plate. Thus, the granitic rocks of the ED are geochemically diverse and their origin and tectonic evolution are still controversial.

Author details

Gaafar A. El Bahariya
Geology Department, Tanta University, Tanta, Egypt

*Address all correspondence to: gbahariya@yahoo.com; gaafar.elbahariya@science.tanta.edu.eg

References

[1] Pitcher, W. S., 1997. The Nature and Origin of Granite. Chapman & Hall, DOI 10.1007/978-94-011-5832-9, 387 p.

[2] Streckeisen, A.L., 1976. To each plutonic rock its proper name. Earth Science Review **12**: 1-33.Sylvester, P. J., 1989. Post - collisional alkaline granites. Journal of Geology **97**: 261-280.

[3] Chappell, B. W. and White, A. J. R., 1974. Two contrasting granite types. Pacific Geology **8**: 173-174.

[4] Loiselle, M.C. and Wones, D. R., 1979. Characteristics and origin of anorogenic granites. Geol. Soc. Am. Abstract, Programs. **11**: 468

[5] Shand, S. J., 1927. The Eruptive Rocks. John Wiley. New York.

[6] Maniar, P.D. and Piccoli, P.M., 1989. Tectonic discrimination of granitoids. Geological Society of America Bulletin **101**: 635-643.

[7] Whalen, J.B., Currie, K.L. and Chappell, B.W., 1987. A-type granites: geochemical characteristics, discrimination and petrogenesis. Contributions to Mineralogy and Petrology **95**: 407-419.

[8] Pearce J.A, Harris N.B.W and Tindle A.G., 1984. Trace element discrimination diagrams for the tectonic interpretation of granitic rocks. Journal of Petrology **25**: 956-983.

[9] Harris N, Pearce J.A, Tindle A.G., 1986. Geochemical Characteristics of Collision-Zone Magmatism. In: Coward MP, Ries AC (eds) Collision tectonics, vol 19, Special Publications Geological Society, London, pp 67-81. https://doi.org/10.1144/GSL.SP.1986.019.01.04

[10] Sylvester, P. J., 1989. Post - collisional alkaline granites. Journal of Geology **97**: 261-280.

[11] Barbarin, B., 1990. Granitoids: main petrogenetic classifications in relation to origin and tectonic setting. Geological Journal **25**: 227-238.

[12] Picher, W.S., 1997. The Nature and Origin of Granite, Chapman & Hall, 387p

[13] Winter, J.D., 2001. An Introduction to Igneous and Metamorphic Petrology. Prentice Hall. 697p

[14] Gill, R., 2010. Igneous Rocks and Processes: A Practical Guide. John Wiley & Sons-Blackwell, ISBN 978-1-4443-3065-6, 428p

[15] Frost, B.R, Barnes, C.G, Collins, W.J, Arculus, R.J, Ellis D.J, Frost C.D.; 2001. A geochemical classification for granitic rocks. Journal of Petrology **42**: 2033-2048.

[16] Nedelec, A. and Bouchez, J.L., 2015. Granites: petrology, structure, geological setting, and metallurgy. Oxford, Ltd, Croydon, ISBN 978-0-19-870561-1, 335p

[17] Harris, N., Ayres, M. and Massey, J.; 1995. Geochemistry of granitic melts produced during the incongruent melting of muscovite – implications for the extraction of Himalayan leucogranite magmas, Journal of Geophysical Research **100**: 15767-15777.

[18] Eby, G.N., 1990. The A-type granitoids: a review of their occurrence and chemical characteristics and speculations on their petrogenesis. Lithos **26**: 115-134.

[19] El-Gaby S, List FK and Tehran R.; 1988. Geology, evolution and metallogenesis of the Pan-African Belt in Egypt. In: El Gaby S, Greiling RO, editors. The Pan African Belt of Northeast Africa and Adjacent Areas. Braunschweig/Wiesbaden: Vieweg & Sohn, 17-68.

[20] El Bahariya G.A and Abd El-Wahed M.A., 2003. Petrology, mineral chemistry and tectonic evolution of the northern part of Wadi Hafafit area, Eastern Desert of Egypt. In: 3rd international conference on the geology of Africa. Assiut University, 2, 201-231.

[21] El Bahariya G.A.; 2008a. Geology and petrology of Neoproterozoic syntectonic anatectic migmatites around Wadi Abu Higlig, Hafafit region, Eastern Desert, Egypt. Egyptian Journal of Geology 5225-5254.

[22] El Bahariya G.A.; 2009. Geology and petrogenesis of Neoproterozoic migmatitic rock association, Hafafit Region, Eastern Desert, Egypt: Implications for syntectonic anatectic migmatites. Lithos **113**: 465-482.

[23] Andresen A, El-Rus MMA, Myhre PI, Boghdady GY., 2009. Corfu F. U-Pb TIMS age constraints on the evolution of the Neoproterozoic Meatiq Gneiss Dome, Eastern Desert, Egypt. International Journal of Earth Sciences **98**: 481-497.

[24] El Bahariya, G. A., 2019. Geochemistry and Tectonic Setting of Neoproterozoic Rocks from the Arabian-Nubian Shield: Emphasis on the Eastern Desert of Egypt, In: Applied Geochemistry with Case Studies on Geological Formations, Exploration Techniques and Environmental Issues, Mazadiego, L. F., Garcia, E. D.M., Barrio-Parra, F. and Izquierdo-Díaz, M; (eds) IntechOpen, DOI: 10.5772/intechopen.82519

[25] Akaad MK, Noweir AM, Abu El Ela AM, El Bahariya GA., 1997. The Um Esh olistostromal melange, Qift-Quseir region, central Eastern Desert, Egyptian Journal of Geology **41**: 465.

[26] El Bahariya GA., 2006. Petrology, mineral chemistry and metamorphism of two pan-African ophiolitic metagabbro occurrences, Central Eastern Desert, Egypt. Egyptian Journal of Geology **50**: 183-202.

[27] El Bahariya G.A., 2007. Geology, compositional variation and petrogenesis of possible MORB-type ophiolitic massive and pillowed metabasalts from the Pan-African belt, Eastern Desert, Egypt. Egyptian Journal of Geology **51**: 41-59.

[28] El Bahariya G.A.; 2018a Geology, geochemistry, and source characteristics of Neoproterozoic arc-related clastic metasediments, Central Eastern Desert, Egypt. Arabian Journal of Geoscience **11**: 87.

[29] El Bahariya G.A.; 2018b. Classification of the Neoproterozoic ophiolites of the Central Eastern Desert, Egypt based on field geological characteristics and mode of occurrence. Arabian Journal of Geoscience **11**: 313.

[30] El-Ramly M.F., 1972. A new geological map for the basement rocks in the Eastern and Southwestern Deserts of Egypt. Ann. Geol. Surv. Egypt **2**: 1-18.

[31] Stern R.J and Hedge C.E; 1985. Geochronologic constraints on late Precambrian crustal evolution .7in the Eastern Desert of Egypt. American Journal of Science **285**: 7-127.

[32] Bentor Y.K., 1985. The crustal evolution of the Arabo-Nubian massif with special reference to the Sinai Peninsula. Precambrian Research **28**: 1-74.

[33] Stern R.J., 1979. Late Precambrian crustal environments as reconstructed from relict igneous minerals, Central Eastern Desert of Egypt. Ann. Geol. Surv. Egypt **9**, 9-13.

[34] Hassan M.A and Hashad A.H., 1990. Precambrian of Egypt. In: Said R (ed) The geology of Egypt. Balkema, Rotterdam, 201-245.

[35] Hussein A.A, Ali M.M and El-Ramly M.F; 1982. A proposed new classification of the granites of Egypt. Journal of Volcanological and Geothermal Research **14**: 187-198.

[36] Asran, A.M., 2021. Rock Succession in Egyptian Nubian Shield. In: Hamimi Z., Arai S., Fowler AR., El-Bialy M.Z. (eds) The Geology of the Egyptian Nubian Shield. Regional Geology Reviews. Springer, Cham. http://doi-org-443.webvpn.fjmu.edu.cn/10.1007/978-3-030-49771-2-6, 53-81.

[37] Akaad MK, Noweir AM., 1980. Geology and lithostratigraphy of the Arabian Desert orogenic belt of Egypt Lat. 25° 35′ and 26° 30′ N. Bull. Inst. Appl. Geol. King Abdul Aziz Univ. Jeddah, 127-135.

[38] Akaad MK, Noweir A, Koth H., 1979. Geology and petrographically of the granite association of the Arabian Desert, orogenic belt of Egypt between lat. 25° 35′ and 26° 30′. Delta J. Sci. **3**: 107-151.

[39] Liégeois J.P and Stern R.J., 2010. Sr–Nd isotopes and geochemistry of granite–gneiss complexes from the Meatiq and Hafafit domes, Eastern Desert, Egypt: no evidence for pre-Neoproterozoic crust. Journal of African Earth Science **57**: 31-40.

[40] Loizenbauer J, Wallbrecher E, Fritz H, Neumayr P, Khudeir A.A and Kloetzli U., 2001. Structural geology, simple zircon ages and fluid inclusion studies of the Meatiq metamorphic core complex: Implications for Neoproterozoic tectonics in the Eastern Desert of Egypt. Precambrian Research **110**: 357-383.

[41] Saleh G.M, Dawood Y.H and Abd El-Naby H. H., 2002. Petrological and geochemical constraints on the origin of the granitoid suite of the Homret Mikpid area, south Eastern Desert, Egyptian Journal of Mineralogical and Petrological Sciences **97**: 47-58.

[42] El-Nisr S.A and El-Sayed M.M., 2002. The role of fractional crystallization and assimilation in the evolution of the zoned Mukhattata pluton, Eastern Desert, Egypt. Chemie der Erde **62**: 216-236.

[43] Gharib M.E and Ahmed A.H., 2012. Late Neoproterozoic volcanics and associated granites at Wadi Ranga, south Eastern Desert, Egypt: a transition from subduction related to intra-arc magmatism. Lithos **155**: 236-255.

[44] Abd El-Naby, H.H., 2021. The Egyptian Granitoids: an up-to-date Synopsis In: Hamimi Z., Arai S., Fowler AR., El-Bialy M.Z. (eds) The Geology of the Egyptian Nubian Shield. Regional Geology Reviews. Springer, Cham. http://doi-org-443.webvpn.fjmu.edu.cn/10.1007/978-3-030-49771-2_6, 239-294.

[45] Finger F, Dorr W, Gerdes A, Gharib M, Dawoud M; 2008. U-Pb zircon ages and geochemical data for the monumental granite and other granitoid rocks from Aswan, Egypt: Implications for the geological evolution of the Western margin of the Arabian Nubian shield. Mineralogy and Petrology **93**: 153-183.

[46] Noweir A.M, Sewifi B.M and Abu El Ela A.M., 1990. Geology, petrography, geochemistry and petrogenesis of the Egyptian younger granites. Qatar Univ. Sci. Bull. **10**: 363-393.

[47] El Bahariya, G.A.,Abu anbar,M.M. and El Galy, M.M., 2008b. Petrology and geochemistry of Um Rus and Samadi granites, central Eastern Desert, Egypt: Implications for I-type granites of variable magma sources. Ann. Geol. Surv. Egypt 1-15.

[48] Farahat E.S, Mohamed H.A, Ahmed A.F, El Mahallawi M.M; 2007. Origin of I- and A-type granitoids from the Eastern Desert of Egypt: Implications for crustal growth in

the northern Arabian-Nubian shield. Journal of African Earth Science **49**: 43-58.

[49] Basta F.F, Maurice A.E, Bakhit B.R, Azer M.K, El-Sobky A.F., 2017. Intrusive rocks of the Wadi Hamad Area, North Eastern Desert, Egypt: Change of magma composition with maturity of Neoproterozoic continental island arc and the role of collisional plutonism in the differentiation of arc crust. Lithos **288-289**: 248-263.

[50] El-Bialy, M. Z.; 2020. Precambrian Basement Complex of Egypt. In Hamimi et al. (eds), The geology of Egypt. Springer, Nature Switzerland AG, 37-72.

[51] Azer, M.K. and Asimo, B.D., 2021. Petrogenetic Evolution of the Neoproterozoic Igneous Rocks of Egypt. In: Hamimi Z., Arai S., Fowler AR., El-Bialy M.Z. (eds) The Geology of the Egyptian Nubian Shield. Regional Geology Reviews. Springer, Cham. http://doi-org-443.webvpn.fjmu.edu.cn/10.1007/978-3-030-49771-2_6, 343-372

[52] Petro, W.L. Vogel, T.A. and Wilbard, J.T., 1979. Major element chemistry of plutonic rock suites from compressional and extensional plate boundaries. Chemical Geology **26**: 217-235.

[53] Pearce J.A and Gale G.E., 1977. Identification of ore deposition environment from trace element geochemistry. Geol. Soc. London Spec. Publ. **7**: 4-24.

[54] Lundmark A.M, Andresen A, Hassan M.A, Augland L.E and Boghda, G.Y. 2012. Repeated magmatic pulses in the East African Orogen in the Eastern Desert, Egypt: An old idea supported by new evidence. Gondwana Research **22**: 227-237.

[55] Hassan, M, Abu-Alam, T.S, Hauzenberger, C, Stüwe, K.; 2016. Geochemical signature variation of Pre-, Syn- and Post-shearing intrusives within the Najd Fault System of Western Saudi Arabia. Lithos **263**: 274-291.

[56] Maurice AE, Basta FF and Khiamy AA., 2012. Neoproterozoic nascent island arc volcanism from the Nubian Shield of Egypt: magma genesis and generation of continental crust in intra-oceanic arcs. Lithos **132-133**: 1-20.

[57] Moghazi A.M, Mohamed F.H and Kanisawa S., 1999. Geochemical and petrological evidence of calc-alkaline and A-type magmatism in the Homrit Waggat and El-Yatima areas of eastern Egypt. Journal of African Earth Science 2535-2549.

[58] Ali KA, Zoheir BA, Stern RJ, Andresen A, Whitehouse MJ, Bishara WW., 2016. Lu-Hf and O isotopic compositions on single zircons from the North Eastern Desert of Egypt, Arabian- Nubian shield: Implications for crustal evolution. Gondwana Research **32**: 181-192.

[59] Khalil A.E.S, Obeid M.A, Azer M.K and Asimow P.D., 2018. Geochemistry and petrogenesis of post-collisional alkaline and peralkaline granites of the Arabian-Nubian Shield: a case study from the southern tip of Sinai Peninsula Egypt. International Geological Review **60**: 998-1018.

[60] Moghazi A.M.; 2002. Petrology and geochemistry of Pan-African granitoids, Kab Amiri area, Egypt—Implications for tectonomagmatic stages in the Nubian shield evolution. Mineralogy and Petrology **75**: 41-67.

[61] Wilson, M., 1994. Igneous petrogenesis. Chapman & Hall, London, p 466.

[62] Pearce J.A and Peate D.W., 1995. Tectonic implications of the composition of volcanic arc magmas. Ann. Rev. Earth Planet. Sci. 23, 251-285.

[63] El-Bialy M.Z, Streck M.J; 2009. Late Neoproterozoic alkaline magmatism in the Arabian-Nubian Shield: the post collisional A-type granite of Sahara-Umm Adawi pluton, Sinai Egypt. Arabian Journal of Geosciences **2**: 151-174.

[64] El-Bialy M.Z, Omar M.M; 2015. Spatial association of Neoproterozoic continental arc I-type and post-collision A-type granites in the Arabian-Nubian Shield: The Wadi Al-Baroud Older and Younger Granites, North Eastern Desert, Egypt. Journal of African Earth Sciences **103**: 1-29.

[65] Abdelfadil K.M, Obeid M.A, Azer M.K, Asimow P.D., 2018. Late Neoproterozoic adakitic lavas in the Arabian-Nubian shield, Sinai Peninsula Egypt. Journal of Asian Earth Sciences **158**: 301-323.

[66] Mohamed F.H and El-Sayed M.M., 2008. Post-orogenic and anorogenic A-type fluorite-bearing granitoids, Eastern Desert, Egypt: petrogenetic and geotectonic implications. Chemie der Erde **68**: 431-450.

[67] Moghazi A.M, Hassanen M.A, Mohamed F.H and Ali S., 2004. Late Neoproterozoic strongly peraluminous leucogranites, South Eastern Desert, Egypt: petrogenesis and geodynamic significance. Mineralogy and Petrology **81**: 19-41.

[68] Avigad D, Gvirtzman Z., 2009. Late Neoproterozoic rise and fall of the northern Arabian-Nubian shield: the role of lithospheric mantle delamination and subsequent thermal subsidence. Tectonophysics **477**: 217-228.

[69] Bonin B., 2007. A-type granites and related rocks: Evolution of a concept, problems and prospects. Lithos **97:** 1-29.

[70] Ali B.H, Wilde S.A, Gaber, M.M., 2009a. Granitoid evolution in Sinai, Egypt, based on precise SHRIMP U-Pb zircon geochronology. Gondwana Research **15**: 38-48.

[71] Grebennikov AV; 2014. Erratum to "A-type granites and related rocks: petrogenesis and classification". Russian Geology and Geophysics **55**: 1353-1366.

[72] Dobretsov N.L., 2003. Mantle plumes and their role in the formation of anorogenic granites. Geology and Geophysics **44**: 1243-1261.

[73] Abu-Alam TS and Stüwe K., 2009. Exhumation during oblique transpression: the Feiran-Solaf region, Egypt. Journal of Metamorphic Geology **27**: 439-459.

[74] Andresen A, El-Rus MMA, Myhre PI, Boghdady GY., 2009. Corfu F. U-Pb TIMS age constraints on the evolution of the Neoproterozoic Meatiq Gneiss Dome, Eastern Desert, Egypt. International Journal of Earth Sciences **98**: 481-497.

[75] Ali KA, Andresen A, Stern RJ, Manton WI, Omar SA, Maurice AE., 2012. U-Pb zircon and Sr-Nd-Hf isotopic evidence for a juvenile origin of the c 634 Ma El-Shalul Granite, Central Eastern Desert, Egypt. Geological Magazine **149**: 783-797.

[76] Zoheir, B, Goldfarb. R, Holzheid. A, Helmy H and El Sheikh. A., 2019. Geochemical and geochronological characteristics of the Um Rus granite intrusion and associated gold deposit, Eastern Desert, Egypt. Geosci Front. https://doi.org/10.1016/j.gsf.2019.04.012

[77] Ali KA, Stern RJ, Manton WI, Kimura J-I, Khamees HA., 2009b. Geochemistry, Nd isotopes and U-Pb SHRIMP dating of Neoproterozoic volcanic rocks from the Central Eastern Desert of Egypt: new insights into the _750 Ma crust-forming event. Precambrian Research **171**: 1-22.

[78] Andresen A, Abu El-Enen MM, Stern RJ, Wilde SA, Ali KA., 2014. The

Wadi Zaghra metaconglomerates of Sinai, Egypt: new constraints on the Ediacaran tectonic evolution of the northernmost Arabian-Nubian Shield. International Geological Review **56**: 1020-1038.

[79] Morag N, Avigada D, Gerdesb A, Belousovac E and Harlavand Y., 2011. Crustal evolution and recycling in the northern Arabian-Nubian Shield: new perspectives from zircon Lu–Hf and U-Pb systematics. Precambrian Research **186**: 101-116.

Chapter 3

Microstructure Features in Paleo and Neoproterozoic Granitic Rocks, Southeastern Region of Brazil

Leonardo Gonçalves and Cristiane Castro Gonçalves

Abstract

In this section we present the characteristic microstructures of granitic rocks and mylonites exposed in the southeastern region of Brazil, belonging to the geological context of the Neoproterozoic Araçuaí orogen and the Paleoproterozoic Mineiro belt. The studied rocks include most biotite and biotite hornblende tonalite and granodiorite and minor granite showing similar signatures to magnesian, calc-alkalic, metaluminous to peraluminous magmas. As a representative of continental or oceanic magmatic arcs, these granitoids are composed of plagioclase, K-feldspar, quartz, biotite, hornblende, garnet, muscovite, and sillimanite as major phases and pyroxene, apatite, epidote, allanite, titanite, zircon, monazite, rutile, magnetite, ilmenite, hematite, Fe-Ti oxides, pyrite, chalcopyrite, pyrrhotite, and graphite as accessory minerals. Typical magmatic textures include concentric and oscillatory zoning and exsolution, while isolated deformation microstructures and biotite replacing hornblende represent late-magmatic features. On the other hand, wide-spread undulose extinction, mechanical twinning, deformation bands, and recrystallized grains register crystal-plastic deformation under a variable temperature condition. Pyroxene and garnet fish and quartz ribbons are microstructures that record the temperature peak, under granulite facies in mylonites. Indeed, unshaped quartz grains and extremely lobated phase boundaries indicate the partial melting process.

Keywords: microstructure, granitic and mylonitic rocks, magmatic arcs, Araçuaí orogen, Mineiro belt

1. Introduction

Plate tectonic movement has been known at least for the last 50 years as a key mechanism responsible for remodeling the Earth surface to what we observe nowadays. It is linked to processes that occur in the most external layer of the planet and also to those acting in its inner parts including the terrestrial magnetic field created during the rotation of the liquid lower mantle against the solid core [1]. Someone studying the evolution of the Earth and trying to understand why intelligent life is present only here thus might keep in mind the role played by plate tectonics. Many researchers have attributed the success of the Earth as a living mechanism due to processes involved in the plate movements, such as those triggered by consumption and creation of crust in subduction zones and mid-oceanic ridges, respectively, capable of providing the fuel for life (oxygen and water, among other gases). Importantly, these processes have no longer been documented in other planets in the solar system [2] and where there is consumption or production of crust is the place quartz-feldspathic rocks can be found. In subduction zones, magnesian, calc-alkalic granitic rocks are predominantly found, while the formation of plagiogranite is linked to mid-oceanic ridge areas.

The lithosphere—the skin layer of the Earth—comprising the upper mantle and the whole crust, both of continental and oceanic origin, plays a critical role for the evolution of the continents. The geological processes involved in this hard rheological structure that moves from time to time are very unique and essential to constrain and reconstruct the geological history of a region. Particularly, regarding the continental crust that covers around 30% of the Earth surface today, become very important the study of its main composite, the granitic rocks. Essentially composed of quartz and feldspar minerals, these rocks could tell us their complete history, from the initial crystallization to metamorphic and post-deformational events, and thus reveal the behavior of a major part of the lithosphere in a geological area ([2] and references therein).

In this sense, we present the main characteristic microstructures of granitoids and mylonites exposed in the southeastern region of Brazil. The granitic rocks, belonging to the Paleoproterozoic Mineiro belt and the Neoproterozoic Araçuaí orogen (**Figure 1**), regions which could be used as two natural laboratories, provide preserved igneous as well as deformational features that show how

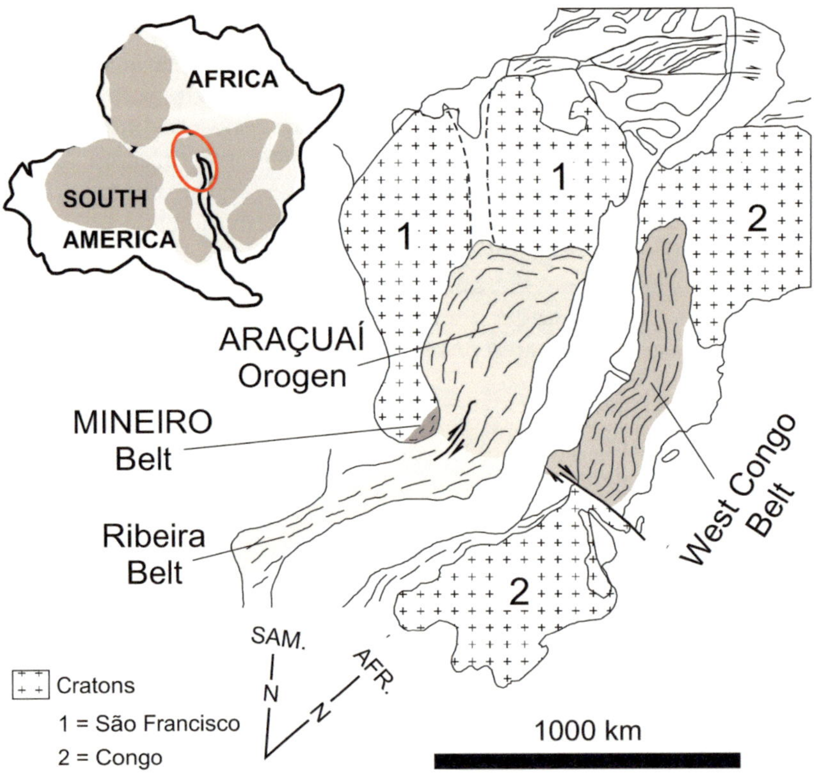

Figure 1.
Upper left image showing the geotectonic scenario of West Gondwana (modified from [3]). Dark gray, cratons; light gray, orogenic belts. Relative positions of the Paleoproterozoic Mineiro belt and Araçuaí orogen presently exposed along south American and African margins of the Atlantic (modified from [4]).

small structures tell a big history, involving, for example, the deformation of the lithosphere in orogenic systems.

2. Geological context

2.1 The Paleoproterozoic Mineiro belt

The Mineiro belt represents a segment of a Paleoproterozoic accretionary orogen preserved in the southern part of the São Francisco Craton, southeastern Brazil. It forms a region of approximately NE–SW trending with 180-km-long and 50-km-wide dimensions [5]. This region consists mainly of orthogneisses and metavolcanic sedimentary rocks intruded by Paleoproterozoic granitic rocks and records metamorphic conditions ranging from greenschist to amphibolite facies [6]. Geochronological data available in the literature spans its evolution from Siderian to Rhyacian times, ca. 2.46 to 2.09 Ga (see [7] for a comprehensive review).

Regarding the granites exposed in the region, they consist mainly of biotite hornblende tonalite and biotite trondhjemite, with common occurrence of dioritic and/or tonalitic enclaves. They show a magnesian, metaluminous to slightly peraluminous character, and many of them have TTG-like affinity [8, 9]. Indeed, these granitic bodies show chemical and isotopic characteristics similar to rocks formed in a subduction-related magmatic environment, being interpreted as both oceanic and continental magmatic arcs, which were responsible for building this orogenic system [7, 9].

2.2 The Neoproterozoic Araçuaí orogen

The Araçuaí-West Congo orogenic system is developed between the São Francisco and Congo cratons during the Brasiliano-Pan-African event [10]. The Brazilian counterpart, or Araçuaí, represents a segment of this Neoproterozoic collisional orogen located eastward to the São Francisco Craton, central eastern Brazil. It forms an area of approximately N-S trending with 700-km-long and 400-km-wide dimensions that curves to NE–SW in its southern part toward the Neoproterozoic Ribeira belt [11]. The whole area encompasses rift-related volcanic and sedimentary units, ophiolite slices, arc-related volcanic-sedimentary rocks, and granites that record regional metamorphism not higher than amphibolite facies conditions (see [12] for a comprehensive review).

Regarding the granites, they mark the pre-, syn-, and post-collisional stages of the Araçuaí orogen evolution and evolved from the Ediacaran to the Cambrian times, between ca. 630 and 480 Ma [13]. Composed mostly of biotite hornblende tonalites, hornblende biotite granodiorites, two mica, and garnet sillimanite granites, with local occurrence of mafic and intermediate enclaves, they show slightly peraluminous signature, with minor metaluminous terms. With magnesian and calc-alkaline characteristics, these granitic rocks show chemical and isotopic features compatible with subduction-related magmatism and peraluminous collisional type granites, such as those from Himalayan chain. In addition, some of them, mostly the younger ones, present mantle-like signature interpreted to be formed during mantle decompression or extensional orogenic collapse [13–15].

3. Microstructures and their implications

Granitoids or granitic rocks are defined as felsic plutonic rocks, composed of quartz, plagioclase, and K-feldspar as major phases, and vary from tonalite (Ca-rich member) to K-feldspar granite, including the Na-rich term, trondhjemite, and also Si-poor members. These rocks can be produced by a combination of magmatic processes, with mantle contribution or exclusively related to crustal sources, and thus could reveal a complex geological history in a region. Such a conundrum must be kept in mind when using granitoid and mylonite features to unreveal the geological processes of an area.

Macro- and microstructures show the interaction with adjacent rocks, regarding emplacement mechanisms, internal crystallization and magma mixing processes, and also post-crystallization metamorphism, deformation, or even partial melting. For example, characterizing the microstructures in granites is the way to recognize how one of the main constituents of the lithosphere is formed and behaves through the geological time. Keeping in mind the understanding of Earth dynamics based on rock rheology, it is the way to better constrain the mechanical behavior of the lithosphere and its variations with depth, which should be closely linked to plate tectonics [2]. Thus, we show representative microstructures preserved in the granitic and mylonitic rocks from the Paleoproterozoic Mineiro belt and the Neoproterozoic Araçuaí orogen, interpreted to be formed during collisional and subduction-related events, thus building either continental or oceanic magmatic arcs.

These rocks are composed, in general, of plagioclase, K-feldspar, quartz, biotite, hornblende, garnet, muscovite, and sillimanite as major phases and pyroxene, apatite, epidote, allanite, titanite, zircon, monazite, rutile, magnetite, ilmenite, hematite, Fe-Ti oxides, pyrite, chalcopyrite, pyrrhotite, and graphite as accessory minerals and have been largely investigated in previous studies (see [8, 9, 14, 16, 17] for details). The characteristic microstructures of quartz and feldspars, the main phases and the most studied minerals among those observed, are shown as the record of magmatic, late-magmatic, deformation, and partial melting processes. Pyroxene, hornblende, and garnet microstructures are presented as features that allow constraining thermodynamic boundary conditions for the recorded processes, from igneous to metamorphic rocks.

Feldspar features may represent key ways to better understand the magmatic and post-magmatic features [18–22], among which may be highlighted the different zoning patterns, exsolution, corrosion features, and those regarding deformation processes. Even a normal zoning, with grains showing Ca-rich cores and Na-rich rims, can show a more complex magmatic process when the Na enrichment from the core is not progressive, and there is an abrupt transition to a Na-rich boundary (**Figure 2A**). It should not represent a progressive cation exchange during a continuous crystallization process (fractioning) and may be related to the partial melting process if the Ca content in the grain core is in disagreement with what is expected for the An content in plagioclase from granites [23]. In this case, the high An content core may regard the magma source. It possibly reflects a low-temperature granite, when the temperature condition that led to the melting process was not high enough to consume the high An plagioclase from the source, in this case, possibly a mafic rock. Instead of normal zoning, euhedral plagioclase grains from the Alto Maranhão Suite, Mineiro belt, show an oscillatory changing in their composition evidencing cycles of Ca input in the system. Although it may be confirmed by microchemical analysis, the occurrence of Ca-rich second phases highlights such zoning pattern (**Figure 2B**). Besides, coarse plagioclase grains are, sometimes, rounded, showing corroded borders (**Figure 2C**). These features reflect the magma mingling processes, which, in that case, is supported by macro- and microstructures related to the occurrence of dioritic enclaves [8, 17].

Perthite and anti-perthite (**Figure 2D, E**) are widespread features and can be observed in granites from both geological regions. Besides the compositional

Figure 2.
Cross-polarized optical photomicrographs showing typical igneous textures—Granitic rocks. (A, B) Normal concentric zoning in plagioclase: (A) abrupt transition from a Ca-rich core, with Ca content anomalously high, to boundary progressively Na richer; (B) cyclical interchange between Ca- and Na-rich strips. (C) Coarse rounded plagioclase grain surrounded by biotite. Exsolutions in feldspar: Perthitic (D) and anti-perthitic (E) grains. (F) Myrmekite texture records the interaction between plagioclase and K-feldspar. (A, D–F), Araçuaí orogen; (B, C), Mineiro belt.

implications, occurrence of K (perthite) and Na (anti-perthite) rich terms [24], these features reveal crystallization and cooling process conditions. The coexistence of K and Na feldspar indicates subsolvus crystallization condition that should reflect low-temperature and/or high-H_2O pressure [24, 25], possibly under eutectic point. The exsolutions record condition below the solvus temperature into the immiscibility gap of alkali-feldspars. If perthite and anti-perthite coexist in the case of a subsolvus granite, otherwise it is a hypersolvus granite. Once plagioclase and K-feldspar coexist and share boundaries, under tension, a reaction along these phase boundaries is induced liberating silica (SiO_2) [26]. In this case vermicular quartz appears to have a myrmekite texture, which is common in a variety of crystallization or even post-crystallization condition (**Figure 2F**).

Considering the stress state at the late stages of crystallization, when the melt fraction is much lower than the crystal volume, one could observe myrmekite and a bunch of reactions among mineral phases or between minerals and the residual melt, as exemplified by the biotite nuclei in hornblende crystals (**Figure 3A**). At this point localized deformation features representing crystal-plastic deformation related to the current crystallization process also become common. Quartz always portrays this final crystallization stage, showing undulose extinction or deformation bands (**Figure 3B**) that can evolve to subgrain formation. Initial and localized feldspar deformation can be observed through the formation of mechanical twinning in plagioclase grains (**Figure 3C**).

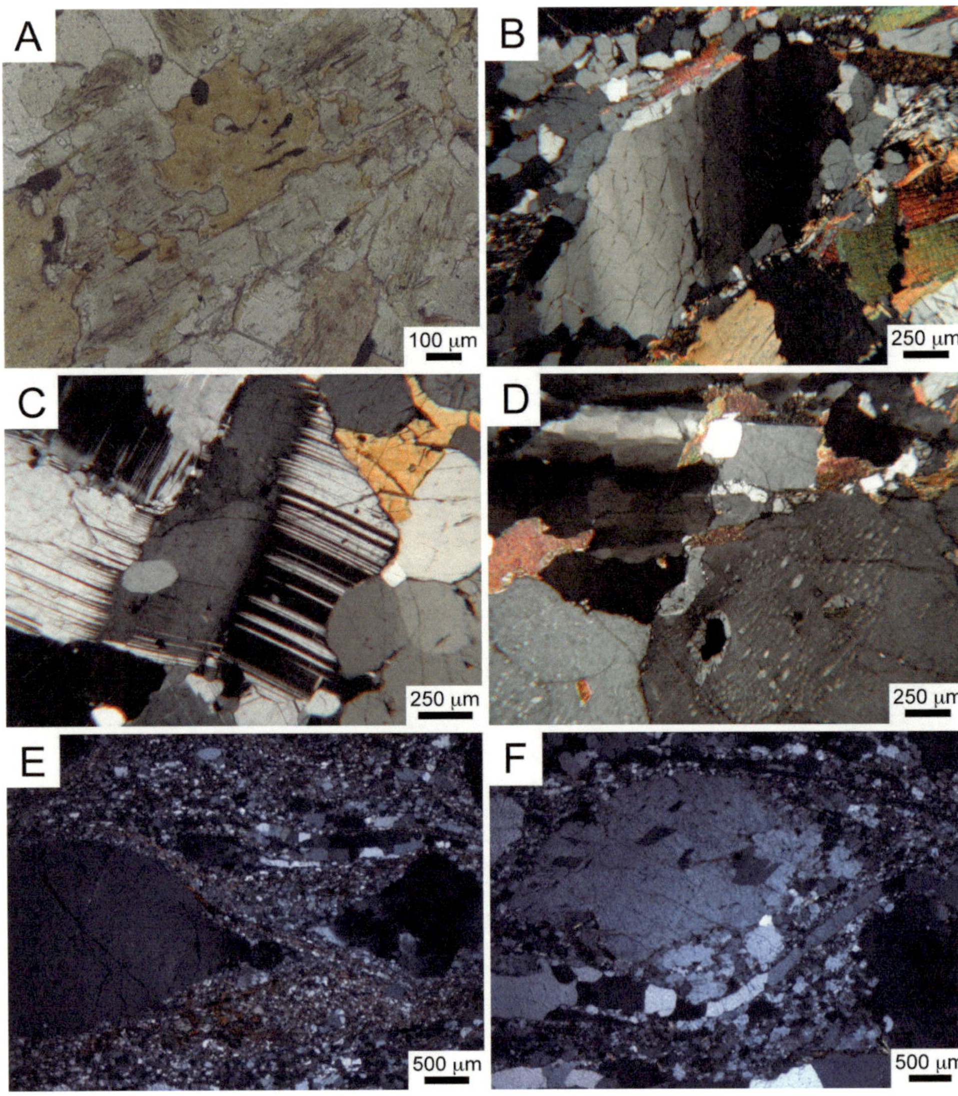

Figure 3.
Optical photomicrographs showing late-magmatic (A–D) and deformational (E, F) features, slightly deformed granites and mylonites, respectively. (A) Biotite core in hornblende grain (plane-polarized light). (B) Quartz grain with deformation bands. (C) Mechanical twinning in a kinked plagioclase grain. (D) Quartz grain showing undulose extinction in a chessboard pattern (upper left corner) and perthitic feldspar (lower right corner). (E, F) feldspar fish surrounded by a quartz-feldspathic recrystallized matrix: K-feldspar (E) and anti-perthitic feldspar (F) porphyroclasts surrounded by recrystallized grains and quartz ribbons (B–F: Cross-polarized photomicrographs). (A) Mineiro belt; (B–F) Araçuaí orogen.

In a post-crystallization condition, under an imposed stress, the localized crystal-plastic deformation features that record grain interaction during the late crystallization stage become the main microstructures, widespread through the aggregates. In a progressive process, quartz grains develop subgrains, under high temperature showing a chessboard pattern [27] (**Figure 3D**). Progressively, new recrystallized grains are formed, defining a matrix surrounding strong porphyroclast grains (**Figure 3E**), feldspar, hornblende, pyroxene, and garnet. If this matrix is not weak enough or its proportion and distribution have not achieved a critical value to localize deformation, the porphyroclasts will also accommodate deformation. Feldspar grains initially showing mechanical twinning will develop undulose

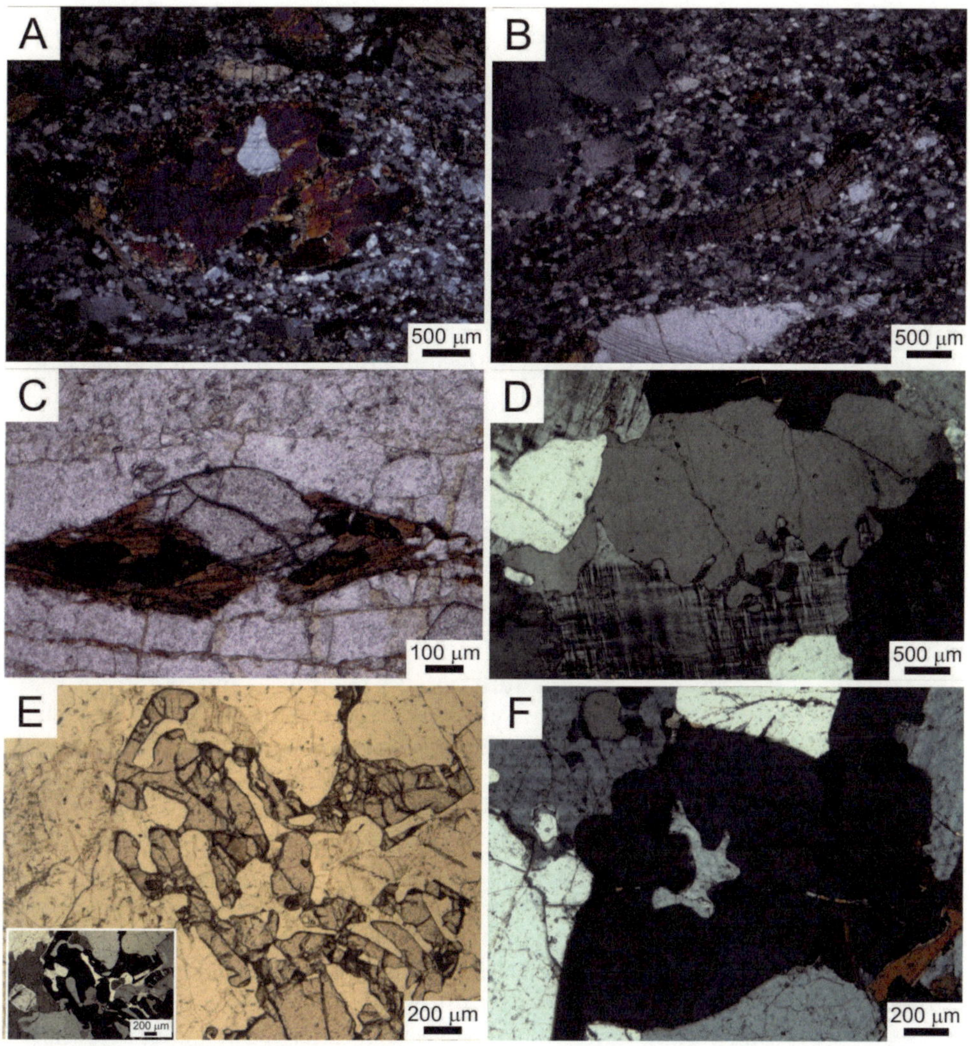

Figure 4.
Optical photomicrographs recording high-temperature deformation (A–C; mylonites) and partial melting process in granites (D–F). (A, B) plastically deformed pyroxene porphyroclasts: Pyroxene fish (A) and twisted pyroxene grain (B) surrounded by recrystallized quartz-feldspathic matrix and feldspar porphyroclasts (cross-polarized photomicrographs). (C) Garnet fish partially replaced by biotite and surrounded by quartz ribbons (plane-polarized light). (D) Strongly lobated quartz-microcline interphase boundary (cross-polarized light). (E) Interdigitated garnet and quartz grains showing, as seen in the previous image, highly lobated interphase boundaries (plane-polarized light for the main image and cross-polarized light for the image in the inset, lower left corner). (F) Unshaped quartz grain within a feldspar grain (center) (cross-polarized light). (A–F) Araçuaí orogen.

extinction, subgrains, and progressively new recrystallized grains, which will compound the quartz-feldspathic matrix (**Figure 3E**). At this point, the porphyroclast shape may reflect the deformation pattern and stress field, mainly if they are asymmetrical, as feldspar fish, when they reflect the vorticity of the deformation mechanism (**Figure 3E, F**). Considering the increasing temperature, quartz grains, with low misorientation, coalesce and form ribbons, parallel to the rock foliation and surrounding the remnant porphyroclasts (**Figure 3E, F**).

The highest deformation temperature is registered by the crystal-plastic deformation of the high-temperature minerals, in the case of such granitic mylonites, pyroxene and garnet. The Araçuaí orogen is cut by bundles of shear zones in which sheared granites are found from greenschist to granulite facies conditions [28]. **Figure 4A–C** shows the microstructures that record high-temperature deformation, typical of granulite facies, when crystal-plastic deformation condition of pyroxene is achieved. Pyroxene fish with strain shadows and recrystallized tails (**Figure 4A**) and/or twisted pyroxene grains (**Figure 4B**) are surrounded by a recrystallized quartz-feldspathic matrix in which quartz ribbons are recognized. The same is observed for garnet, whose grains many times are replaced by biotite (**Figure 4C**).

If the temperature is high enough to overcome the solidus temperature for the granite, it melts. In this case microstructures regarding the partial melting process are seen. Strongly lobated phase boundaries, melt pools, and features related to mineral reactions should be recognized, as well as perithetic mineral phases [29, 30]. This is the case of granites outcropping in the intracontinental part of the Rio Doce Arc—Araçuaí Orogen [31]. These granites show strongly lobated phase boundaries shared by quartz and feldspars (**Figure 4D**), remains of garnet which are interdigitated with quartz grains (**Figure 4E**) and anhedral quartz grains with very irregular boundaries filling triple junctions of feldspar grains, along their boundaries or as unshaped inclusions, regarding melt pools (**Figure 4F**).

4. Concluding remarks

From this quick view of microstructure features in granites and mylonites, there is no doubt that recognizing and quantifying them is a critical step to better understand and define magmatic, late-magmatic, and deformational processes. There is no substantial petrological study without a consistent description and analysis of the microstructural pattern(s) in the considered aggregates. The identified relations among different mineral phases and grain boundary and phase boundary geometries are fundamental to conduct petrology studies based on mineral chemistry, P–T–t trajectories, geothermobarometry, or even geochronology. It is not possible to conduct a petrology project if the aggregate is not known, for example, the mineral phases, their internal features, and how they interact. Besides, the understanding of rock rheology necessarily passes by the definition of processes responsible for the aggregate formation and its stabilization through time. Talking about typically igneous and sheared granites, the mechanical behavior of these quartz-feldspathic rocks represents the way to understand how the crustal part of the lithosphere is formed and accommodates stress through time.

Invariably, geoscientists trying to reproduce in laboratory what happens at large scale in nature need to use microstructural observations to establish comparisons between the experimental and natural microstructures and solve geologic problems. Researchers looking deep into the microstructure features of granites allow the characterization of processes such as fractional crystallization in a magmatic chamber, chilling mechanisms, mineral reactions, and ultimately how these rocks behave under stress. That is possible because the grain-scale mechanisms are unveiled. As a consequence a better knowledge about the lithosphere is achieved, and questions such as why its deformation is localized in narrow zones in their boundaries may be better discussed, as well as what are the mechanisms controlling the plate movements. Specifically, the presented microstructures show that in the Araçuaí orogen, granites and mylonites register from typical crystallization to partial melting processes, some of them showing features of high-temperature deformation. On the other hand, in the Mineiro belt, granites show typical magmatic to late-magmatic features, including an important process of magma mingling.

Based on the observed features, some processes can be depicted:

i. Magmatic differentiation through concentric zoning in plagioclase crystals (**Figure 2A**)

ii. Magma mingling due to oscillatory zoning in plagioclase grains (**Figure 2B, C**)

iii. Cooling processes associated with perthite and anti-perthite textures (**Figure 2D, E**)

iv. Reactions between K-feldspar and plagioclase grains recorded by myrmekitic texture (**Figure 2F**)

v. Late-magmatic stages during magmatic crystallization by localized deformational features (bands in quartz, mechanical twinning in feldspar) and reaction between minerals and final melts (e.g., biotite replacing hornblende) (**Figure 3A–D**)

vi. Ductile deformational processes by crystal-plastic deformation features such as undulose extinction, chessboard texture, quartz ribbons, recrystallized quartz feldspathic matrix, porphyroclasts of garnet, feldspar, and pyroxene with asymmetrical shapes sometimes showing recrystallized tails (**Figures 3E, F** and **4A–C**)

vii. Partial molting process due to the occurrence of quartz pools, interdigitated phase boundaries seen between quartz and feldspars, and also quartz and garnet with high density of inclusions, as well as unshaped inclusions (**Figure 4D–F**)

Acknowledgements

The authors thank everybody working at the Laboratory of Microscopy and Microanalyses (LMic-DEGEO-UFOP) for all sorts of facilities and the excellent work environment. We are also grateful to Carolina Ribeiro, Jessyca Carneiro, Marcos Pimenta, Reginaldo Vieira, and Wilker Soares for providing part of the micrographs that illustrate this chapter. We would like to acknowledge the IntechOpen for the invitation and opportunity and for financially supporting this publication. We also want to thank Dr. Gaafar El Bahariya for the editorial handling and suggestions that improved the manuscript and Kristina Kardum for the service management (CCG acknowledges the CNPq - Brazilian Research Council - grant: 429072/2016-3, for the financial support).

Author details

Leonardo Gonçalves* and Cristiane Castro Gonçalves
Department of Geology, Federal University of Ouro Preto, Ouro Preto - MG, Brazil

*Address all correspondence to: leonardogeologo@hotmail.com; leonardo.goncalves@ufop.edu.br

References

[1] Kearey P, Klepeis KA, Vine FJ. Global Tectonics. 3rd ed. Chichester: Wiley-Blackwell; 2009. p. 482

[2] Bürgmann R, Dresen G. Rheology of the lower crust and upper mantle: Evidence from rock mechanics, geodesy, and field observations. Annual Review of Earth and Planetary Sciences. 2008;**36**:531-567. DOI: 10.1146/annurev.earth.36.031207.124326

[3] DeWit M, Jeffery M, Bergh H, Nicolaysen L. Geologic Map of Sectors of Gondwana. Tulsa: AAPG and University of Witwatersrand; 1988

[4] Porada H. Pan-African rifting and orogenesis in southern to equatorial Africa and eastern Brazil. Precambrian Research. 1989;**44**:103-136. DOI: 10.1016/0301-9268(89)90078-8

[5] Teixeira W, Ávila CA, Dussin IA, Correa Neto AV, Bongiolo EM, Santos JO, et al. A juvenile accretion episode (2.35-2.32Ga) in the Mineiro belt and its role to the Minas accretionary orogeny: Zircon U-Pb-Hf and geochemical evidences. Precambrian Research. 2015;**256**:148-169. DOI: 10.1016/j.precamres.2014.11.009

[6] Ávila CA, Teixeira W, Bongiolo EM, Dussin IA, Vieira TAT. Rhyacian evolution of subvolcanic and metasedimentary rocks of the southern segment of the Mineiro Belt, São Francisco Craton, Brazil. Precambrian Research. 2014;**243**:221-251. DOI: 10.1016/j.precamres.2013.12.028

[7] Moreira H, Seixas L, Storey C, Fowler M, Lasalle S, Stevenson R, et al. Evolution of Siderian juvenile crust to Rhyacian high Ba-Sr magmatism in the Mineiro Belt, southern São Francisco Craton. Geoscience Frontiers. 2018;**9**:977-995. DOI: 10.1016/j.gsf.2018.01.009

[8] Seixas LAR, Bardintzeff J-M, Stevenson R, Bonin B. Petrology of the high-Mg tonalites and dioritic enclaves of the ca. 2130 Ma Alto Maranhão suite: Evidence for a major juvenile crustal addition event during the Rhyacian orogenesis, Mineiro Belt, Southeast Brazil. Precambrian Research. 2013;**238**:18-41. DOI: 10.1016/j.precamres.2013.09.015

[9] Seixas LAR, David J, Stevenson R. Geochemistry, Nd isotopes and U-Pb geochronology of a 2350 Ma TTG suite, Minas Gerais, Brazil: Implications for the crustal evolution of the southern São Francisco craton. Precambrian Research. 2012;**196-197**:61-80. DOI: 10.1016/j.precamres.2011.11.002

[10] Pedrosa-Soares AC, Noce CM, Wiedemann CM, Pinto CP. The Araçuaí–West Congo orogen in Brazil: An overview of a confined orogen formed during Gondwanaland assembly. Precambrian Research. 2001;**110**:307-323. DOI: 10.1016/S0301-9268(01)00174-7

[11] Alkmim FF, Marshak S, Pedrosa-Soares AC, Peres GG, Cruz SCP, Whittington A. Kinematic evolution of the Araçuaí–West Congo orogen in Brazil and Africa: Nutcracker tectonics during the Neoproterozoic assembly of Gondwana. Precambrian Research. 2006;**149**:43-64. DOI: 10.1016/j.precamres.2006.06.007

[12] Gonçalves L, Alkmim FF, Pedrosa-Soares A, Gonçalves CC, Vieira V. Mint: From the plutonic root to the volcanic roof of a continental magmatic arc: A review of the Neoproterozoic Araçuaí orogen, southeastern Brazil. International Journal of Earth Sciences (Geol Rundsch). 2018;**107**:337-358. DOI: 10.1007/s00531-017-1494-5

[13] Pedrosa-Soares AC, De Campos C, Noce CM, Silva LC, Novo T, Roncato J,

et al. Late Neoproterozoic–Cambrian granitic magmatism in the Araçuaí orogen (Brazil), the eastern Brazilian Pegmatite Province and related mineral resources. Geological Society, London. Special Publications. 2011;**350**:25-51. DOI: 10.1144/SP350.3

[14] Gonçalves L, Farina F, Lana C, Pedrosa-Soares AC, Alkmim F, Nalini HA Jr. New U–Pb ages and lithochemical attributes of the Ediacaran Rio Doce magmatic arc, Araçuaí confined orogen, southeastern Brazil. Journal of South American Earth Sciences. 2014;**52**:129-148. DOI: 10.1016/j.jsames.2014.02.008

[15] Tedeschi M, Novo T, Pedrosa-Soares AC, Dussin I, Tassinari C, Silva LC, et al. The Ediacaran Rio Doce magmatic arc revisited (Araçuaí-Ribeira orogenic system, SE Brazil). Journal of South American Earth Sciences. 2016;**68**:167-186. DOI: 10.1016/j.jsames.2015.11.011

[16] Santos-Dias JC, Gonçalves L, Gonçalves CC. Contrasting oxygen fugacity of I- and S-type granites from the Araçuaí orogen, SE Brazil: An approach based on opaque mineral assemblages. Mineralogy and Petrology. 2019;**113**:667-686. DOI: 10.1007/s00710-019-00670-2

[17] Vieira RR, Gonçalves CC, Gonçalves L. Evidências da colocação sintectônica de plutons revelada por estudos de campo, petrográficos, microestruturais e de química mineral: Estudo de caso da Suíte Alto Maranhão (2130 Ma), Cinturão Mineiro, Sudeste do Brasil. Anuário do Instituto de Geociências – UFRJ. 2020;**43**(1):376-396. DOI: 10.11137/2020_1_376_396

[18] Vernon RH. K-feldspar megacrysts in granites–phenocrysts, not porphyroblasts. Earth-Science Reviews. 1986;**23**:1-63. DOI: 10.1016/0012-8252(86)90003-6

[19] Paterson SR, Vernon RH, Tobisch OT. A review of criteria for the identification of magmatic and tectonic foliations in granitoids. Journal of Structural Geology. 1989;**11**:349-363. DOI: 10.1016/0191-8141(89)90074-6

[20] Paterson SR, Fowler TK Jr, Schmidt KL, Yoshinobu AS, Yuan ES, Miller RB. Interpreting magmatic fabric patterns in plutons. Lithos. 1998;**44**:53-82. DOI: 10.1016/S0024-4937(98)00022-X

[21] Vernon RH. Review of microstructural evidence of magmatic and solid-state flow. Visual Geosciences. 2000;**5**(2):1-23. DOI: 10.1007/s10069-000-0002-3

[22] Zak J, Verner K, Tykova P. Multiple magmatic fabrics in plutons: An overlooked tool for exploring interactions between magmatic processes and regional deformation? Geological Magazine. 2008;**145**(4):537-551. DOI: 10.1017/S0016756808004573

[23] White AJR, Chappell BW. Petrographic discrimination of low- and high-temperature I-type granites. Resource Geology. 2004;**54**:215-226. DOI: 10.1111/j.1751-3928.2004.tb00203.x

[24] Klein C, Dutrow B. The Manual of Mineral Science. 23rd ed. Chichester: John Wiley & Sons; 2007. p. 704

[25] Winter JD. An Introduction to Igneous and Metamorphic Petrology. 2nd ed. New Jersey: Prentice Hall; 2010. p. 702

[26] Phillips ER. Myrmekite–one hundred years later. Lithos. 1974;**7**:181-194. DOI: 10.1016/0024-4937(74)90029-2

[27] Kruhl JH. Prism- and basal-plane parallel subgrain boundaries in quartz: A microstructural geothermobarometer. Journal of Metamorphic Geology. 1996;**14**:581-589. DOI: 10.1046/j.1525-1314.1996.00413.x

[28] Silva CMT, Alkmim FF, Pedrosa-Soares AC. Geometria e evolução do feixe de zonas de cisalhamento Manhuaçu-Santa Margarida, Orógeno Araçuaí, MG. Revista da Escola de Minas. 2009;**62**(1):23-34. DOI: 10.1590/S0370-44672009000100005

[29] Viegas G, Archanjo CJ, Vauchez A. Fabrics of migmatites and the relationships between partial melting and deformation in high-grade transpressional shear zones: The Espinho Branco anatexite (Borborema Prince, NE Brazil). Journal of Structural Geology. 2013;**48**:45-56. DOI: 10.1016/j.jsg.2012.12.008

[30] Melo MG, Lana C, Pedrosa-Soares AC, Frei D, Alkmim FF, Alkmin LA. Two cryptic anatetic events within a syn-collisional granitoid from the Araçuaí Orogen (southeastern Brazil): Evidence from the polymetamorphic Carlos Chagas batholith. Lithos. 2017;**277**:51-71. DOI: 10.1016/j.lithos.2016.10.012

[31] Gonçalves L, Alkmim FF, Pedrosa-Soares AC, Dussin IA, Valeriano CM, Lana C, et al. Granites of the intracontinental termination of a magmatic arc: An example from the Ediacaran Araçuaí orogen, southeastern Brazil. Gondwana Research. 2016;**36**:439-458. DOI: 10.1016/j.gr.2015.07.015

Chapter 4

Deformation Features and Structures in Some Igneous and Metamorphic Rocks: A Case Study of Central African Fold Belt in Cameroon

Benjamin Ntieche, Pauline Wokwenmendam Nguet, Zakari Nchouwet, Mahomed Aziz Mounjouohou and Daouda Mfepat

Abstract

The first phase is marked by the S1 foliation. The second phase is marked by fold (F_2), lineation (L_2), and boudins (B_2). The third phase is marked by subvertical foliation (S_3), shear (C_3), and lineation (L_3). The fourth phase is essentially a brittle phase. Granites present mostly magmatic deformation features; meanwhile, granite mylonites and gneisses present submagmatic to nonmagmatic deformation features. Mylonitization occurred during a ductile-brittle transition phase. Coarse-grained granites emplaced in the lower crustal level, while protomylonites and mylonites occurred in the middle crust and ultramylonites in the upper crustal level. Fine-grained granite was filtered and channeled through the middle crust shear zone areas to be settled on an upper crustal level as irregular spot within the ultramylonites and gneisses. Granites and granites mylonites were syntectonically emplaced during the D_3 sinistral shearing phase.

Keywords: Mgbanji, granitoids, deformation, ductile-brittle, syntectonically, shearing

1. Introduction

Igneous, metamorphic, and sedimentary rocks generally host structures or features characterizing their condition or their emplacement environment. For igneous and metamorphic rocks, the deformation features provide information on the condition of the emplacement, the tectonic environment, and or the protolith nature. The Central African Fold belt to which belongs the Mgbanji area is situated between the Congo-Saõ Francisco and the West African cratonic masses, and it is mostly composed of Neoproterozoic plutonic and Pan-African metamorphic rocks [1–3]. The Mgbanji granites and their host gneisses present several deformation features testifying multiple deformation phases underwent by the area and the complex rock emplacement mechanism.

In this chapter, we present structural and petrographic data derived from systematic geological mapping, detailed petrographic descriptions, and structural characterization carried out in outcrops and rock thin sections coupled with kinematic studies in order to determine the petrography, the deformation phases, the origin, and the emplacement condition of the rocks.

2. Geological setting

The Mgbanji area belongs to the eastern part of the Adamawa-Yade Domain in the Central African Fold Belt (CAFB) in Cameroon (**Figure 1**). The CAFB was formed as a result of the convergence of the West African and the Congo-São Francisco cratons as well as the Sahara Metacraton during the assembly of West Gondwana [5, 6]. The CAFB is a major collisional belt that underlies the region from the West African craton to East Africa [3, 7]. It underlies parts of Nigeria, Cameroon, Chad, and Central African Republic, and extends further eastward to Sudan and Uganda [7]. Also known as the Oubanguide Fold Belt or the Pan-African Equatorial Fold Belt (PANEFB: [8]), the Central African Fold Belt in Cameroon belongs to the orogenic belt of Central Africa. It is a mega-chain-oriented E-W with a length of more than 5000 km over a 300-km width.

Three lithological domains have been identified in the CAFB in Cameroon, namely the Adamawa-Yade Domain (AYD), the Yaounde Domain (YD), and Northwestern Cameroon Domains (NWC) [3, 9]. Moreover, in [1, 10–12], the PANEFB (or CAFB) is subdivided into three (03) distinct geodynamic domains separated by four main shear and fault zones. They are from North to South:

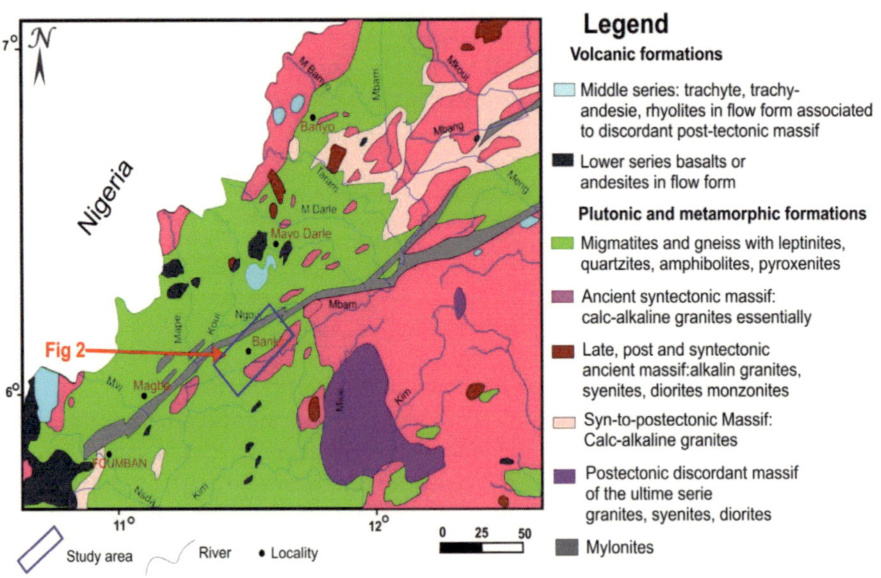

Figure 1.
Geological map of the western Adamawa Yade domain (modified from [4]).

the North Cameroon Domain, the Center Cameroon Domain, and the South Cameroon Domain.

The AYD [3, 9] or the Center Cameroon Domain [10] to which belongs our study area is the largest domain of the CAFB in Cameroon. It extends from the north of Bafia to the south of Poli. It is separated from the North by the Tchollire-Banyo fault (TBF), and its southern boundary is marked by the Sanaga fault. It displays syn to post-tetonic granitoids (leucogranites, granites, and syenites). These granitoids intrude high-grade gneisses that represent a Palaeoproterozoic basement, which was likely dismembered during the Pan-African assembly of western Gondwanaland [3, 9, 13]. According to [1, 10–12, 14], the geological formations of this domain are polycyclic and have been affected by three (03) deformation phases accompanied by metamorphic recrystallization in the amphibolite facies. The first phase D_1 is marked by a schistosity S_1, which carries a mineral line L_1; shear planes C_1; and the folds P_1. The metamorphism associated with this phase is characterized by the amphibolite facies assemblages from low to medium level. The second phase D_2 corresponds to a phase of transcurrent tectonics. It presents shear C_2, folds P_2, axial plane schistosity S_2, and lineation L_2. The associated metamorphism is characterized by assemblages of the amphibolite facies of high degree. Phase D_3 is a brittle tectonic phase, characterized by shear C_3 and large folds P_3.

Few geological works have been done in Mgbanji area. However, the early works taking into account the study area are mainly those done by [15–20]. Those studies demonstrated the presence of granites, gneisses, and mylonites on Foumban-Bankim shear zone that includes our study area. Those granitoids are igneous in nature and of calc-alkaline character. They are I-type and pre- to syn-tectonically emplaced [16, 19, 20]. Ntiéche et al. [20] demonstrated that the Magba area, adjacent to the Mgbanji area, underwent four phases of deformation under a transpressive tectonic regime.

3. Methods

For the field part of the work, a global positioning system (GPS), camera, clinometer, geological hammers, and map were used. This phase involved mapping, collection of samples, and investigation of structural elements on outcrops. A total of 30 samples were collected in the field. To assess the trend of structures, dip and strike measurements were taken. Other field assessment and tectonic features helped to determine the geotectonic setting of the rocks.

The laboratory phase involved the preparation of thin sections for petrographic study of selected samples. Structural analyses and petrographic studies were done by observing thin sections under the microscope. The deformation history and kinematic analysis of the whole area were deduced from the field measurements. Foliation and lineation, fault and shears trajectories along with meso-to-microscopic criteria of coaxial, or noncoaxial strain (e.g., symmetry or asymmetry of shear bands, boudins, and tails around porphyroclasts, fault, and folds) were studied.

4. Field relationship

Three main rock types are recorded in the Mgbanji area. They are granites, granite mylonites, and gneisses (**Figure 2**). Granites are mostly coarse-to-fine grained cropping in the hill top and the valley (**Figure 3(a)**). They are cropped in the form of slab or dome of varying size (1–50 m long). They are gray or pink in color and host epidote veins in some outcrop (**Figure 3(b)**). Some outcrops present biotite flakes molding feldspar phenocrysts.

Medium- and fine-grained granites are mostly mylonitized and situated at the hill top or in the river bed. They are characterized by mantled and fractured feldspar porphyroclasts preferentially oriented.

Few coarse-grained granites outcrops show magmatic of stage (pre-full crystallization fabric (pfc)) marked by oriented feldspar porphyroclasts, while the mylonitized one show mostly solid-state deformation fabric.

Fine-grained granites display millimetric-sized quartz and feldspar with biotite flakes sparsely dotted in the rock.

Mylonitized granites are well represented in the study area. They occur as slabs or domes at the hill top and valley. Depending on the degree of crushing, granites mylonites can be classified into protomylonite, mylonites, and ultramylonites (**Figure 3(c)–(e)**). Some outcrops present banding or foliation marked by the

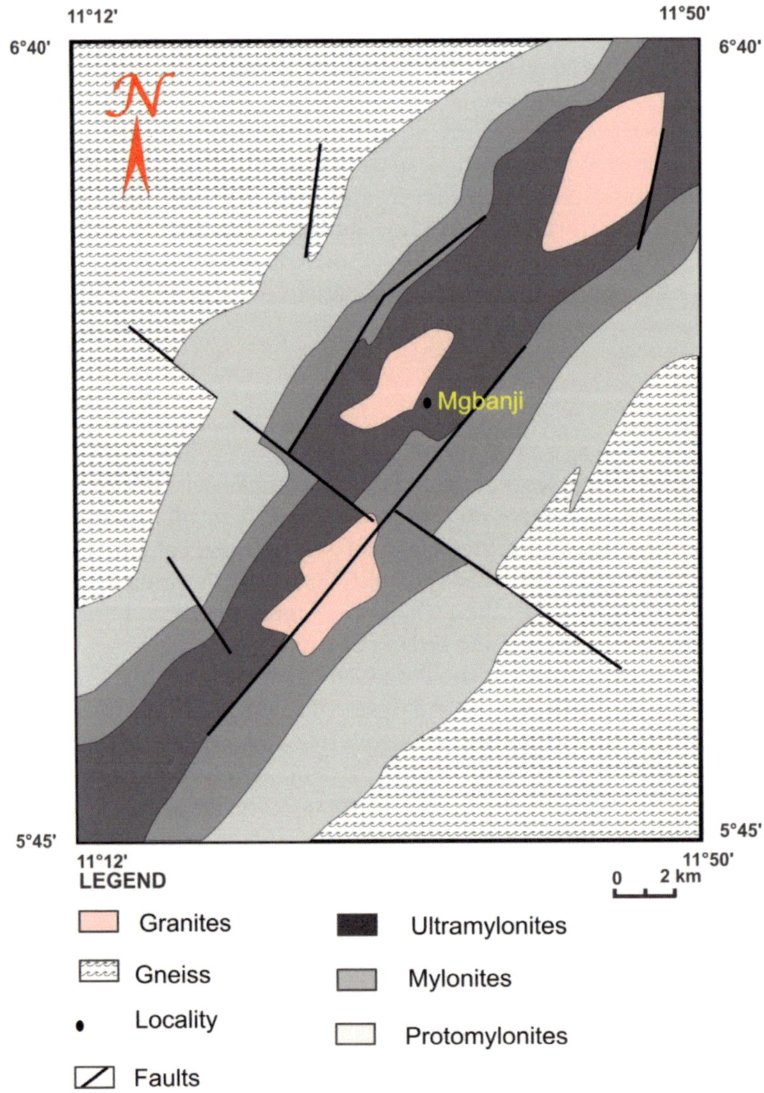

Figure 2.
Geological map of the study area.

Figure 3.
Field photographs. (a) Fine-grained granite outcrop. (b) Coarse-grained granite outcrop. Note aligned and oriented euhedral K-feldspar megacrysts perpendicular to the epidote vein. (c) Protomylonite outcrop showing aligned deformed quartz and feldspar grains. (d) Mylonite outcrop with reduced quartz and feldspar grain size. (e) Ultramylonite outcrop presenting cataclasis marked by the reduction of the grain size to a pasty matrix. (f) Gneiss outcrop presenting foliation marked by alternation of amphibolites and quartzofeldspathic layers.

alternation of dark mafic and clear felsic bands. Clear bands present oriented sigma "σ" and delta "Θ" porphyroclasts, giving the sinistral sense of shear to the rock. They also present mineral lineation marked by aligned biotite, quartz, and feldspar porphyroclasts (**Figure 3(c)** and **(d)**). In some outcrops amphibolites injections are notice in the granite mylonite. The fine-grained mylonitized granites present some chlorite and epidote vein enrichments. The mylonitic foliation is outlined by aligned biotite and quartz ribbons. In some places, the progressive increase in strain is marked by the presence of the protomylonite, mylonite, and ultramylonite subzones within the sheared granite.

Gneisses occur in a slab form and mostly confined to the hill slopes. They present foliation highlighted by alternating dark mafic and clear quartzofeldspathic bands (**Figure 3(f)**). In some places, lineations are characterized by the alignment of biotite flakes, quartz ribbons, and feldspar porphyroclasts.

Some gneiss outcrops present migmatitic features such as the combination of melted and unmelted gneissic components.

5. Microscopic features

5.1 Granites

Granites present coarse- to fine-grained texture and consist of quartz, feldspar, biotite pyroxene, and amphibole as principal minerals, while apatite, sphene, epidote, and opaques are the accessory mineral phase (**Figure 4(a)**). Sericite, perthite, myrmekite, and chlorite are the secondary minerals.

Quartz is fine- to coarse-grained and mostly anhedral and present in few places, tilled and imbricated porphyrocrysts. Quartz phenocrysts are associated to plagioclase and K-feldspar, while microcrysts are filling phenocrysts interstices. Some quartz grains develop from plagioclase porphyrocryst suggesting the incipient metasomatism during granite formation (**Figure 4(b)**). Plagioclases are predominantly subhedral and present twins. Some plagioclase laths are partially altered into sericite. K-feldspars are subhedral and mostly transformed into perthite. They are associated to quartz and plagioclase and present opaque inclusions in places. Biotites are subhedral and partially altered into chlorite. They fill quartz and feldspar interstices along with euhedral sphene (**Figure 4(a)**). Pyroxene and apatite are rare in the rock and occur respectively as subtabular and acicular crysts associated to amphibole and biotite.

5.2 Granites mylonites

Granites mylonites present granoclastic, cataclastic, and mylonitic texture (**Figure 4(c)**). They consist of quartz, K-feldspar, plagioclase, biotite, amphibole, pyroxene and microcline as principal mineral, chlorite and sericite as secondary minerals, and sphene and opaques as accessory minerals. The rock is marked by weakly to well-developed fabric characterized by elongated quartz and feldspar grains.

Quartz is in the form of polycrystalline ribbon preferably oriented and forming the clear band with feldspar. Some thin sections present also pasty or flame quartz molding feldspar and amphibole clasts. Some quartz and feldspar yield to brittle failure characterized by crystal fragmentation or shearing. In places, crystal underwent grain refinement through recrystallization. Some microclasts present waving or curved form testifying the deformation intensity (**Figure 4(c)**). Plagioclase and K-feldspar are subhedral to anhedral and present in places seritization reaction (**Figure 4(d)**). Plastic deformation is marked on feldspar by deformation lamellar, kinking or folding and the sigma "σ" and delta "Θ" porphyroclast giving the sense of shear to the rock. Mylonitic texture is developed in some thin sections and characterized by clear quartz and feldspar bands and dark ferromagnesian bands. In some places, subrounded feldspar and quartz are aligned in the clear band giving the ocellar mylonitic texture to the rock. Amphibole and biotite present flaky form or fish-like structure testifying the deformation of the rock. In some places,

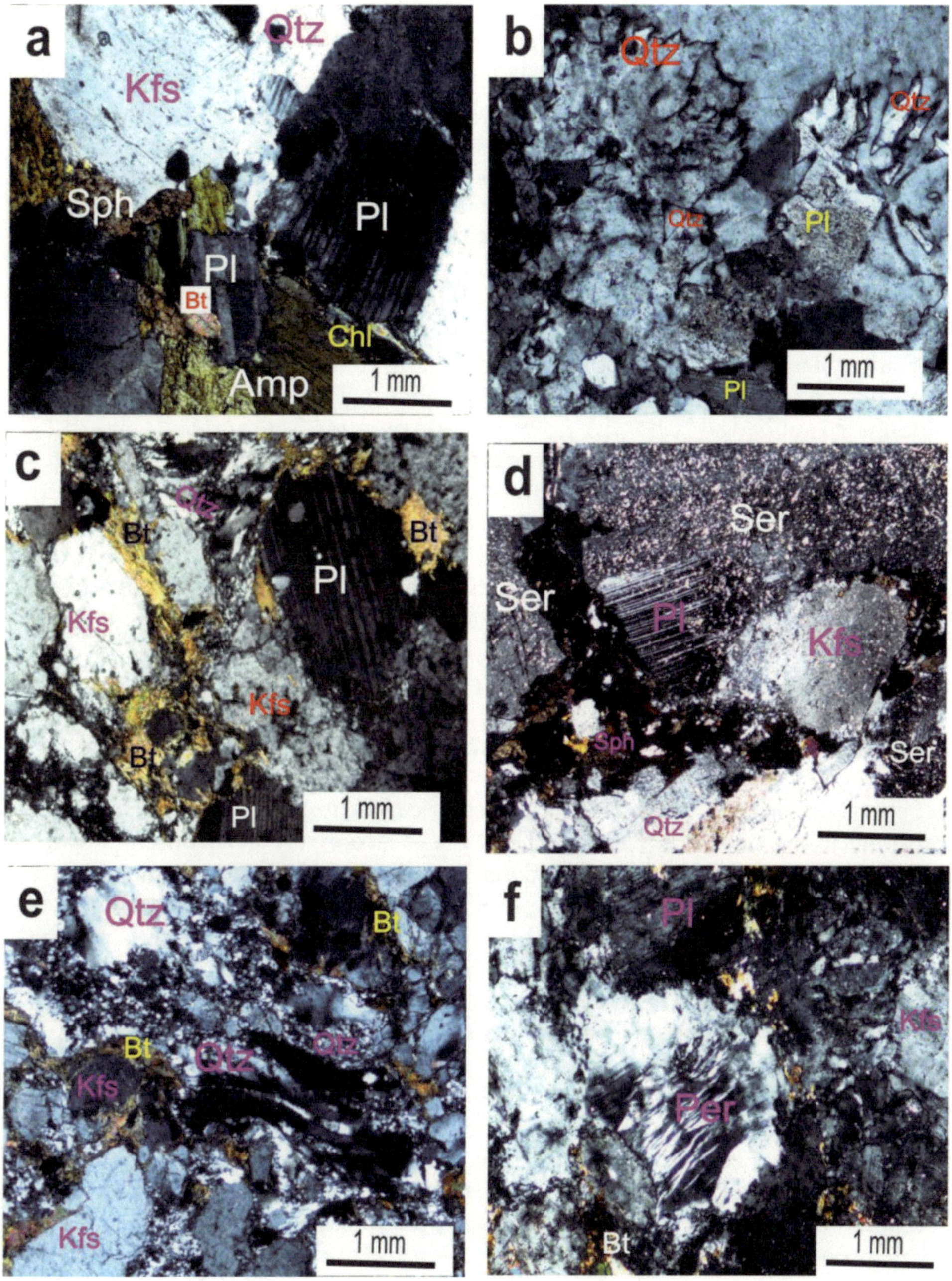

Figure 4.
Photomicrographs of mineral textures and reactions (cross-polarized light). (a) Coarse-grained texture marked by euhedral to subhedral feldspar, quartz, amphibole, and biotite. This is a typical evidence of magmatic structure in the rock. (b) Quartz replacing plagioclase in granite suggesting the incipient metasomatism by fluids derived from evolved crustal rocks. (c) Mylonitic texture marked by K-feldspar and plagioclase molded by biotite and quartz. Note the curved or waving quartz at the photo top and quartz grain embayment by plagioclase. (d) Sericitization reaction on plagioclase. Also note early sphene crystals trapped within the microfracture fillings in granite mylonite. (e) Grano-lepidoblastic texture on gneiss. Note the quartz and feldspar porphyroblasts in the matrix of fine-grained-elongated and recrystallized quartz with undulose extinction. (f) Perthitization reaction on K-feldspar in gneiss.

amphibole or biotite are altered into chlorite. Opaques and zircon are located in the mineral interstices or are present as inclusions in the center or the margins of some biotite minerals.

5.3 Gneisses

Gneisses present various textures. The principals are granoblastic and lepidoblatic textures (**Figure 4(e)**). Quartz, plagioclase, orthoclase, amphibole, and biotite are principal minerals; chlorite and perthite are secondary minerals, while sphene and epidote are the accessory phases.

Quartz is the most abundant mineral in the rock and is represented by two generations. The first generation is the porphyroblasts and the second generation is the microblasts (**Figure 4(e)**). The porphyroblasts are subhedral to anhedral and present boundary relationship with plagioclase and K-feldspar. The microblasts form the matrix and also fill the feldspar and hornblende interstices (**Figure 4(e)**). Both microblasts and porphyroblasts are generally confined and oriented in the clear microbands with some feldspar porphyroblasts. Plagioclases are in the form of elongated subhedral laths with polysynthetic twining. Some crystals present quartz crystal inclusions. The most represented feldspar is orthoclase. They are subhedral in some thin section, and mantled in the form of delta "Θ" porphyroblast. Perthitization reaction is common on feldspar porphyroblasts (**Figure 4(f)**). Amphiboles are anhedral and occur mostly in the form of amphibole fish. They are associated with biotite and sphene. Biotites are in the form of elongated flake in the dark microbands along with amphibole and sphene. Some flakes are totally or partially altered into chlorite.

6. Deformation features and kinematic indicators

Several deformations features and kinematic indicators are observed on Mgbanji granitoids in the Central African Fold Belt in Cameroon, either on outcrops scale or under the microscope. They are foliations, lineations, folds, faults, joints, delta "Θ," and sigma "σ" porphyroclasts, shears, boudins, kinking, micas, and amphibole fish.

On the outcrop scale, gneisses are of the rock types presenting all the deformation features observed on the study area. They are foliation, lineations, shears, joints, faults, and folds. The foliation is marked by the alternation of clear quartzofeldspathic and dark ferromagnesian bands. There are three different foliation generations on gneisses. The first is represented by parallel E-W-oriented horizontal bands (**Figure 5(a)**). The second is rare and is N-S oriented and results from the reorientation of the first foliation generation. It is mostly parallel to the isoclinals fold axes (**Figure 5(b)**). The third foliation generation is subvertical and oriented NE-SW (**Figure 5(c)** and **(d)**). It originated from the reorientation of the second generation foliation. Lineations follow mostly the foliation trend. Shears are mostly present on the third foliation generation and are mostly sinistral. Joints and faults are chronologically the last structures on all the rock types of the study area. Some joints are dry, while others are filled with quartzofeldspathic materials (**Figure 5(d)**).

Granites mylonites present several deformation features. Protomylonites and mylonites present some plastic deformation features, while ultramylonites present only brittle features such as faults, joints, and shears. Protomylonites present lineations marked by feldspar and quartz-oriented grains (**Figure 5(e)**). Foliation consists of alternation of quartzofeldspathic and ferromagnesian bands. S-C structures are in places seen on protomylonites and mylonites. Folds are rarely seen on ultramylonites but are common on protomylonites and mylonites. Fault and shears are present on all the mylonitic rocks.

Granites present lineations, faults, joints, and "false" boudins. Linenations consist of aligned feldspar porphyroclast on the outcrop, generally oriented

Figure 5.
Photomicrographs showing field deformation structures. (a) Foliation S_1 on gneiss; (b) F_2 folded asymmetric boudins B_2 on gneiss; (c) S_3 subvertical foliation on gneiss; (d) S_3 vertical gneissic foliation perpendicular to the J_4, quartzofeldspathic filled joint; (e) L_2 and L_3 lineations on granite mylonite; (f) F_4 fault on granite. Note also the "false" boudin B_4 made of quartzofeldspathic material encircling a portion of granite rock.

NE-SW. Faults are mostly normal faults. In some places, they are reduced to dry joints characterized by undisplaced component and empty fracture. Some joints are filled of quartzofeldspathic materials which are rarely boudinated. False boudins are made up of concentric quartzofeldspathic layers surrounding the granite material (**Figure 5(f)**). Depending on crosscutting relationship, faults and joints are the last deformation features on all the rock types in the study area.

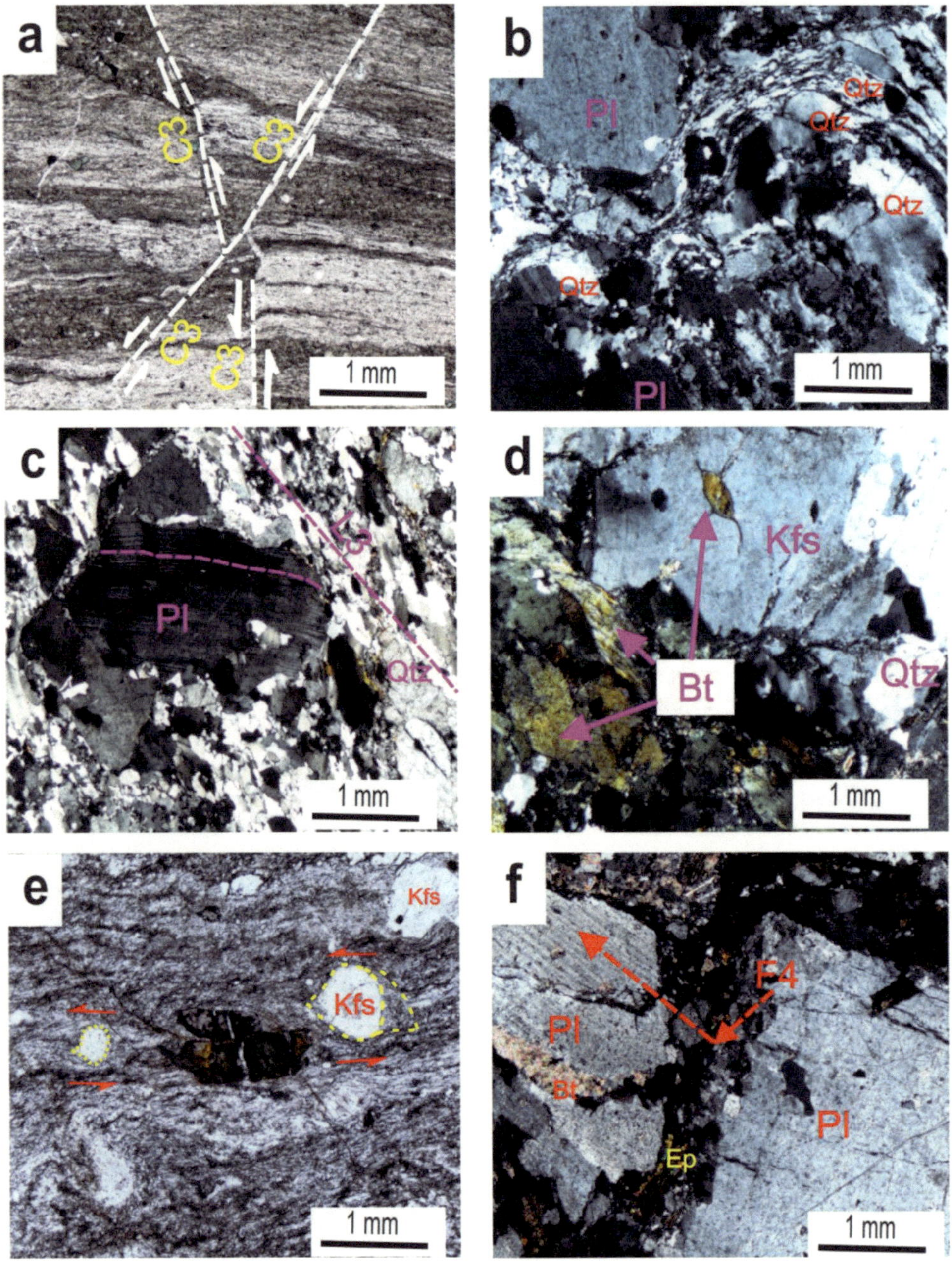

Figure 6.
Photomicrographs presenting microstructures and kinematic indicators ((b), (c), (d), and (f) = cross polarized light; (a) and (e) = plane polarized light). (a) Microfoliation marked by alternated clear quartzofeldspathic and dark ferromagnesian band crosscut by C_3 conjugated sinistral shear on granites ultramylonite; (b) stream quartz with undulatory extinction molding plagioclase and filling their cracks. Note also deformation lamellae in plagioclase; (c) kink deformation on plagioclase. The subeuhedral outline of the plagioclase phenocryst is oblique to aligned recrystallized quartz grain (lineation L3) giving the S-C fabric to the section. The plagioclase phenocryst is recrystallized along its margins, and present visible twins. The shape fabric and the intracrystalline deformation features can therefore be interpreted as submagmatic. This granite mylonite is syntectonically emplaced; (d) granite mylonite present embayed biotite fish on feldspar porphyroclasts; (e) "Θ" and "σ" porphyroclasts give the sinistral shear sense to the rocks. Also note the quiet pressure shadow around porphyroclast on the right; and (f) normal microfault on plagioclase megacryst. Note also the biotite mineral filling the plagioclase crack and epidote trapped in the fault fracture.

On thin sections, deformation structures are common. They are foliations, lineations, kink deformation on plagioclase, undulatory extinction on quartz, and deformation lamellae on feldspar, micas and amphibole fish, delta "Θ" and sigma "σ" porphyroclasts, and faults.

Foliation consists of alternation of quartzofeldspathic and ferromagnesian bands on mylonites. Bands are in place crosscut by conjugated sinistral shears (**Figure 6(a)**). Mylonites present flame and pasty quartz with undulatory extinction molding residual quartz grains. They also present feldspar porphyroclast with crystal deformation lamellae (**Figure 6(b)**). Mineral lineation is marked in mylonite by aligned recrystallized quartz crystals along with few biotite flakes. The feldspar porphyroclasts are in places recrystallized along their margins. They show kink deformation marks and visible polysynthetic twinning. The feldspar porphyroclast orientation is oblique to the recrystallized quartz mineral lineation, giving the microscopic S-C foliation to the rock section (**Figure 6(c)**). Biotite fish crystals are also present on mylonite. They are embayed on feldspar porphyroclasts and also seen at the feldspar edge (**Figure 6(d)**).

Granite mylonites in general host delta "Θ" and sigma "σ" porphyroclasts giving the shear sense of deformation to the outcrop. On ultramylonites, those porphyroclasts are located in the crushed matrix resulted from cataclasis (**Figure 6(e)**). They show a sinistral sense of shear as the conjugated shears are early described. The last deformation microstructure present in the Mgbanji granitoids is fault. On granite and granite mylonites, they are present on several minerals. The most faulted mineral is plagioclase. It presents normal microfault with sinistral sense of movement. The fault crack is more or less filled with epidote and sphene crystals which are a continuous phase in the rock matrix (**Figure 6(f)**).

7. Discussion

7.1 Deformation phases

Depending on crosscutting relationship, four deformation phases have been identified on studied rocks.

The first phase is exclusively represented on gneisses and is marked by the horizontal foliation (or banding) (S_1) and lineation (L_1). The foliation consists of the alternation of clear quartzofeldspathic and dark bands. They are mostly subhorizontal and oriented E-W (**Figure 5(c)**). Lineation is marked by randomly oriented quartz feldspar and biotite minerals. It is mostly parallel to foliation.

The second phase of deformation is recorded on gneisses and granite mylonites. That phase is characterized by folding (P_2), lineation (L_2), and boudin (B_2). Folds (P_2) are isoclinals with fold axes oriented principally N-S. The lineations are parallel to P_2 fold axes. Boudins (B_2) are complete and asymmetric. They are made of ferromagnesian layer (mostly amphibolites) and are folded in places (**Figure 5(b)**).

The third phase of deformation is represented on gneisses and granite mylonites. It seems to be the major phase on the field. It presents more deformation features compared to the other phases. Those features are resulting from the reorientation of previous phase's features. It hosts foliation (F_3), lineation (L_3), and shear (C_3). The foliation (S_3) is mostly subvertical and consists of alternated quartzofeldspathic and ferromagnesian bands oriented NE-SW (**Figure 5(c)** and **(d)**). Lineations (L_3) are hosted by foliation (S_3) and oriented in the same direction. Shears (C_3) consist of the fracturation and displacement of foliation (S_3). The shear axes are mostly oriented NW-SE or NE-SW. They are observed both on outcrop and on thin sections. On thin section, D_3 deformation phase is marked by conjugated shears (C_3), lineation (L_3), and sinisterly oriented "Θ" and "σ" feldspar porphyroclasts (**Figure 6(a), (c),** and **(e)**).

The fourth phase of deformation is present on all the three rock types of the study area, but it is mostly highlighted on granite. It is mostly a brittle phase. It is marked by dry and filled joints. The filled joints are made of quartzofeldspathic materials and may be qualified as a late D_3 phase. The filled joints crosscut the foliation (S_3) and are oriented E-W to NE-SW (**Figure 5(d)**). In some places, filled joints are circular and surround the granite material as a "false" boudin (B_4) (**Figure 5(f)**). The dry joints are chronologically the last feature on the study area of rocks. They are characterized by fractures classified as diaclase or fault (F_4) depending respectively on the tilling (diaclase) or the displacement (fault) of the component (**Figure 5(f)**). This chronology has been also mentioned in the recent study from the Central Cameroon Shear Zone [20].

7.2 Rock emplacement mechanism

Rocks within deformation areas are generally slightly, moderately, or highly strained and may exhibit very nice deformation microstructures. Deformed rocks present matrix with reduced grain size surrounding elongated faulted or sheared grains or grain aggregates. Many questions are steel a debate nowadays. Among them, we have the question of the identification of the presence or the absence of melt during deformation in rocks. Those questions are important for the problem of the melt extraction, the interpretation of plutonic magma ascent, and the mechanisms of emplacement. To have a good understanding of those questions, it is important to use macroscopic, mesoscopic, and microscopic structural criteria.

Coming to the present study, field relationships, mesoscopic and microscopic observations, and the structural data testify that Mgbanji granitoids were emplaced during the brittle-ductile deformation events occurring under the influence of an active strike-slip shear zone.

Conjugated mechanism may have acted simultaneously during the studied granitoid emplacement. They are cataclasis (microcracks, microfaults), diffusive mass transfer by solution (interpenetrating grain, pressure shadow), intracrystalline plasticity (deformation twin, odulatory extension, kink band, subgrain, grain shape fabric and ribbon, new grain), and solid-state diffuse mass transfer transformation (reaction rim, corona, relic mineral). All these mechanisms can be regrouped into magmatic, nonmagmatic (solid state), or submagmatic (few melt) deformation and can help to qualify the shear zone as magmatic, submagmatic, or nonmagmatic.

It is important to mention that nonmagmatic microstructures are those formed during deformation without any melt present. Such a deformation has previously been called subsolidus or solid-state deformation. Submagmatic deformation is characterized by the presence of few melt during deformation, while the magmatic deformation refers to the total presence of melt.

On Mgbanji granitoids, magmatic microstructures are only seen on granite. They are coarse-grained texture containing euhedral feldspar, amphibole, biotite, and sphene minerals (**Figure 4(a)**) and the presence of magmatic myrmekitic textures [21, 22]. In contrast to granites, granite mylonites (protomylonites, mylonites, and ultramylonites) present nonmagmatic to submagmatic deformation microstructures.

The nonmagmatic structures on Mgbanji granitoids are represented by plagioclase deformation lamellae, pressure shadows filled by epidote around megacrysts, and ribbon grains formed by intracrystalline plasticity in quartz, embayed, or overgrown grain boundaries [23, 24]. Granite mylonites contain new recrystallized quartz grains with higher frequencies of contacts between like grains (**Figure 7(b)**). This feature has been also revealed in [25] and is justified by the fact that new grains preferentially nucleate on the same phase.

The submagmatic structures on studied granite mylonites are highlighted by the tectonically induced myrmekites (**Figure 7(d)**) formed in volumes perpendicular to the shortening direction [21, 22].

The Mgbanji granite mylonites present elongated ribbons of quartz grains showing undulatory extinction. Quartz presents sutured margins on ribbons and some areas of small strain free polygonal grains. The new quartz or feldspar grains are in places concentrated in a mantle that partly surrounds the older mega feldspar grains, giving the core-and-mantle structure to the mineral (**Figure 7(a)**). These features are characteristic of syntectonic emplacement of the granite mylonites. Reaction rims of altered mineral around feldspar grains are present on granite mylonite. This texture may surely testify to DMT (**Figure 7(e)**).

The high frequency of undulose extinction in quartz within granites mylonites suggests that intracrystalline plasticity occurs during submagmatic deformation (**Figure 7(b)**). This assumption is mentioned here because quartz subgrain formation or recrystallization has been taken as an indicator of submagmatic deformation in rock where an overall igneous texture is preserved without any nonmagmatic deformation [26]. Plagioclase and calcite microcracks present biotite, quartz, and sphene or epidote trapped within the fillings (**Figure 7(a), (c),** and **(f)**). In some places, microcracks are intragranular and filled with pasty quartz suggesting that the plagioclase crystal was in contact with melt (**Figure 7(c)**). The microcracks filling is petrographycally continuous with groundmass phase of the rock.

Other submagmatic microstructures are the deformation twins and bent twins (kinking) in granite mylonites plagioclases. Such deformation features are known to be formed by intracrystalline plasticity in submagmatic phase (**Figures 6(c)** and **7(d)**).

Paradoxically, some authors demonstrated that the cataclasis of the crystal is not only occurring exclusively during the non-magmatic deformation but can also occur during sub-magmatic events [22, 27, 28]. Also, it has been proved that S-C fabrics may form either in submagmatic or nonmagmatic deformation [29].

On the field scale, the melt was localized in pressure shadows at the ends of boudins, and also faults and joints on Mgbanji granitoids were filled [30]. This observation coupled to the microscopic observation of some kinematic indicators ("σ" and "δ" porphyroclasts) indicate that the rock was syntectonically emplaced [31]. The presence of quartz replacing plagioclase in some granite mylonite samples suggests the possible incipient metasomatism by fluids derived from evolved crustal rocks (**Figure 4(b)**) [32].

Considering all abovementioned features, the granitoids of the Mgbanji area seem to be emplaced during plastic- to solid-state deformation stage.

It has been also demonstrated that intracrystalline plasticity and solid-state DMT are strongly controlled by temperature but relatively insensitive to pressure. That is why cataclasis is generally attributed to the upper crust (because of the low pressures at that level). Following that rule, crystal plasticity is supposed to occur in the lower crust or in the lithospheric mantle due to the high temperatures in deep earth areas.

The brittle-plastic transition (cataclasis—plasticity) results to intermediate deformation regime of semi-brittle behavior characterized by microfractures

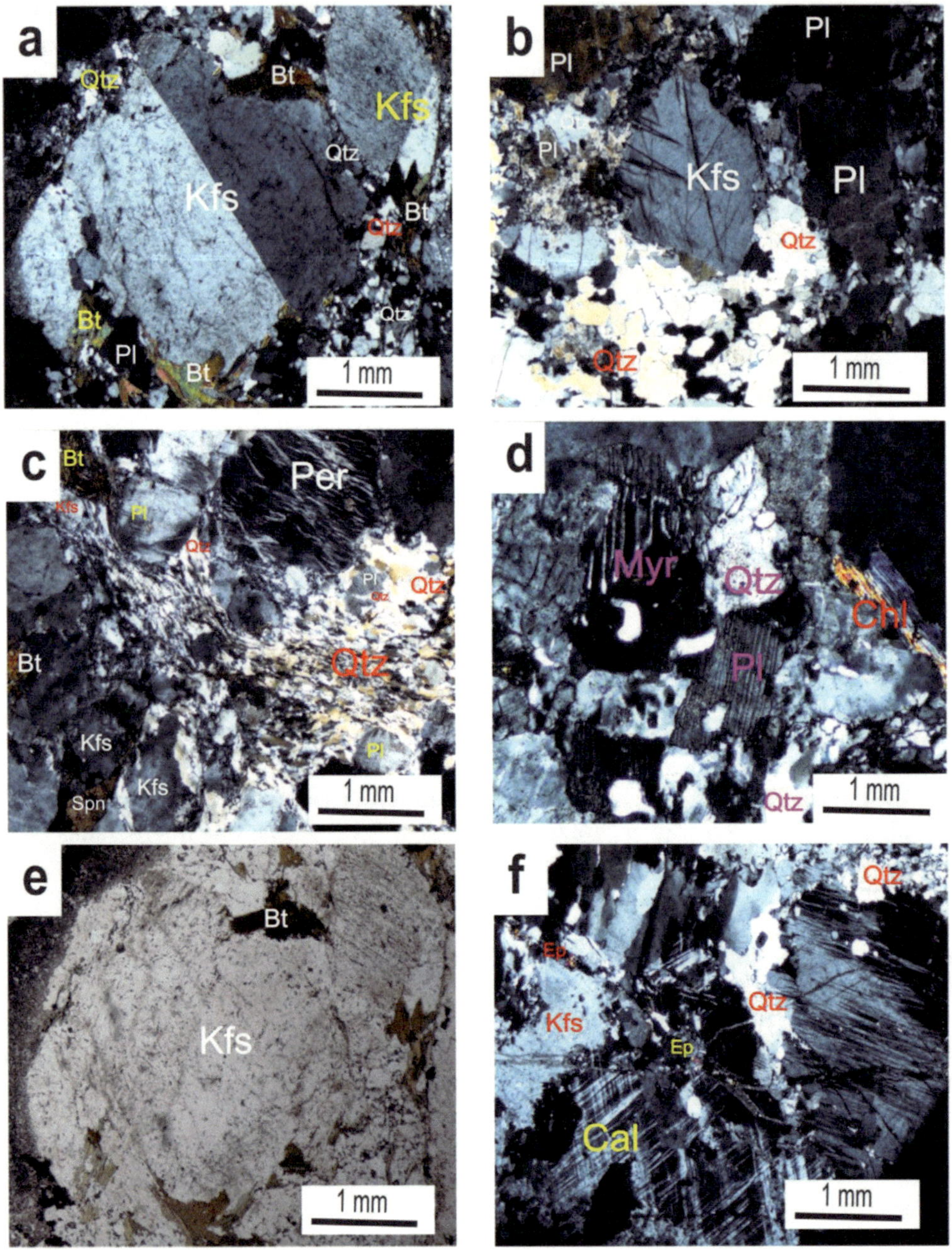

Figure 7.
Photomicrographs presenting microstructures ((a), (b), (c), (d), and (f) = cross-polarized light; (e) = plane polarized light). (a) Megacryst of feldspar (orthoclase) from granite mylonites presenting microcracks filled with biotite and quartz. The biotite and quartz are the same phases observed in the groundmass. Also note the core-mantle structure on plagioclase. (b) Typical feature of dynamic recrystallization in quartz involving grain boundary migration. Note the sutured margins on quartz ribbons and also domains of small strain free polygonal grains. Also note the K-feldspar presenting a leave like shape with unfilled microcracks (c) perthite exsolution and stream quartz molding K-feldspar porphyroblasts. Note the K-feldspar microcryst presenting microcrack filled with pasty quartz. (d) Tectonically induced myrmekites characterized by their grow in volumes perpendicular to the shortening direction. Also note exsolution lamellae on plagioclase. (e) Reaction rims of altered minerals around K-feldspar megacryst. This texture testifies to the transformation by diffusive mass transfer (DMT). Also note the biotite embayment and (f) calcite microveins filled with epidote and quartz.

interaction with intracrystalline plasticity [33, 34]. Two transitions in deformation mechanism, namely the brittle-semi-brittle transition and semi-brittle-plastic transition are registered with increasing depth respectively in the upper crust and in the middle crust levels. But, even though it has been proved that high stresses may also cause cataclasis at greater depths or higher temperatures, the mechanism we suggest for the Mgbanji magma emplacement operates on three different crustal levels, namely lower, middle, and upper crustal levels (**Figure 8**).

In the lower crustal level, the granitic magma may have originated from the in situ crustal melting [35]. That level favored the formation of coarse-grained granite in the study area.

The middle level is characterized by the presence of mylonites and protomylonites, presenting submagmatic and mylonitic structures, typical of syntectonic magmatism. The presence of subvertical lineation on mylonites and protomylonites may indicate that the middle level zone reacted as feeder zones acting as melt collectors (**Figure 8**) [36].

The presence of both submagmatic and solid-state deformation features demonstrates that the strike-slip event may have been the active phenomenon during and after the total crystallization of the magma.

The magmatic to mylonitic fabric of the Mgbanji granitoids are in general parallel to the solid-state deformation features encountered in the country rocks, demonstrating that the Mgbanji granites seems coeval with the transcurrent tectonics in the Cameroon Central Shear Zone [20].

The upper crustal level or subsurface display spots of fine-grained granites (with homogeneous appearance in the field and few or no solid-state deformation features) surrounded progressively by ultramylonites, mylonites, and protomylonites (**Figure 8**). The fine-grained granite may be the outcome of tectonically driven solid-liquid partitioning, operating as filter-press, in the middle crust. At upper crustal levels, real magmatic fabrics are not common because of the higher temperature contrasts with the surrounding rocks, facilitating rapid magma cooling.

Magma intrusion may have been forced to flow toward local dilatational, low pressure sites, such as the boudin necks [37].

Field and microscopic observations of kinematic indicator point to the emplacement of the Mgbanji granitoids during the D_3 sinistral deformation within an active strike-slip shear zone.

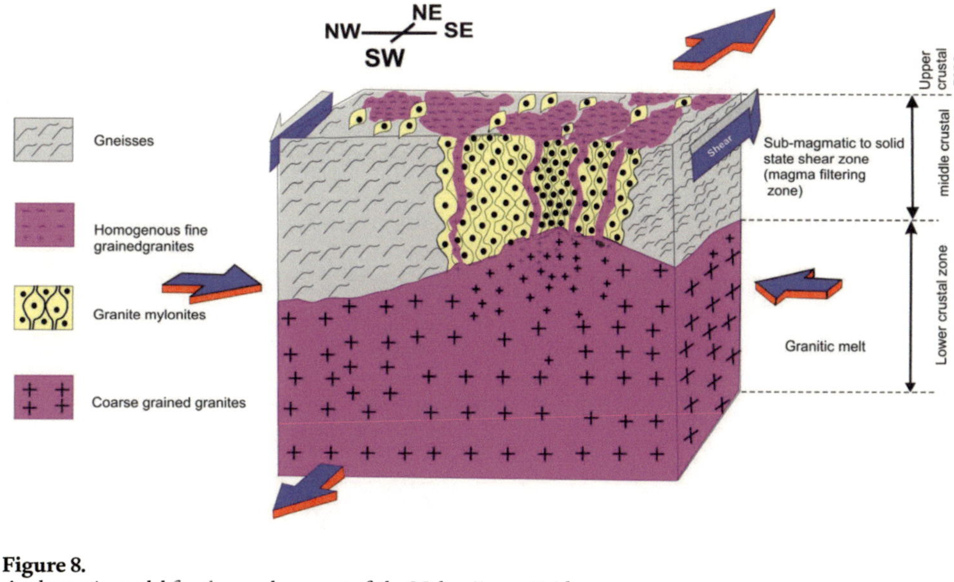

Figure 8.
A schematic model for the emplacement of the Mgbanji granitoids.

8. Conclusion

The Mgbanji area is made up of three main rock types, namely granite, granite mylonites, and gneisses. Field, mesoscopic and microscopic observations show coarse- to fine-grained texture for granites, granoclastic, cataclastic, and mylonitic texture for granite mylonites and granoblastic and lepidoblastic textures for gneisses.

Four deformation phases have been recorded in the study area's rocks. The first phase is marked by the foliation (S_1) which is mostly recognized on gneisses. The second phase is marked by fold (F_2), lineation (L_2), and boudins (B_2) on gneisses and granite mylonite. The third phase is marked by subvertical (S_3) foliation, shears (C_3), and lineations (L_3) recognized on granites mylonites and gneisses. The fourth phase is essentially a brittle phase and was recognized on all the three rock types of the study area.

Granites present mostly magmatic deformation features, while granite mylonites and gneisses present submagmatic to nonmagmatic deformation features. Mylonitization occurred on a brittle-ductile transition phase. Coarse-grained granites emplaced in the lower crustal level, while protomylonite and mylonites occurred in the middle crust and ultramylonites in the upper crust. Fine-grained granite was filtered and channeled through the middle crust shear's areas to be settled at the upper crustal level as irregular spot within the mylonites and gneisses. Deformation structures demonstrated that the studied area is a composite ductile-brittle shear zone. Granites and granites mylonites were syntectonically emplaced during the D_3 sinistral shearing phase.

Author details

Benjamin Ntieche[1*], Pauline Wokwenmendam Nguet[2], Zakari Nchouwet[3], Mahomed Aziz Mounjouohou[4] and Daouda Mfepat[3]

1 Geology Laboratory, Higher Teacher Training College, University of Yaounde I, Cameroon

2 National Institute of Geological and Mining Research, Cameroon

3 Faculty of Science, University of Yaounde I, Cameroon

4 Department of Earth Sciences, University of Maroua, Cameroon

*Address all correspondence to: ntiechebenjo@yahoo.fr

References

[1] Ngako V, Affaton P, Njonfang E. Pan-African tectonics in northwestern Cameroon: Implication for the history of western Gondwana. Gondwana Research. 2008;**14**:509-522

[2] Bouyo Houketchang M, Toteu SF, Deloule E, Penaye J, Van Schmus WR. U–Pb and Sm–Nd dating of high-pressure granulites from Tchollir é and Banyoregions: Evidence for a Pan-African granulite facies metamorphism in northcentral Cameroon. Journal of African Earth Sciences. 2009;**54**:144-154

[3] Van Schmus WR, Oliveira EP, Da Silva Filho AF, Toteu SF, Penaye J, Guimaraes IP. Proterozoic Links between the Borborema Province, NE Brazil, and the Central African Fold Belt. London: Geological Society, Special Publications 294 (No. 1); 2008. pp. 69-99

[4] Dumort JC. Geological map of Cameroon at a scale of 1:500,000, East of Douala sheet with explanatory note. Cameroon: Report of the Directorate of Geology and Mines. 1968. p. 69

[5] Castaing C, Feybesse JL, Thieblemont D, Triboulet C, Chevremont P. Palaeogeographical reconstructions of the Pan-African/ Brasiliano Orogen; closure of an oceanic domain or intracontinental convergence between major blocks? Precambrian Research. 1994;**69**:327-344

[6] Abdelsalam MG, Liegeois J-P, Stern RJ. The Saharan Metacraton. Journal of African Earth Sciences. 2002;**34**:119-136

[7] Toteu SF, Van Schmus WR, Penaye J. The Precambrian of Central Africa: Summary and perspectives. Journal of African Earth Sciences. 2006;**44**:7-10

[8] Nzenti JP, Barbey P, Macaudiere J, Soba D. Origin and evolution of the late Precambrian high-grade Yaoundé gneisses (Cameroon). Precambrian Research. 1988;**38**:91-109

[9] Toteu SF, Penaye J, Djomani YP. Geodynamic evolution of the Pan-African belt in Central Africa with special reference to Cameroon. Canadian Journal of Earth Sciences. 2004;**41**:73-85

[10] Nzenti JP, Barbey P, Bertrand JML, Macaudière J. The pan-African chain in Cameroon: Let's look for suture and model. In: S.G.F., 15th Earth Sciences Meeting. Nancy, France; 1994

[11] Ngnotué T, Nzenti JP, Barbey P, Tchoua FM. The Ntui-Betamba high-grade gneisses: A northward extension of the Pan-African Yaounde gneisses in Cameroon. Journal of African Earth Sciences. 2000;**31**:369-381

[12] Njanko T, Nedelec A, Affaton P. Synkinematic high-K calc-alkaline plutons associated with the Pan-African Central Cameroon shear zone (W-Tibati area): Petrology and geodynamic significance. Journal of African Earth Sciences. 2006;**44**:494-510

[13] Tanko NEL, Nzenti JP, Njanko T, Kapajika B, Nédélec A. New U–Pb zircon ages from Tonga (Cameroon): Coexisting Eburnean-Transamazonian (2.1 Ga) and Pan-African (0.6 Ga) imprints. Comptes Rendus Geosciences. 2005;**337**:551-562

[14] Toteu SF, Van Schmus WR, Penaye J, Michard A. New U–Pb and Sm–Nd data from north-Central Cameroon and its bearing on the pre-Pan African history of Central Africa. Precambrian Research. 2001;**108**:45-73

[15] Njonfang E. Contribution to the study of the relationship between the "Cameroon line" and the Adamawa direction: 1-Petrology and Geochemistry of associated tertiary

magmatic formations. 2-Petrology Geochemistry and Structure of pan-African granitoids in the Foumban-Bankim shear zone (West Cameroon and Adamaoua) [PhD thesis]. Cameroon: University of Yaounde I; 1998. p. 392

[16] Njonfang E, Ngako V, Kwekam M, Affaton P. The Foumban-Bankim calc-alkaline ortho-gneisses: Evidence of an internal area of active sheared pan-African margin. Reports of the Academy of Sciences Paris. 2006;**338**:606-616

[17] Njonfang E, Ngako V, Moreau C, Affaton P, Diot E. Restraining bends in high temperature shear zones: The "Central Cameroon Shear Zone," Central Africa. Journal of African Earth Sciences. 2008;**52**:9-20

[18] Ntiéche B. Petrographic and geo-environmental study of the Magba granitoids West Cameroon [Master thesis]. University of Yaounde I; 2009. p. 53

[19] Ntiéche B, Mohan MR, Moundi A, Mounjouohou MA. Petrogenesis and geochemical characterization of the granitoids of the Magba shear zone West Cameroon Central Africa. Open Journal of Geology. 2016;**6**:812-839

[20] Ntiéche B, Mohan MR, Moundi A. Granitoids of the Magba Shear Zone, West Cameroon, Central Africa: Evidences for emplacement under transpressive tectonic regime. Journal of the Geological Society of India. 2017;**89**:33-46

[21] Simpson C. Deformation of granitic rocks across the brittle-ductile transition. Journal of Structural Geology. 1985;**7**:503-511

[22] Hibbard MJ. Deformation of incompletely crystallised magma systems: Granitic gneisses and their tectonic implications. Journal of Geology. 1987;**951**:543-561

[23] Dell'Angelo LN, Tullis J, Yund RA. Transition from dislocation creep to melt-enhanced diffusion creep in fine-grained granitic aggregates. Tectonophysics. 1987;**139**:325-332

[24] Park Y, Means WD. Direct observation of deformation processes in crystal mushes. Journal of Structural Geology. 1996;**18**:847-858

[25] Ashworth JR, Schneider H. Deformation and transformation in experimentally shock-loaded quartz. Physics and Chemistry of Minerals. 1985;**11**:241-249

[26] Bouchez JL, Gleizes G. Two-stage deformation of the Mont-Louis-Andorra granite pluton (Variscan Pyrenees) inferred from magnetic susceptibility anisotropy. Journal of the Geological Society. 1995;**152**:669-680

[27] Bouchez JL, Delas C, Gleizes G, Nedelec A, Cuney M. Submagmatic microfractures in granites. Geology. 1992;**20**:35-38

[28] Karlstrom KE, Miller CF, Kingsbury JA, Wooden JL. Pluton emplacement along a ductile thrust zone, Pine mountains, southeastern California: Interaction between deformational and solidification processes. Bulletin Geological Society of America. 1993;**105**:213-230

[29] Blumenfeld P, Bouchez J-L. Shear criteria in granite and migmatite deformed in the magmatic and solid states. Journal of Structural Geology. 1988;**10**:361-372

[30] Quick JE, Sinigoi S, Negrini L, Demarchi G, Mayer A. Synmagmatic deformation in the underplated igneous complex of the Ivrea-Verbano zone. Geology. 1992;**20**:613-616

[31] Wickham SM. The segregation and emplacement of granitic magmas. Journal of the Geological Society of London. 1987;**144**:281-297

[32] Weeksteen G. The volcanic massifs of the Bamoun Area. Cameroon: Cameroon Geological Service. Annual Report; 1954

[33] Carter NL, Kirby SH. Transient creep and semibrittle behaviour of crystalline rocks. Pure and Applied Geophysics. 1978;**116**:807-839

[34] Kirby SH, Kronenberg AK. Deformation of clinopyroxenite: Evidence for a transition in flow mechanisms and semibrittle behaviour. Journal of Geophysical Research. 1984;**89**:3177-3192

[35] Moyen J-F, Nédélec A, Martin H, Jayananda M. Syntectonic granite emplacement at different structural levels: The Closepet granite, South India. Journal of Structural Geology. 2003b;**25**(4):611-631

[36] Vigneresse JL, Bouchez JL. Successive granitic magma batches during pluton emplacement: The case of Cabeza de Araya (Spain). Journal of Petrology. 1997;**38**(12):1767-1776

[37] D'Eramo F, Tubía JM, Pinotti L, Vegas N, Coniglio J, Demartis M, et al. Granite emplacement by crustal boudinage: Example of the Calmayo and El Hongo plutons (Córdoba, Argentina). Terra Nova. 2013;**25**(5):423-430

Chapter 5

The Importance of Mechanical Transport, Rock Texture, and Mineral Chemistry in Chemical Weathering of Granites: The Melechov Massif, Czech Republic

Václav Procházka, Miroslav Žáček, Petr Sulovský, Tomáš Vaculovič, Lenka Rukavičková and Dobroslav Matějka

Abstract

Data of 41 or more elements in superficial as well as drill-core samples of the peraluminous Lipnice and Melechov granites, located several kilometers apart in northern Moldanubian Batholith, are evaluated. Weathering of both granite types proceeded in virtually identical time and environment, but it shows very different patterns. In the weathered Lipnice granite, Al_2O_3 slightly increased, loss on ignition (LOI) increased strongly, and contents of all other major elements except for Fe are lower (however, reconcentration of K, Mg, and Ti in secondary phases is possible). In the relatively coarse-grained and more acidic Melechov granite, the depleted major elements are Si, Fe, Ti, Mn, and Mg. Strongly increased Al in half of weathered samples is independent on the moderate increase of LOI and relatively small changes of Na, Ca, K, and P contents. These samples are relatively poor in quartz, which is the result of fossil weathering, mechanical mineral separation, and erosion processes. In the Lipnice granite, however, chemical weathering dominated over mechanical fractionation due to a more compact character of the rock (as well as of biotite and plagioclase). Regarding trace elements, enrichment in Ga and loss of U are the only changes documented in both granite types (in different proportions however). The rare-earth element (REE) fractionation is generally weak, but in the Lipnice granite, two processes are proven: (i) dissolution of apatite which has an M-type lanthanide tetrad effect in the fresh rock and (ii) formation of positive Ce anomaly.

Keywords: granite weathering, temperate climate, trace elements, apatite, lanthanide tetrad effect, grain size

1. Introduction

Knowledge of chemical weathering processes is important for pedology, sedimentology, hydrogeochemistry, environmental chemistry, and petrology. Chemical weathering is most frequently studied in magmatic and high-grade metamorphic rocks which are, by their nature, thermodynamically unstable in superficial conditions. At the same time, chemical composition of the fresh rock is an important fingerprint of magma origin and differentiation. Several elements relatively conservative at weathering, like Sc, Th, and rare-earth elements (REEs), have been also used to assess the source composition of ancient sediments [1].

Chemical weathering of rocks is a complicated interplay of alteration and decomposition of primary minerals including removal of ions in solution, formation of secondary minerals, and removal of the solid weathering products. Ideal conditions for bedrock decomposition differ from ideal conditions for erosion. Therefore, in geological time, environmental changes may lead to the formation of complex profiles, where fossil weathering can be documented [2].

While the research of weathering crusts in the Bohemian Massif was mainly dedicated to economic clay deposits, fossil weathering and small-scale clay formation have been independently documented in various rocks [3]. Pivec [4] described kaolinization in the Říčany granite which probably took place in a warm and humid climate in the Cretaceous. The study of low-temperature fracture fillings in a 100-m-deep borehole in the Kouty granite of the Melechov Massif (MM) [5, 6] documented an intense downward transport of supergene clay enriched in fine-grained resistant accessory minerals prevalently in the lower Cretaceous. In this way, also clay minerals of hydrothermal origin could be concentrated. Štemprok [7] showed that hydrothermal kaolinite in granites and greisens occurs in the whole profile of a 1.5-km-deep borehole at Cínovec.

The Melechov Massif is a good representative of the granite body in highland areas of the Bohemian Massif, with thin soil cover and low rate of recent chemical weathering. However, influence of weathering was documented down to ca. 50 m of depth [6, 8]. In this chapter we will focus on the comparison of chemical weathering of Lipnice and Melechov granites, which proceeds in nearly the same time and space but shows very different patterns due to differences not only in rock and mineral chemistries but also in properties relevant for hydrology, mechanical transport, and mineral separation.

2. Geological setting

The study area is located in the northwestern part of the Bohemian-Moravian Highlands in the Czech Republic (**Figure 1a**). The granites of MM represent the northernmost surface body of Moldanubian Batholith. The massif was formed during the Variscan orogeny (330–300 Ma) in high-grade metamorphic rocks of Moldanubian Unit (prevalently biotite paragneisses and migmatites). The relatively older granites of Lipnice and Kouty types outcrop in the outer part of MM, prevalently on SE (see **Figure 1b**). The central part is formed by the younger Melechov granite type and its derivative, the Stvořidla type. For additional geological and geochemical information, see [9–14].

Similar to other crystalline units of Bohemian Massif, the area was largely peneplenized during the Mesozoic and then uplifted as late as during Pliocene and Pleistocene [15]. Thus, the rocks may have been affected by supergene alterations more than 100 million years ago (even in warmer and more humid climate), as also indicated by Pb isotope ages of fracture fillings [5, 6]. The present climate is temperate with a mean annual temperature of 7 °C and humid, with maximum rainfall in summer; the annual precipitation is roughly 750 mm. Erosion base represented by rivers Sázava and Želivka is located in altitudes 350–380 m. The region has been little industrialized and belongs to the relatively unpolluted ones within central Europe. However, acidification also influenced the area at the end of the twentieth century [16].

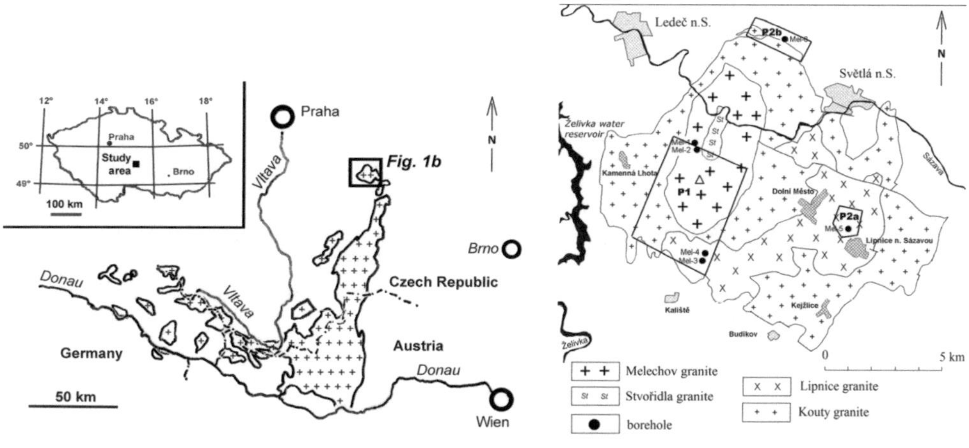

(a) (b)

Figure 1.
(a) Situation of the Melechov Massif in the Czech Republic and in the Moldanubian Batholith (delimitation of the batholith from Mísař et al. [15]). (b) Granites of the Melechov Massif [10] with localization of boreholes and polygons of geochemical mapping. Triangle indicates the top of Melechov (715 m).

The Lipnice granite is fine-grained, with a grain size mostly below 2–2.5 mm. U-Th-Pb monazite ages of this granite range according to the electron microprobe dating from 308 ± 13 to 315 ± 23 Ma [17]. The major minerals are represented by quartz, oligoclase (locally albitized), K-feldspar (mainly microcline), biotite (with dominant annite or siderophyllite component [18, 19]), and muscovite (less abundant than biotite). K-feldspar partly replaces plagioclase, and domains unusually rich in K-feldspar are common, possibly formed by recrystallization of poikilitic microcline observed in some gneisses nearby. In addition, nodules rich in sillimanite and both micas are also abundant, representing probably restite from a material similar to paragneisses present within the area. The significant accessory minerals are ilmenite, fluorapatite (in the following "apatite" only), monazite (Ce), zircon, rutile/anatase, and locally secondary fluorite or pyrrhotite [19–21].

The delimitation of the Lipnice granite boundary in the south and east is difficult and controversial. Also more acidic intrusive and vein rocks appear frequently in the southwest of the granite body [13, 14]. Nevertheless between Lipnice nad Sázavou and Dolní Město, the granite is remarkably homogeneous [18, 21, 22], regardless of the ubiquitous presence of small restite nodules. Microscopic alterations of biotite (with formation of chlorite, muscovite/illite, TiO_2 phase, and K-feldspar) and of plagioclase, i.e., mainly formation of sericite (illite), together with rock fracturation and porosity were recently investigated in detail in the Mel-5 borehole in the Lipnice granite [8].

The Melechov granite to alkali-feldspar granite is relatively coarse-grained (with typical grain size ~5 mm) and more fractionated than the Lipnice type. The major minerals are quartz, K-feldspar (prevalently microcline), albite (prevalently in perthite with An < 5, however magmatic albite and oligoclase also occur), muscovite and biotite (siderophyllite; less abundant than muscovite). Abundant apatite is the dominant accessory mineral. Zircon and monazite are less abundant, whereas primary ilmenite and rutile are scarce [20]. Tourmaline (schorl) is distributed irregularly; xenotime occurs rarely [19].

Chemically the differentiation of Melechov granite is manifested by higher content of SiO_2 and P_2O_5 and lower content of elements which are compatible in peraluminous granites (mainly Mg, Ti, Fe, Ca, Zr, LREE, Th). In addition the

(a) Major elements												
Samples		**SiO_2**	**TiO_2**	**Al_2O_3**	**Fe_2O_3t.**	**MgO**	**MnO**	**CaO**	**Na_2O**	**K_2O**	**P_2O_5**	**LOI**
Lipnice granite												
Fresh, Mel-5 (16.5–150 m), n = 18	Mean	69.87	0.54	14.93	2.49	0.78	0.044	1.06	2.91	5.46	0.301	1.21
	St. dev.	0.24	0.03	0.12	0.05	0.05	0.004	0.12	0.06	0.15	0.017	0.05
Weathered, P2a low-Al group, n = 20	Mean	70.76	0.44	15.15	2.33	0.67	0.027	0.79	2.65	5.09	0.290	1.72
	St. dev.	0.70	0.05	0.27	0.33	0.12	0.007	0.21	0.22	0.16	0.033	0.49
Weathered, P2a high-Al group, n = 26	Mean	67.82	0.49	17.36	2.69	0.74	0.028	0.70	2.35	5.06	0.271	2.38
	St. dev.	1.46	0.08	0.52	0.55	0.13	0.006	0.14	0.34	0.26	0.040	0.93
Fresh, Mel-3 (36.4–106.5, 175–175.5 m), Mel-4 (22–168.4 m), n = 30	Mean	69.79	0.48	14.99	2.53	0.83	0.039	1.10	2.95	5.43	0.256	1.35
	St. dev.	0.41	0.04	0.13	0.12	0.05	0.003	0.07	0.09	0.17	0.010	0.17
Weathered, P1, n = 16	Mean	67.21	0.47	17.14	2.65	0.76	0.034	0.77	2.54	5.12	0.280	2.99
	St. dev.	2.97	0.08	1.83	0.54	0.15	0.010	0.25	0.48	0.33	0.075	1.38
Melechov granite												
Fresh, Mel-1 (37–199.5 m), Mel-2 (22–193 m), n = 44	Mean	72.98	0.14	14.66	1.26	0.25	0.045	0.74	3.51	4.54	0.41	1.20
	St. dev.	0.71	0.01	0.43	0.19	0.08	0.011	0.14	0.52	0.26	0.02	0.43
Weathered, P1 low-Al group, n = 25	Mean	70.44	0.09	16.49	1.00	0.20	0.032	0.71	4.03	4.68	0.48	1.81
	St. dev.	1.16	0.02	0.62	0.14	0.03	0.006	0.12	0.29	0.20	0.08	0.53
Weathered, P1 high-Al group, n = 29	Mean	65.32	0.10	21.51	1.04	0.20	0.030	0.68	3.97	4.62	0.47	1.95
	St. dev.	1.06	0.02	0.89	0.13	0.03	0.007	0.11	0.30	0.28	0.07	0.52

(b) Trace elements and element ratios (average)						
	Lipnice granite			**Melechov granite**		
Samples / **Element**	**Mel-5 fresh**	**P2a weathered —low Al**	**P2a weathered —high Al**	**Mel-1, Mel-2 fresh**	**P1 weathered —low Al**	**P1 weathered —high Al**
Ba	528.4	514.4	486.6	168.1	174.7	172.3
Be	3.8	4.2	4.2	3.8	4.3	4.1
Co	3.9	3.6	4.5	1.0	1.0	0.9
Cs	8.0	7.4	8.6	24.5	25.2	24.9
Ga	23.0	23.9	27.1	21.4	22.3	25.1
Hf	6.5	6.5	6.4	2.0	2.3	2.2
Nb	9.1	8.5	9.2	14.6	13.0	13.7
Rb	322.3	327.9	333.4	296.7	313.6	314.6
Sn	5.2	6.4	7.4	19.1	24.1	25.8

(b) Trace elements and element ratios (average)						
	Lipnice granite			Melechov granite		
Samples / Element	Mel-5 fresh	P2a weathered —low Al	P2a weathered —high Al	Mel-1, Mel-2 fresh	P1 weathered —low Al	P1 weathered —high Al
Sr	109.0	103.8	96.7	92.9	98.0	94.6
Ta	0.81	0.75	0.82	3.5	2.96	3.12
Th	42.9	46.1	43.9	3.9	4.1	4.5
U	9.9	6.6	6.1	13.5	5.0	5.3
V	24	24	30	5.5	4	4
W	1.4	1.9	1.6	48	4.3	4.6
Zr	224.5	213.5	210.6	54	55.6	53.7
Y	15.7	15.5	15.7	11.4	13.1	12.7
La	56.7	52.3	52.6	9.1	9.9	9.7
Gd	5.45	5.03	5.19	2.10	2.22	2.16
Lu	0.17	0.16	0.15	0.09	0.10	0.09
REE total	297	281	271	51.6	53.2	51.9
K/Rb	141	130	127	128	125	123
Ca/Sr	70.1	53.6	50.8	60	51.8	52.1
Al/Ga	3437	3367	3512	3638	3930	4573
Th/U	4.37	7.37	7.21	0.31	0.87	0.88
Zr/Hf	34.4	32.9	33.1	26.6	24.2	24.8
Nb/Ta	11.4	11.8	11.3	4.23	4.5	4.4
Eu/Eu*	0.247	0.235	0.239	0.43	0.443	0.43
Ce/Ce*	1.06	1.09	1.09	1.05	0.99	1.00
Pr/Pr*	1.11	1.01	1.00	1.02	0.97	0.97
Y/Ho	32.6	34.0	33.9	32.8	35.5	34.6

Primary data are from [13, 21]. Note that borehole samples of "fresh" Melechov granite include also hydrothermally altered rock which is relatively common. Weathered samples from boreholes were excluded.
$Eu/Eu^* = 2Eu_N/(Sm_N + Gd_N)$.
$Ce/Ce^* = 3Ce_N/(2La_N + Nd_N)$.
$Pr/Pr^* = 3Pr_N/(2Nd_N + La_N)$.

Table 1.
Statistics of elements in boreholes (Mel-1 to Mel-5) and in soil base samples of polygons (P1, P2a).

Melechov type has significantly lower fluorine content (0.08–0.14 wt.%) than the Lipnice type (usually 0.20–0.25 wt.%) and somewhat lower content of K (see also **Table 1a**; **Figure 2**).

In comparison with the Lipnice type, the Melechov granite is generally more affected by subsolidus alterations [13, 14]: chloritization; baueritization with formation of secondary Ti, Fe, and Zn oxides [23]; sericitization of plagioclase (with formation of secondary apatite); probably kaolinization of feldspars; and locally carbonatization. Samples affected by such alteration processes (except for carbonatization) exhibit weaker chemical and mineralogical changes (including coloration by ferric pigments) during supergene weathering, whose influence is sometimes difficult to be distinguished from alteration at higher temperature [24].

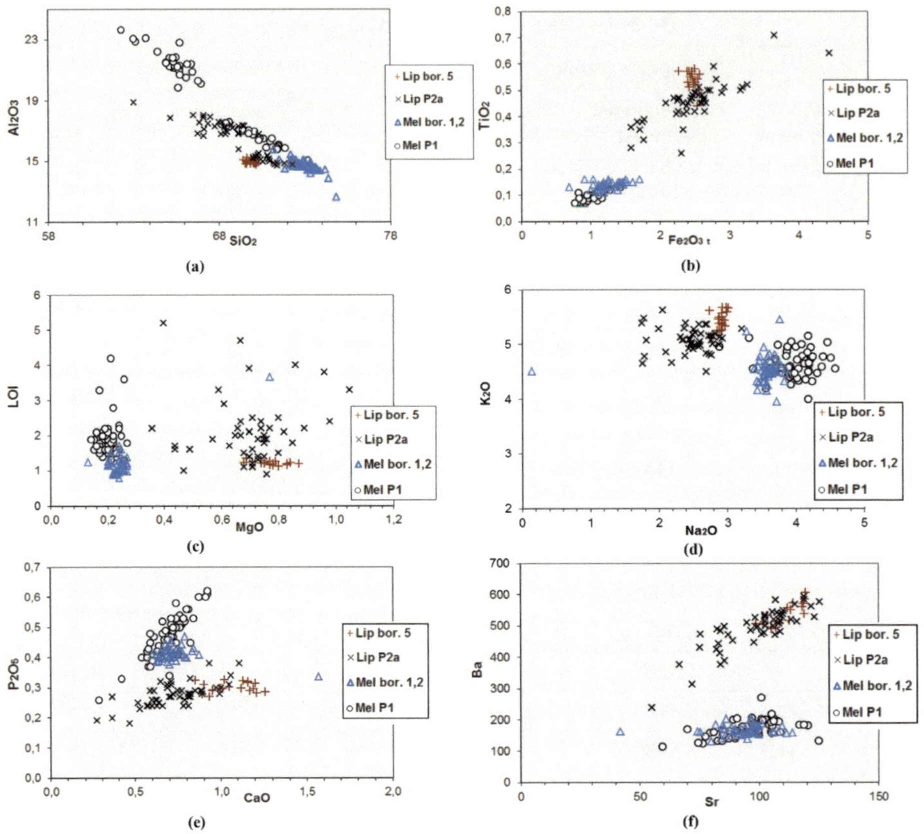

Figure 2.
(a–f) Plots of selected pairs of chemical parameters (major components in wt. %, Sr and Ba in ppm) in Lipnice and Melechov granites from boreholes Mel-5, Mel-1, and Mel-2 (including fresh, altered, and weathered samples) and from soil cap in polygons P2a and P1.

3. Material and methods

Since the 1990s, the Melechov Massif has been the subject of intensive investigation, leaded by the Radioactive Waste Repository Authority of the Czech Republic (RAWRA = SÚRAO). The results include numerous whole-rock analyses of granites from cores of boreholes 100–200 m deep and from shallow drill holes. These data are completed in unpublished reports [13, 21] and with respect to weathering partly discussed in [25–27].

The fresh granites are represented by core samples taken in intervals of 5–10 m from boreholes Mel-1 and Mel-2 (Melechov type) and Mel-3 to Mel-5 (Lipnice type). However, due to complicated geology and petrology in boreholes Mel-3 and Mel-4, which were intentionally drilled in inhomogeneous environment [13, 14], we present results from these two boreholes only marginally. Borehole samples displaying visible indices of weathering have been included in graphs (**Figure 2**) but not in the statistical processing (**Table 1**). We also used nine analyses of generally very slightly weathered Lipnice granite from outcrops in polygon P2a, analyzed under the same conditions [22]. The major elements were analyzed in the labs of the Czech Geological Survey (ČGS) in Prague and trace elements in the ACME Laboratories, Vancouver, Canada (by ICP-OES and ICP-MS).

As for weathered granite, analyses of samples taken from the lower part of shallow drill holes (soil cap) in the polygons P1 (which also includes the boreholes

Mel-1 to Mel-4) and P2a (which includes borehole Mel-5) were used. These samples were collected in profile grid making a regular network (interval between profiles is ca. 750–900 m, sampling step on profile 150 m). Samples for chemical analyses were collected from the maximum attainable depth, this varying from 0.6 to 2.8 m (mostly close to 2 m in Lipnice granite and 1.5 m in Melechov granite)—see [21] for details. Analyses of saprolites containing pegmatite or quartz veins were excluded from the data processing in the presented study.

The polygon P1 extends from contact of the Melechov and Stvořidla granites to exocontact of the Lipnice granite with paragneisses (the gneisses/migmatites sampled in this area seem to be chemically very similar to the Lipnice granite [21]). Between the Lipnice and Melechov granites outcrops Kouty granite, compositionally and texturally largely transitional between the Lipnice and Melechov types; this rock, however, is represented by relatively few samples and so is not considered here. The area is dominated by the Melechov hill (715 m) and covered prevalently by a managed spruce forest. The major soil type in the forest is dystric cambisol, locally podzolic [28]. Especially in the southern part of P1 (i.e., mainly on Lipnice and Kouty granites), there are also agricultural fields and meadows.

The polygon P2a north of Lipnice nad Sázavou represents Lipnice granite (only locally with acidic vein intrusions) and is dominated by the Holý vrch hill (620 m). Almost all samples were taken in the forest (prevailing spruce) and only few samples in abandoned quarries. The dominant soil type is lithic cambisol, in places also pseudogleys occur [28].

Samples from soil cap were analyzed in the ACME Laboratories for both major and trace elements, the methods applied for the presented trace elements being the same as in case of samples from boreholes. The fact that major elements have not been determined under the same conditions in samples of boreholes and of soil cap implies that subtle differences between fresh and weathered rocks have to be treated carefully; however it plays no role in comparison of behavior of the Lipnice and Melechov granite types.

Regarding the distribution of trace elements in rock-forming minerals of the fresh granite, mainly data from the previous studies [19, 20] are considered. In addition, new trace element analyses of apatite are presented. REE, Y, Sr, Th, and U in apatite have been determined by LA-ICP-MS spot analyses in polished sections of rocks at the Department of Chemistry of Faculty of Science, Masaryk University, Brno.

Instrumentation for the LA-ICP-MS consists of a laser ablation system UP 213 (New Wave, USA) and an ICP-MS spectrometer Agilent 7500 CE (Agilent, Japan). A commercial Q-switched Nd:YAG laser ablation device works at the wavelength of 213 nm. Helium was used as the carrier gas. For measurements we used hole drilling mode for the duration of 40 seconds for each spot. Laser ablation was performed with a spot diameter of 25 μm, laser fluence 4.5 J cm^{-2}, and repetition rate 10 Hz. The isotopes were measured with integration time 0.1 s / isotope. Optimization of LA-ICP-MS parameters (gas flow rates, sampling depth, electrostatic lens voltages of the MS) was performed with the glass reference material NIST SRM 612 with respect to maximum S/N ratio and minimum oxide formation (ThO^+/Th^+ count ratio 0.2%, U^+/Th^+ count ratio 1.1%).

4. Results

4.1 Composition of fresh and weathered rocks

Several interesting facts not related to weathering, especially some vertical compositional gradients of fresh rocks, have been also found. Here, we present this information only to distinguish the influence of weathering processes from original granite inhomogeneity. The chemical contrasts between Lipnice and Melechov granite types (including the ratios of isovalent elements Zr/Hf and Nb/Ta and the Eu/Eu* ratio) are mostly not significantly affected by weathering; however in case of few elements, they were enhanced (Na, P) or smoothened up to reversed (Si, K, U).

4.1.1 Major elements

Core samples of Lipnice granite from the borehole Mel-5 have very low variability regarding major as well as trace elements (see also **Table 1**), and the results show no relation to rock pigmentation and microscopic alteration patterns (chloritization, sericitization) which were investigated in detail [8] in this borehole (except for the uppermost weathered sample, representing the "brown granite"). In Mel-3 and Mel-4, the situation is more complicated: in addition to the typical Lipnice granite, also magmatic vein rocks and a relatively bright, more acidic variety of Lipnice granite were found [13]. In addition, hydrothermally altered granites occur (in contrast to Mel-5). We excluded such samples from the data processing; however it is obvious that even the "fresh" Lipnice granite from boreholes Mel-3 and Mel-4 is less representative than that from Mel-5.

Similarly, the polygon P1 is very inhomogeneous in comparison with P2a [21], and in addition redistribution and contamination of weathered material have been more intensive there [27]. For these reasons and due to the fact that the Lipnice granite is represented by relatively few samples in P1, we evaluated the weathering of Lipnice granite mainly by comparison of borehole Mel-5 and polygon P2a.

Comparing borehole samples taken from various depths, iron oxidation is observed near the surface [25], which is also supported by several slightly weathered samples taken from outcrops in P2a [22]. However, Fe_2O_3 and FeO were not determined in samples from soil cap singly, what precludes the application of Fe_2O_3/FeO ratio as otherwise a powerful indicator of chemical weathering in our study.

The weathered Lipnice as well as Melechov granites are enriched in Al_2O_3 and H_2O (which due to low C and S content represents the majority of loss of ignition), especially at the expense of SiO_2. Two distinct groups of weathered granite can be distinguished according to Al content, most notably in the Melechov type: $Al_2O_3 \leq 17.6\%$ and $Al_2O_3 \geq 19.8\%$ (**Figure 2a**). As shown by [24, 26, 27], the second group is enriched in small particles, which include secondary minerals, detrite of plagioclase, micas and chlorite, and accessory minerals, and is relatively depleted in quartz (the original overlying quartz-rich eluvia were mostly eroded; nevertheless in places, sandy eluvium was preserved and locally used as a building material). Such mineral fractionation was most effective in the coarse-grained Melechov granite, where plagioclase had been already intensively affected by subsolidus alteration and small particles could be transported through a skeleton formed by quartz and K-feldspar [27].

In the Lipnice granite, two groups with different enrichment in Al can be observed in polygon P2a, but they are not so contrasting (Al_2O_3 = 14.57–15.82 wt. % and 16.66–19.11 wt. %, respectively; in the first group, SiO_2 is not lower than in borehole). The most contrasting single parameter of chemical weathering intensity is hydration (expressed as LOI), reaching higher values than in the Melechov type (**Figure 2c**).

The behavior of K is complicated: its content seems to be slightly decreased in weathered Lipnice granite. However, there is no trend of ongoing K removal with weathering intensity (**Table 1**; **Figure 2d**). One of the possible explanations is formation of illite. In weathered Melechov granite, no systematic shift of K content is observed.

Both Ca and Na exhibit very different behaviour in the two granites. In weathered Melechov type, Ca is comparable and Na even higher than in the fresh rock (**Figure 2d** and **e**). In contrast, in the Lipnice type, Ca and Na are strongly depleted during weathering. Mg is slightly depleted in weathered granites of both types; however note that in the Melechov granite, this could be related rather to the original magma inhomogeneity than to the weathering (see **Figure 3d**). As for Mn, it is removed by weathering in both granites (**Table 1**).

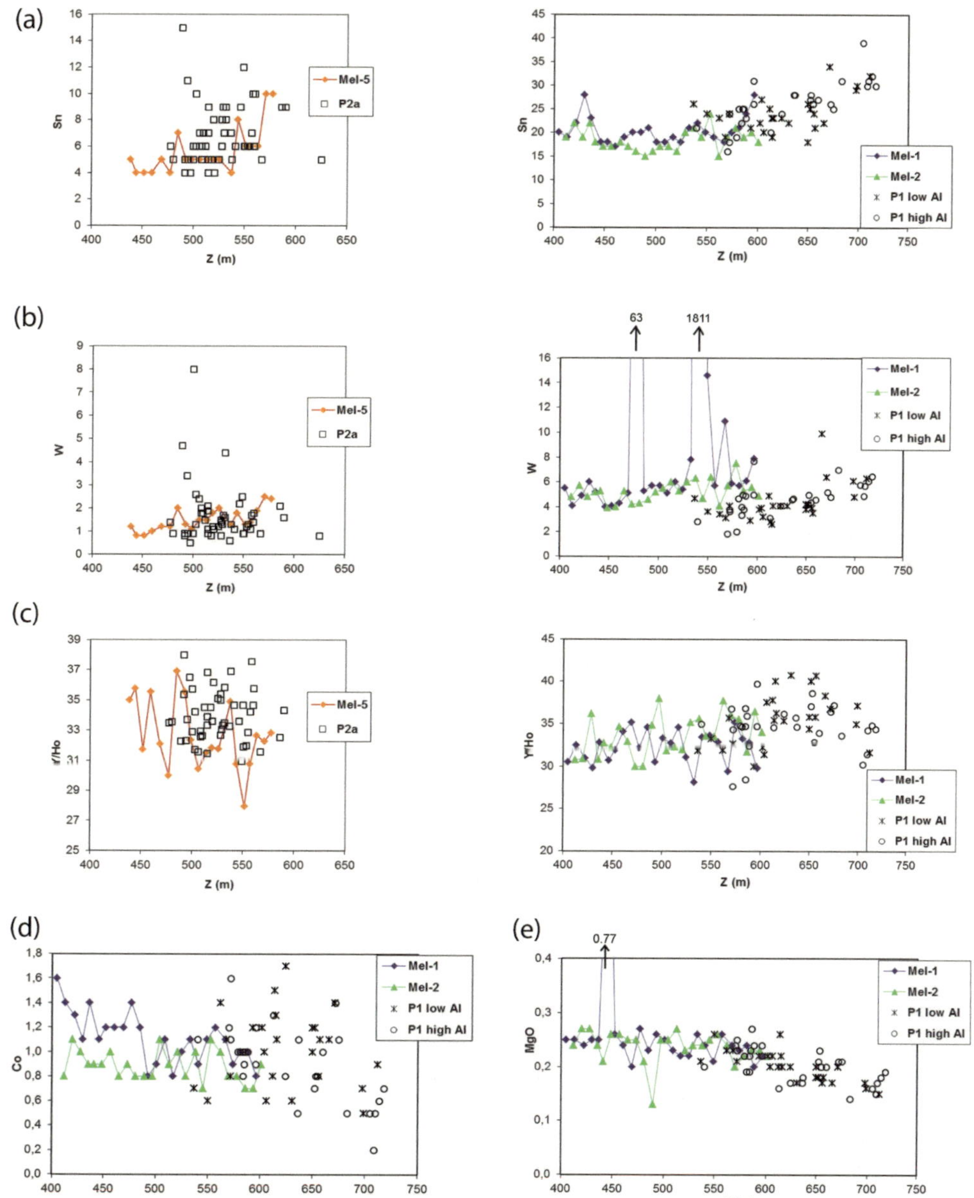

Figure 3.
Relationships of selected element concentrations and ratios in granites of boreholes and polygons (soil cap) to altitude. (a)–(c) Sn (ppm), W (ppm), and Y/Ho ratio in Lipnice and Melechov granites. (d) Co (ppm), (e) MgO (wt. %) in Melechov granite.

Phosphorus, slightly depleted in Lipnice granite in P2a only and possibly enriched in weathered Melechov granite, has in all weathered granites positive correlation with calcium (**Figure 2e**), missing in the fresh rock. This indicates importance of apatite. Negative correlation of P and LOI especially in weathered Lipnice type is an indication for apatite dissolution. Another indirect evidence for apatite dissolution and phosphorus mobility is the common occurrence of P-rich limonite in eluvia and low-temperature fracture fillings ([6] and unpublished data of V. Procházka).

Iron content decreased in the weathered Melechov granite but not in Lipnice granite (**Figure 2b**), where the correlation of Fe and LOI (R = 0.68) indicates the possibility of passive (re)concentration of Fe in weathered rock. Note that Fe content (especially Fe_2O_3) is also significantly lower in borehole Mel-2 than in Mel-1 [13], perhaps as a result of subsolidus alterations of Melechov granite. The behavior of Ti is similar to that of Fe; however weathered Lipnice granite in polygon P2a is mostly significantly depleted in Ti relatively to the fresh rock (**Figure 2b**).

The total carbon content was not measured in boreholes; in soil cap samples, it is usually smaller than 0.2 wt. %, the peak value being 0.36% in P1 and 0.95% in P2a.

It follows that mainly elements concentrated in plagioclase are depleted in the Lipnice granite whereas elements originally concentrated mainly in biotite (or its alteration products) are depleted in both granite types, except for Fe in Lipnice granite.

4.1.2 Trace elements

Only elements showing significant fractionation during weathering at least in one granite type are presented here.

Sr, Ba. Strontium and barium are significantly depleted in weathered Lipnice granite but not in weathered Melechov granite (**Figure 2f**). Despite the removal of Sr, the mean Ca/Sr ratio of weathered Lipnice granite is significantly lower than that of fresh rock (**Table 1b**).

Co. There is no systematic trend of Co content in Lipnice granite. In Melechov type, however, Co has a decreasing trend with altitude (as a compatible element), and it cannot be excluded that weathering leads to enrichment in Co (**Figure 3e**). The Co content in borehole Mel-2 is mostly lower than in Mel-1.

Ga. Gallium was passively concentrated in chemically weathered rocks similarly to Al. However, the Al/Ga ratio in weathered Melechov granite is significantly higher than in fresh rock. In weathered Lipnice granite, such a systematic trend is not observed; only the variability of Al/Ga ratio is much higher.

Sn. Tin content is at average higher in weathered Melechov granite than in the fresh rock (**Figure 3a**; **Table 1b**). This may reflect vertical differentiation trend only, which however is not apparent in boreholes Mel-1 and Mel-2. Possible explanation is stronger chemical fractionation in apical part of the Melechov granite intrusion (according to [29], the original contact was not far above the present top of Melechov hill). Interestingly, Mg in the Melechov granite exhibits opposite behavior to Sn (**Figure 3d**). In the Lipnice granite, Sn also has an increasing trend with altitude, complicating the evaluation of possible weathering influence (**Figure 3a**).

V. In weathered Lipnice granite, V content is higher than in boreholes. The concentration of V by Fe-oxyhydroxides is one of the possible explanations. In Melechov granite, vanadium was mostly below detection limit (<5 ppm), the exception being borehole Mel-1. It follows that weathering led rather to removal of V from the Melechov type; however the effect of older alterations could be similar.

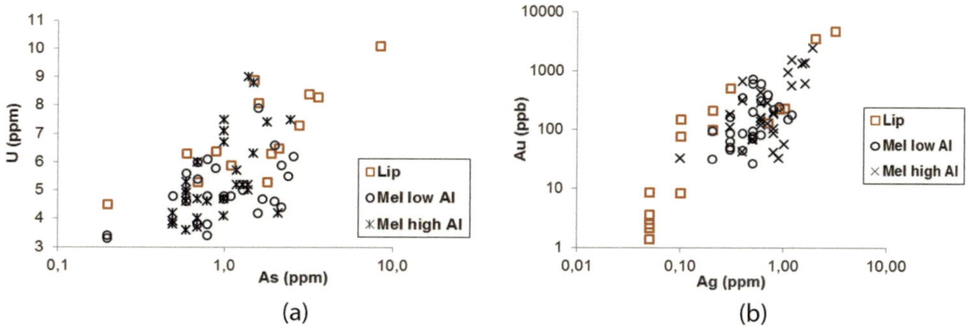

Figure 4.
(a) Relation of As and U content (in ppm) in Lipnice and Melechov granites in polygon P1 (samples of Melechov granite divided into groups with lower and higher Al content). (b) Gold and silver in granite eluvia of Lipnice and Melechov types in P1.

W. Of the elements analyzed, W is the most differentiated one by vertical fractionation. Its content increases with altitude in boreholes and even in P1, probably reflecting fluid-dominated upward transport of incompatible elements. After distinguishing this vertical trend, it is obvious that weathered Melechov granite is significantly depleted in W and weathered Lipnice granite possibly too (**Figure 3b**); in both cases, however, mean values are biased by outliers.

U. In Lipnice granite, the U contents both in the samples from shallow pits and even from outcrops are significantly lower than in the borehole samples. The removal of U from Melechov granite was yet more intensive than from the Lipnice type. As shown by mineral chemistry data and mass balance calculations [19, 30], the rock-forming accessory minerals (monazite, zircon, apatite; in the Melechov type also xenotime) contain at most 70% of U in the Lipnice granite and < 50% U in Melechov granite (in fresh rocks), the rest being obviously in an unstable phase. One possibility is uraninite which was found scarcely [20]; perhaps more important is uranium bound to Fe-oxyhydroxides (see also [5]) and along grain boundaries [31].

In the soil cap, there is correlation of U and As, missing in boreholes and suggesting formation of secondary uranium arsenate, in both granite types in polygon P1 (**Figure 4a**; in P2a, As was not analyzed).

Unlike U, no systematic shift of Th concentration was observed. Therefore, Th/U ratio increased during weathering of both granites (**Table 1b**). Thorium is concentrated predominantly in monazite whose Th/U ratio is higher than that of respective whole rock [19, 20, 30].

Au, Ag. Both elements have been systematically determined only in eluvia of P1 polygon. Nevertheless the high concentrations, especially of Au in Al-rich eluvia, cannot be explained by their high content in the original granite, and probably not only passive concentration during weathering but also supergene contamination was important [27]. In six samples from boreholes Mel-1 and Mel-2, the peak Au content (measured by ET-INAA) is 4.5 ppb (V. Procházka & J. Mizera, unpublished data), i.e., by 1–3 orders of magnitude lower than in eluvia of Melechov granite (**Figure 4b**).

REE, Y. REEs are generally little affected by weathering. However, the most weathered (high-Al) group of Lipnice granite shows some depletion in total REE, and evaluation of their mutual ratios revealed several trends. The comparison of variability of individual REEs in boreholes and in eluvia shows that the variation coefficient (the mean/standard deviation ratio) in weathered Melechov granite has a distinct minimum at Eu (**Table 2**). It follows that in weathered Melechov granite, the content of feldspars (the major reservoir of Eu^{2+}) in individual samples is more stable than that of the main carriers of trivalent REEs, including Sm and Gd—monazite and apatite. The fact that similar situation is not observed in the Lipnice granite could be related to more intense weathering of feldspars and to more homogeneous distribution of monazite in the Lipnice type (in Melechov granite, the most of monazite is bound to large apatite crystals [19]).

In "fresh" Melechov granite of borehole Mel-2, the REE distribution including elevated Eu/Eu* ratio is very similar to weathered Melechov granite, the weathered granite having only somewhat higher total REE (**Figure 5b**). This shows that the effect of surface weathering on REE distribution of Melechov granite was very similar to the effect of former alteration processes, which were more intensive in Mel-2 than Mel-1. Note that the lower REE content in Melechov granite is associated with relatively greater analytical uncertainty.

To display subtle changes at a relatively low degree of weathering, we normalized REE in weathered samples by the average value of REE in boreholes (**Figure 5**).

The Y/Ho ratios of weathered granites seem to be somewhat elevated, which is however partly masked by vertical fractionation (**Figure 3c**).

In the Lipnice granite of outcrops (in P2a), there are relative minima of Ce and Pr, resembling the W type of tetrad effect. The appearance of only first tetrad can be related to REE distribution in apatite (see Section 4.2.). In more weathered samples (soil cap) of P2a, the minimum at Ce gradually disappears, and rather a positive cerium anomaly is formed. The W-type tetrad effect and positive Ce anomaly partially mask one another. It can be summarized that some portion of REE controlled by apatite (with M-type tetrad effect) was removed from weathered rocks, but Ce was partly immobilized by oxidation to Ce^{IV}.

The fact that fractionation of La, Ce, Pr, and Nd does not reflect the original granite inhomogeneity is documented by **Figure 6**.

	La	Ce	Pr	Nd	Sm	Eu	Gd	Tb	Dy	Y	Ho	Er	Tm	Yb	Lu
Lip bor. 5	0.05	0.04	0.04	0.05	0.05	0.06	0.07	0.09	0.07	0.06	0.08	0.09	0.13	0.11	0.10
Lip P2a	0.16	0.15	0.15	0.15	0.15	0.14	0.13	0.14	0.13	0.14	0.16	0.17	0.15	0.17	0.17
Mel bor. 1,2	0.12	0.11	0.10	0.13	0.12	0.10	0.13	0.12	0.13	0.10	0.13	0.12	0.28	0.13	0.20
Mel P1	0.20	0.20	0.20	0.21	0.21	0.14	0.22	0.22	0.22	0.20	0.19	0.19	0.22	0.19	0.22

Table 2.
Variation coefficients for individual REE and Y in fresh and weathered granites (bor. = borehole Mel-5 or Mel-1 and Mel-2).

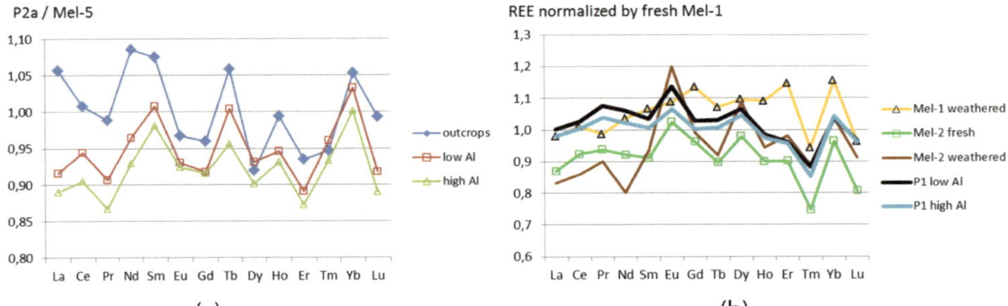

Figure 5.
(a) REE in Lipnice granite of soil base (divided into low-Al and high-Al groups) and small rock outcrops in polygon P2a, normalized by the mean of the borehole Mel-5. (b) REE in various groups of samples of Melechov granite normalized by the mean of fresh Melechov granite from borehole Mel-1.

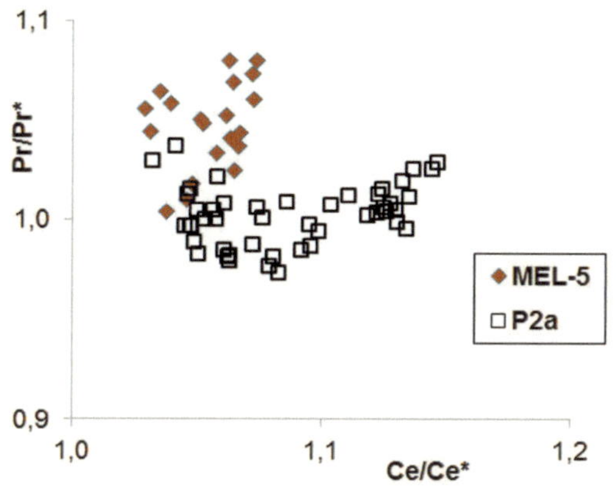

Figure 6.
Plot of cerium anomaly and Pr/Pr (as a manifestation of the first tetrad) in the Lipnice granite in borehole Mel-5 and polygon P2a (soil cap).* $Ce/Ce^* = 3Ce_N/(2La_N + Nd_N)$; $Pr/Pr^* = 3Pr_N/(2Nd_N + La_N)$. *Normalizing values from [32].*

Out of other REEs, the only systematic fractionation in weathered granites is a relative maximum of Yb (**Figure 5a** and **b**).

Other elements concentrated in monazite (Th), zircon (Zr, Hf), and (Fe-)Ti oxides (Nb, Ta) are not significantly affected by weathering, which is also true for Zr/Hf and Nb/Ta ratios. Therefore, it seems that dissolution of primary accessory minerals except for apatite was not significant for REE behavior during weathering. The locally observed alteration of monazite can be attributed to Ca-rich hydrothermal fluids [20].

(a) Lipnice granite, Kopaniny (standardized to assumed Ca content 38 wt. %)													
An. no.	**1**	**2**	**3**	**4**	**5**	**8**	**9**	**10**	**14**	**15**	**16**	**Mean**	**St. dev.**
Sr	128	153	140	122	121	102	129	118	105	124	151	**127**	15
Y	2101	2755	2477	2182	2040	1979	2124	1959	2031	2470	1171	**2117**	367
La	384	486	473	431	330	370	373	399	405	484	222	**397**	70
Ce	1182	1633	1762	1661	1128	1363	1304	1500	1511	1461	842	**1395**	245
Pr	180	267	287	271	166	226	226	209	221	212	142	**219**	41
Nd	970	1353	1203	1352	815	1039	1132	890	956	1033	743	**1044**	183
Sm	376	475	391	475	318	367	417	350	347	386	233	**376**	63
Eu	9	10	9	10	9	8	10	7	7	9	12	**9**	1,5
Gd	310	525	372	502	368	440	420	316	345	396	262	**387**	74
Tb	48	89	66	65	58	70	63	47	61	58	41	**60**	12
Dy	351	605	516	412	378	475	393	330	422	449	294	**421**	81
Ho	67	98	103	79	73	82	74	64	74	80	61	**78**	12
Er	207	271	272	247	212	213	196	188	197	193	145	**213**	34
Tm	32	36	38	34	32	27	26	29	31	26	28	**31**	3,6
Yb	204	241	216	188	198	172	174	187	184	167	138	**188**	25
Lu	27	38	30	27	27	22	30	26	25	25	21	**27**	4,1
Th	21	16	9	9	23	20	25	10	15	10	19	**16**	5,3

(a) Lipnice granite, Kopaniny (standardized to assumed Ca content 38 wt. %)												
An. no.	**1**	**2**	**3**	**4**	**5**	**8**	**9**	**10**	**14**	**15**	**16**	**Mean**	**St. dev.**
U	39	63	72	79	54	52	41	43	64	54	34	**54**	13
Ca/Sr	2974	2489	2707	3117	3143	3720	2936	3220	3628	3055	2514	**3046**	361
Eu/Eu*	0.083	0.061	0.070	0.065	0.079	0.057	0.070	0.063	0.058	0.073	0.150	**0.075**	0.024
Ce/Ce*	1.08	1.08	1.13	1.14	1.15	1.10	1.06	1.24	1.20	1.09	1.11	**1.12**	0.05
Pr/Pr*	1.26	1.38	1.43	1.44	1.26	1.44	1.48	1.24	1.30	1.20	1.48	**1.36**	0.10
Y/Ho	31.1	28.0	24.0	27.6	28.1	24.2	28.9	30.8	27.6	30.8	19.2	**27**	3

(b) Melechov granite, Leštinka (standardized to assumed Ca content 35 wt. %)										
An. no.	**1**	**2**	**3**	**4**	**5**	**6**	**7****| **8**	**9**	**Mean**	**St. dev.**
Sr	186	257	257	153	116	98	143	134	212	**177**	58
Y	252	202	207	1573	946	1740	1686	1876	747	**943**	662
La	201	345	271	242	236	220	536	195	283	**249**	46
Ce	385	766	594	655	603	738	1597	616	730	**636**	113
Pr	46	87	78	86	79	109	205	103	94	**85**	18
Nd	138	247	271	370	302	406	863	469	334	**317**	96
Sm	32	46	54	170	91	154	247	195	95	**105**	58
Eu	12.0	13.7	10.4	12.0	7.3	7.2	6.7	6.6	15.7	**11**	3.1
Gd	28	62	60	178	134	249	361	334	100	**143**	98
Tb	11	9	10	33	40	74	77	73	24	**34**	25
Dy	52	42	49	238	204	409	422	422	152	**196**	144
Ho	7	6	8	43	33	59	69	70	28	**32**	23
Er	13	16	13	94	81	143	142	131	62	**69**	49
Tm	1.7	2.5	1.5	13.4	11.1	17.3	15.6	14.3	7.4	**9**	5.8
Yb	16	19	6	68	65	119	104	82	51	**53**	36
Lu	1.2	0.6	1.2	9.7	7.2	11.4	12.6	8.7	6.2	**6**	4.0
Th	6.8	0.7	2.2	4.3	2.3	4.1	201	25.6	7.6	**7**	7.5
U	19	26	26	205	138	236	333	148	105	**113**	79
Ca/Sr	1881	1363	1363	2290	3020	3571	2447	2620	1648	**2220**	753
Eu/Eu*	1.208	0.786	0.553	0.211	0.202	0.113	0.068	0.079	0.492	**0.455**	0.366
Ce/Ce*	0.92	1.03	0.97	1.09	1.06	1.14	1.16	1.03	1.07	**1.04**	0.06
Pr/Pr*	1.03	1.10	1.07	1.05	1.10	1.30	1.09	1.16	1.14	**1.12**	0.08
Y/Ho	34.1	33.5	25.7	36.9	28.7	29.4	24.5	26.8	27.1	**30.3**	3.8

*******Analyzed spot influenced by monazite (not included in the statistics).*
*Eu/Eu** = $2Eu_N/(Sm_N + Gd_N)$.
*Ce/Ce** = $3Ce_N/(2La_N + Nd_N)$.
*Pr/Pr** = $3Pr_N/(2Nd_N + La_N)$.
Normalizing bulk Earth values from [32].

Table 3.
LA-ICP-MS spot analyses of apatite (elements in ppm).

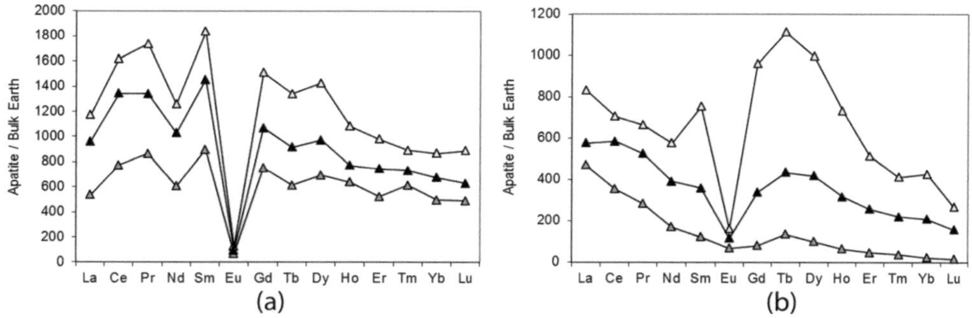

Figure 7.
Bulk Earth normalized [32] REE in apatite (minimum, maximum, and median) as determined by LA-ICP-MS in polished sections of rocks: (a) Lipnice granite, Dolní Město—Kopaniny. (b) Melechov granite, Leštinka.

4.2 REE distribution in apatite

The analyzed trace elements indicate that apatite is a relevant carrier of Y and HREE; however its contribution to whole-rock LREE budget cannot be neglected as well. Results of apatite LA-ICP-MS analyses in polished sections (**Table 3**) are consistent with ICP-MS solution analyses (see [20, 30]). The negative europium anomaly in apatite is deeper than that one in the whole rock. The apatite of Lipnice granite is characterized by strong M-type tetrad effect, forming the first tetrad (whose magnitude is proportional to Pr/Pr* and, in the absence of cerium anomaly, also to Ce/Ce* values) and probably second tetrad (**Table 3a**, **Figure 7a**).

In apatite of Melechov granite (as well as of Kouty and Stvořidla granites), the tetrad effect was already documented before [20]. However, the first tetrad from solution analyses is weak and incomparable to that of apatite in Lipnice type. This is partly due to greater portion of altered and secondary apatite, which does not show tetrad effect (e.g., spot 1 in **Table 3b**). The most spot analyses of apatite in Melechov granite (**Table 3b**, **Figure 7b**) show probably the first three tetrads which, however, are weak.

5. Discussion

5.1 Alteration of primary minerals by chemical weathering

The overall rate of Ca release from the rock is higher than that one of Na, because a large part of the Na budget is relatively better fixed as albite lamellae in perthitic K-feldspar, which is more resistant to weathering than plagioclase; a minor influence of apatite (and perhaps calcite) is possible as well.

Strontium is probably—in contrast to calcium—concentrated in K-feldspar too (see also [33]). This would explain why Ca/Sr ratio decreases at weathering (**Table 1b**). Also note that the Ca/Sr ratio in springwater (~130 at Lipnice granite and ~200 at Melechov granite; **Table 4a**) is significantly higher than in the rocks. White [34] suggested that at the initial stages of weathering, Ca and Sr are released mainly from accessory calcite. According to [35], both calcite and apatite are important sources of Sr in the very early stages of weathering. The loss of Ba and Na shows that the simplest explanation—leaching of Ca and Sr from plagioclase and K-feldspar—is the most likely. Nevertheless the decrease of P_2O_5 in the most weathered samples suggests that apatite contributes to the release of Ca, too. The Sr abundance in apatite is in the range ca. 100–250 ppm (**Table 3**), and so the Ca/Sr ratio of apatite

(a) Statistics of major ions and pH in springs on Melechov and Lipnice granites sampled within the frame of the PADAMOT project [36]; if values below detection limit represented less than half of analyses, they were replaced by 1/2 detection limit

	Substrate rock type	Melechov in the forest (n = 17)		Melechov others (n = 8)		Lipnice (n = 4)	
	Unit	Mean	Median	Mean	Median	Mean	Median
Na^+	mg/l	8.86	9.16	11.41	10.57	8.01	7.67
Mg^{2+}	mg/l	2.97	2.87	4.44	4.81	3.15	2.87
Al	mg/l	0.22	0.2	0.04	0.1	0.1	0.1
K^+	mg/l	1.39	1.38	11.24	5.72	1.63	1.55
Ca^{2+}	mg/l	14.9	13.5	24.9	26	11.9	10.4
Mn^{2+}	µg/l	89.3	25	34.8	9.5	4.4	5
Zn^{2+}	µg/l	41.3	35	18.5	17	9	10
Sr	µg/l	77.8	72.4	117.7	134.6	92.3	81.2
$(NO_3)^-$	mg/l	5.68	2.04	24.8	22.9	20	11.3
F^-	mg/l	0.28	0.19	0.1	0.08	0.08	0.08
$(SO_4)^{2-}$	mg/l	57.6	57.4	47.6	49.8	21.5	22.7
Cl^-	mg/l	2.43	2.21	13	12.5	6.39	4.74
pH		5.32	6.08	6.98	7.04	7.04	7.14
Ca/Sr		192	186	212	193	129	128

DOI: http://dx.doi.org/10.5772/intechopen.91383

(b) Chemistry of springs (all components in mg/l) sampled by L. Rukavičková and co-workers in 2004 within the frame of the RAWRA project; for comparison the discharge of catchment Loukov [16, 44]; values in italics represent 1/2 of detection limit

		Na^+	Mg^{2+}	Al	K^+	Ca^{2+}	$(NO_3)^-$	F^-	$(SO_4)^{2-}$	Cl^-	SiO_2	pH
Object, date	**Method**	**FAAS**	**FAAS**	**FAAS**	**FAAS**	**FAAS**	**HPLC**	**ISE**	**HPLC**	**HPLC**	**FAAS**	**ISE**
Springs on Lipnice granite (n = 5) 27.5. / 22.6.2004	Mean	10.0	4.5	0.20	1.8	17.0	23.2	0.18	48.6	7.1	19.3	6.3
	Median	9.7	4.3	0.20	1.7	18.0	21.0	0.15	40.6	6.6	19.4	6.2
Springs on Melechov granite (n = 24) 27.5. / 22.6.2004	Mean	8.1	2.8	0.42	1.6	14.0	3.4	0.25	56.7	2.7	17.9	5.43
	Median	8.4	2.9	0.27	1.4	14.3	2.3	0.21	57.9	2.3	18.1	5.86
Catchment discharge 1.6.2004		9.8	3.2	0.24	0.5	13.9	*0.15*	0.26	62.7	2.0		5.11
Catchment discharge X.1994–X.2014 (weighed average, monthly sampling)		8.9	3.2	0.36	0.82	13.1	0.29	0.21	59.4	2.1	21.1	5.06

FAAS, flame atomic absorption spectrometry; PMT, photometry; HPLC, high-performance liquid chromatography; ISE, ion-selective electrode.

Table 4.
Springs on Melechov and Lipnice granites sampled within the frame of the (a) PADAMOT project and (b) RAWRA project.

is by 2–3 orders of magnitude higher than that of the whole rock. Thus, apatite weathering can contribute to the lower Ca/Sr ratio in the eluvium as well.

The slight depletion of Lipnice granite in K also indicates some weathering of K-feldspar, because Rb and Cs, concentrated in micas relatively stronger than K (see also analyses of separated micas [37]), are not depleted. However the possibility of selective Rb and Cs concentration (including adsorption) by secondary phases like vermiculite has to be also kept in mind [38–40]. Note that a considerable amount of vermiculite (with minor chlorite) has been proven in concentrates of biotite from eluvia of Lipnice granite [25].

Weathering of feldspars and biotite could also explain the Al/Ga fractionation. It seems that Ga was preferentially removed (in comparison with Al), as documented in the literature [41, 42], but in some places especially in Lipnice granite, it reconcentrated in the eluvium. Ga can be scavenged by Fe-oxyhydroxides [43], as indicated by positive correlation with Fe in P2a.

5.2 Stream water chemistry

Chemical composition of surface water has been systematically monitored in a small catchment Loukov starting with hydrological year 1995. This catchment drains Melechov granite at the eastern slope of Melechov hill. More than 95% is covered by spruce forest [44]. It has been shown [26] that recent mass balance of the catchment explains very poorly the chemical differences between fresh and weathered Melechov granites (especially the behavior of sodium, which is significantly removed in the discharge but not depleted in weathered rock).

Occasionally springs were analyzed at various granite types. The data show that differences between waters draining Lipnice and Melechov granite types are mainly in anions (usually higher sulfate and lower nitrate and chloride in springs on Melechov type), and they can be largely explained by different land use. Sampled springs on Lipnice granite are prevalently in agricultural area, whereas Melechov granite is largely covered by forest, which also enhances atmospheric deposition of sulfur (peaking in the last quarter of the twentieth century). The surface water draining Melechov granite is usually more acidic, and from that reason (as evidenced by negative correlation with pH), it has higher content of Mn, Al, and Zn than springs on Lipnice granite. Obviously the differences in water chemistry cannot explain the different behavior of Ca, Na, and Sr during weathering of Lipnice and Melechov granites because in such case, the content of these cations should be higher in water draining the Lipnice granite. Similarly, concentration of SiO_2 which is depleted mainly in weathered Melechov granite is comparable in discharge from both granite types. On the other hand, the water chemistry data indicate that removal of elements in discharge could explain quite well the observed weathering of Lipnice granite.

Recent accumulation of P and K in the catchment was documented [16]. While some enrichment of the weathered Melechov granite in P is possible (see **Table 1**, **Figure 2**), we should keep in mind that strong retention of both elements can be a short-time phenomenon caused by deposition of dust from agricultural areas and by accumulation of nutrients in biomass (see also [26]) whose volume was increasing during the monitoring period (F. Oulehle, pers.commun.).

For a representative composition of surface water, see **Table 4a** and **b**.

5.3 Erosion and mechanical transport

It was concluded that erosion of a quartz-rich skeleton in the upper part of the weathering profile influenced significantly the present mineral and chemical composition [26, 27]. The Al-rich and Si-poor eluvia represent original lower horizon or domains, relatively depleted in quartz due to enrichment in small grains of weathered and secondary minerals. As indicated by Pb isotope evolution of fracture fillings in granite from borehole PDM-1 (close to Mel-3 and Mel-4), as early as in the Early Cretaceous, there was significant supergene redistribution of U and/or Pb [5, 6]. This event corresponds to the erosion of rocks immediately overlying the granites of Melechov Massif.

Solid-phase physical separation may also explain the question why elements contained in the most chemically resistant minerals (Zr, Hf, Nb, Ta, Th, partly REE; see also [45]) were not passively concentrated in weathered rocks (in case of Nb and Ta even slight depletion cannot be excluded). Also Zr/Hf and Nb/Ta ratios are unaffected. Observations of the relevant accessory minerals (zircon, monazite, rutile, ilmenite) in heavy-mineral concentrates from eluvia showed very weak influence of chemical weathering [21, 25]. However, a significant portion of these minerals in the rock forms very small grains—down to a few μm (zircon and monazite in Melechov type). These very small grains, unless included in other minerals, were easily transported by gravity and water flow away and partly to open fractures deeper in the granite [5, 6].

5.4 Origin of the REE fractionation

The lanthanide tetrad effect in granites and other felsic melts, including experimental ones, has been documented and discussed in numerous publications (e.g., [46–48]). Regarding the tetrad effect in apatite, one possibility is fractionation of monazite ($\pm$ xenotime), which would produce a pattern similar to M-type tetrad effect in the coexisting melt [49, 50]; see also [51]. In monazite within granites of Melechov massif, no fractionation similar to tetrad effect is apparent in EMP data [17, 19]. Nevertheless if monazite crystallized close to apatite, which is true especially in the Melechov granite but partly in the Lipnice granite as well [19], the light REE in apatite would be modified even in the case of only weak tetrad effect in monazite. The difference between apatite composition of Lipnice and Melechov types can be related to the several times higher abundance of monazite in the Lipnice granite. As documented by [48, 52], another important factor can be fluorine content, which is higher in the Lipnice granite as well.

REE fractionation seems to support a hypothesis of formation of P-rich domains in the melt, where monazite and apatite could have influenced one another much more than the remaining melt. Formation of such domains is also supported by conclusions of [19].

As shown by [49], the tetrad effect-like pattern of trivalent REE produced by monazite and xenotime fractionation is more complex. Another important feature is the peak of Yb in the residual melt. Similar Yb peak is observed in weathered Lipnice and Melechov granites (and in altered Melechov granite), when normalized by fresh rocks (**Figure 5**).

Despite the fact that many details of REE fractionation are unanswered, we can sum up that magmatic crystallization of phosphates, possibly with an important role of fluorine, produced complicated REE fractionation among rock-forming minerals, which can be insignificant in whole-rock chondrite-normalized patterns, but it can be enhanced by weathering processes.

Dissolution of apatite whose Y/Ho ratio (**Table 3**) is generally slightly lower than that of whole rock (**Table 1b**) could also lead to slightly elevated Y/Ho ratios of weathered granites (see **Figure 3c**).

6. Conclusions

A unique dataset from the Lipnice and Melechov granites showed that whole-rock analyses of the fresh rock and eluvium combined with the knowledge of element's abundance in primary minerals are a very effective tool for the reconstruction of granite weathering. Combination with borehole data helps to interpret large medium- to small-scale inhomogeneity of weathering processes.

Both granites are depleted in several elements concentrated mainly in biotite: Mg, Ti, Mn (mainly the Lipnice type), and the Melechov type in Fe and V, too. Content of elements concentrated in plagioclase, apatite, and partly K-feldspar (Ca, Na, Sr, P, Ba) decreased in weathered Lipnice granite, but not in Melechov granite. The main factor causing these differences is that superficial chemical weathering of the relatively coarse-grained and permeable Melechov granite was weaker than in Lipnice granite. In addition, the influence of chemical weathering (e.g., on Fe content and REE distribution) could be similar to subsolidus alterations, which had been more intense in the Melechov type. However, in some cases, like carbonatization or U enrichment, the effect of subsolidus alteration is opposite to weathering. In the Lipnice type, especially in the homogeneous P2a polygon (with borehole Mel-5), we were also able to distinguish formation of positive cerium anomaly at weathering and trace the role of secondary phases (Fe-oxyhydroxides, vermiculite) in retention of some elements (Ga). The first stage of Lipnice granite weathering in P2a is characterized mainly by alteration of biotite, not lowering SiO_2 content.

The natural solid-phase separation led to relative depletion of the sampled eluvia in quartz (especially in the Melechov granite) and prevented the most resistant accessory minerals (e.g., zircon) to be passively concentrated in the eluvia. The petrologically important Zr/Hf and Nb/Ta ratios are not significantly affected. Contamination of eluvia by material from quartz veins and other sources led to their enrichment in Au and Ag. A thorough examination of mutual ratios of REE and Y revealed also some influence of apatite dissolution.

The loss of U was significant in both granite types, but more intense in the Melechov type where larger portion of U was allocated to unstable phases or only adsorbed. Gallium was passively concentrated in eluvia of both granites, however stronger in the Lipnice type, possibly thanks to better retention of Fe-oxyhydroxides.

Assessment of weathering behavior of several elements (W, Sn, Co, partly Mg) is complicated by their spatial inhomogeneity in the intrusions, indicated by vertical differentiation in boreholes. In case of strongly incompatible W, we are able to distinguish this vertical trend from obvious removal of W by weathering of the Melechov granite.

Acknowledgements

We thank to Dr. František Woller and Dr. Jiří Slovák from the RAWRA for their help in facilitating the usage of archived data, obtained during the second phase of Melechov test locality project (2004–2006). The work was also supported by the research plan MSM 0021620855 of the Ministry of Education, Youth and Sports of the Czech Republic.

Author details

Václav Procházka[1*], Miroslav Žáček[2], Petr Sulovský[3], Tomáš Vaculovič[4], Lenka Rukavičková[5] and Dobroslav Matějka[6]

1 Česká Geologie, Praha, Czech Republic

2 GEOMIN Ltd., Jihlava, Czech Republic

3 Department of Geology, Faculty of Science of the Palacký University, Olomouc, Czech Republic

4 CEITEC, Masaryk University, Brno, Czech Republic

5 Czech Geological Survey, Praha, Czech Republic

6 Department of Geochemistry, Mineralogy and Mineral Resources, Charles University, Praha, Czech Republic

*Address all correspondence to: vprochaska@seznam.cz

References

[1] Taylor SR, Mclennan SM. The Continental Crust: Its Composition and Evolution. Oxford: Blackwell; 1985. 312 pp

[2] Migoń P, Lidmar-Bergström K. Weathering mantles and their significance for geomorphological evolution of central and northern Europe since the Mesozoic. Earth-Science Reviews. 2001;**56**:285-324

[3] Konta J. Clay Minerals of Czechoslovakia (in Czech). Czechoslovak Academy of Sciences; 1957. 319 pp

[4] Pivec E. Residues of surface kaolinization in granite of Říčany. Čas. Mineral. Geol. 1969;**14**:61-67 (in Czech with English summary)

[5] Procházka V, Zachariáš J, Strnad L. Pb isotopes and model ages of fracture-fillings in granite of the PDM-1 borehole near Rejčkov, Melechov Massif (in Czech with English summary). Ascta Musei Meridionale, Scientiae Naturales. 2015;**55**:4-12

[6] Procházka V, Zachariáš J, Strnad L. Model ages of fracture fillings and mineralogical and geochemical evidence for water-rock interaction in fractures in granite: The Melechov Massif, Czech Republic. Applied Geochemistry. 2018; **95**:124-138

[7] Štemprok M. Petrography and vertical extent of mineralization in the Cínovec granite cupola (in Czech with English summary). Sborník Geologických Věd, Geologie. 1965;**5**:7-106

[8] Staněk M, Géraud Y. Granite microporosity changes due to fracturing and alteration. Solid Earth. 2019;**10**: 251-274

[9] Novotný P. Peraluminous granites of the Melechov massif (in Czech). Zprávy o Geologických Výzkumech. 1985;**18**:31-32

[10] Mlčoch B, Štěpánek P, Breiter K. Brief petrological and petrochemical characteristics of main types of the Melechov massif (in Czech) (MS). Geofond P 86603/2. Praha: Czech Geological Survey; 1995

[11] Matějka D, Janoušek V. Whole-rock geochemistry and petrogenesis of granites from the northern part of the Moldanubian Batholith (Czech Republic). Acta Universitatis Carolinae: Geologica. 1998;**42**:73-79

[12] Woller F, Skopový J. A critical recherche of archived geological information—The Melechov massif (in Czech). Praha: MS RAWRA; 2000

[13] Breiter K. Geological documentation and evaluation of boreholes (in Czech). Geofond P115871. Praha: MS RAWRA; 2006

[14] Breiter K. New results from geological investigation of the Melechov pluton. Zprávy o Geologických Výzkumech. 2006;**2005**:202-205

[15] Mísař Z, Dudek A, Havlena V, Weiss J. Geology of ČSSR I. The Bohemian Massif (in Czech). Praha: SPN; 1983

[16] Oulehle F, Chuman T, Hruška J, Krám P, McDowell WH, Myška O, et al. Recovery from acidification alters concentrations and fluxes of solutes from Czech catchments. Biogeochemistry. 2017;**132**:251-272

[17] Breiter K, Sulovský P. Geochronology of the Melechov granite massif (in Czech with English summary). Zprávy o Geologických Výzkumech. 2005;**37**:16-19

[18] Procházka J, Barnet I, Breiter K, Coubal M, Drábek M, Fottová D, et al. Geological investigation of the test-locality "Melechovský masiv" (in

Czech). Geofond P 93387. MS Czech Geological Survey; 1998

[19] Harlov DE, Procházka V, Förster HJ, Matějka D. Origin of monazite-xenotime-zircon-fluorapatite assemblages in the peraluminous Melechov granite massif, Czech Republic. Mineralogy & Petrology. 2008;**94**:9-26

[20] Procházka V, Matějka D. Rock-forming accessory minerals in the granites of Melechov massif. Acta Universitatis Carolinae, Geologica. 2006;**48**:71-79

[21] Žáček M, Páša J. Geochemical investigation in the Melechov massif area (in Czech). Report for RAWRA, Praha. Jihlava: MS GEOMIN; 2006

[22] Breiter K, Procházka J. 2nd phase of polygon's selection—Detailed on-ground geological research (in Czech). Praha: MS RAWRA; 2004

[23] Procházka V, Uher P, Matějka D. Zn-rich ilmenite and pseudorutile: Subsolidus products in peraluminous granites of the Melechov Massif, Moldanubian Batholith, Czech Republic. Neues Jahrbuch für Mineralogie: Abhandlungen. 2010;**187**(3):249-263

[24] Procházka V, Kašparec I, Knésl I, Matějka D. Small solid particles and chemical weathering of the Melechov granite (in Czech with English summary). Acta Musei Meridionale. 2013;**53**:19-32

[25] Procházka V, Žáček M, Vlček V, Langrová A. Behavior of some elements at weathering of granites of the Melechov massif (in Czech with English summary). Zprávy o Geologických Výzkumech. 2009;**42**:189-195

[26] Procházka V, Žáček M, Fottová D, Matějka D, Štěpánová M. Some relationships between the chemical weathering of Melechov granite and the mass balance of the small catchment Loukov (in Czech with English summary). Acta Musei Meridionale Scientiarum Naturalium. 2011;**51**:5-20

[27] Procházka V, Žáček M, Matějka D. Contamination of weathered Melechov granite (in Czech with English summary). Zprávy o Geologických Výzkumech. 2014;**47**:134-139

[28] Anonymous. Soil maps of the Czech Republic 1:50,000. Czech Geological Survey (web application CZ-00025798-CGS-PUDY50-WMS); 2012–2016

[29] Trubač J, Žák J, Chlupáčová M, Janoušek V. Magnetic fabric and modeled strain distribution in the head of a nested granite diapir, the Melechov pluton, Bohemian Massif. Journal of Structural Geology. 2014;**66**:271-283

[30] Procházka V. Phosphates and accessory oxides in selected granitoids and paragneisses of the Moldanubian area in southeastern and southern Bohemia (in Czech with English summary) [PhD thesis]. Prague: Charles University; 2010

[31] Guthrie VA, Kleeman JD. Changing uranium distributions during weathering of granite. Chemical Geology. 1986;**54**(1–2):113-126

[32] Allégre C, Manhés G, Lewin E. Chemical composition of the Earth and the volatility control on planetary genetics. Earth and Planetary Science Letters. 2001;**185**:49-69

[33] Blum JD, Erel Y, Brown K. $^{87}Sr/^{86}Sr$ ratios of Sierra Nevada stream waters: Implications for relative mineral weathering rates. Geochimica et Cosmochimica Acta. 1994;**58**:5019-5025

[34] White AF, Schulz MS, Lowenstern JB, Vivit DV, Bullen TD. The ubiquitous nature of accessory calcite in granitoid rocks: Implications for weathering, solute evolution, and

petrogenesis. Geochimica et Cosmochimica Acta. 2005;**69**:1455-1471

[35] Erel Y, Blum JD, Roueff E, Ganor J. Lead and strontium isotopes as monitors of experimental granitoid mineral dissolution. Geochimica et Cosmochimica Acta. 2004;**68**:4649-4663

[36] Milodowski AE, Tullborg E-L, Buil B, Gómez P, Turrero M-J, Haszeldine S, et al. Application of mineralogical, petrological and geochemical tools for evaluating the palaeohydrogeological evolution of the PADAMOT study sites. In: PADAMOT Project Technical Report WP2. Harwell, UK: Nirex Ltd.; 2005

[37] Breiter K, Hrubeš M, Mlčoch B, Štěpánek P, Táborský Z. Results of new geologic-petrological studies in the Melechov massif area (in Czech). In: Procházka J, editor. Geological Investigation of the Testing Locality Melechov Massif. Praha: MS RAWRA; 2001

[38] Jeong GY, Cheong CS, Kim J. Rb–Sr and K–Ar systems of biotite in surface environments regulated by weathering processes with implications for isotopic dating and hydrological cycles of Sr isotopes. Geochimica et Cosmochimica Acta. 2006;**70**:4734-4749

[39] Komarneni S, Roy DM. Shale as a radioactive waste repository: The importance of vermiculite. Journal of Inorganic and Nuclear Chemistry. 1979; **41**:1793-1796

[40] Sawhney BL. Selective sorption and fixation of cations by clay minerals: A review. Clays and Clay Minerals. 1972; **20**:93-100

[41] Shiller AM. Enrichment of dissolved gallium relative to aluminium in natural waters. Geochimica et Cosmochimica Acta. 1988;**52**:1879-1882

[42] Shiller AM, Frilot DM. The geochemistry of gallium relative to aluminum in Californian streams. Geochimica et Cosmochimica Acta. 1996;**60**:1323-1328

[43] Hieronymus B, Kotschoubey B, Boulégue J. Gallium behaviour in some contrasting lateritic profiles from Cameroon and Brazil. Journal of Geochemical Exploration. 2001;**72**: 147-163

[44] Fottová D. Hydrochemical budgets in the monitoring network GEOMON. In: Project Report SP/1a6/151/07. Prague: Czech Geological Survey; 2011

[45] Nesbitt HW, Markovics G. Weathering of granodioritic crust, long-term storage of elements in weathering profiles, and petrogenesis of siliciclastic sediments. Geochimica et Cosmochimica Acta. 1997;**61**:1653-1670

[46] Bau M. Controls on the fractionation of isovalent trace elements in magmatic and aqueous systems: Evidence from Y/Ho, Zr/Hf, and lanthanide tetrad effect. Contributions to Mineralogy and Petrology. 1996;**123**:323-333

[47] Irber W. The lanthanide tetrad effect and its correlation with K/Rb, Eu/Eu*, Sr/Eu, Y/Ho, and Zr/Hf of evolving peraluminous granite suites. Geochimica et Cosmochimica Acta. 1999;**63**:489-508

[48] Veksler IV, Dorfman AM, Kamenetsky M, Dulski P, Dingwell DB. Partitioning of lanthanides and Y between immiscible silicate and fluoride melts, fluorite and cryolite and the origin of the lanthanide tetrad effect in igneous rocks. Geochimica et Cosmochimica Acta. 2005;**69**:2847-2860

[49] Tin QD, Keppler H. Monazite and xenotime solubility in granitic melts and the origin of the lanthanide tetrad effect. Contributions to Mineralogy and Petrology. 2015;**169**:1-26

[50] Stepanov AS, Hermann J, Rubatto D. Experimental study of

monazite/melt partitioning with implications for the REE, Th and U geochemistry of crustal rocks. Chemical Geology. 2012;**300-301**:200-220

[51] Yurimoto H, Duke EF, Papike JJ, Shearer CK. Are discontinuous chondrite-normalized REE patterns in pegmatitic granite systems the results of monazite fractionation? Geochimica et Cosmochimica Acta. 1990;**54**(7): 2141-2145

[52] Peretyazhko IS, Savina EA. Tetrad-effects in REE pattern of granitoid rocks: The result of liquid immiscibility in fluorine-rich silicate melts. Doklady Earth Sciences. 2010;**433**:1042-1047

Chapter 6

Changes of Granite Rapakivi under the Biofouling Influence

Dmitry Yu. Vlasov, Elena G. Panova, Marina S. Zelenskaya, Oksana A. Rodina, Alexey D. Vlasov and Katerina V. Sazanova

Abstract

Interdisciplinary study of granite rapakivi biofouling in the natural and anthropogenic environment (St. Petersburg, Vyborg, Southern Finland) was carried out. The biodiversity of microorganisms (cyanobacteria, micromycetes, and organotrophic bacteria) and various types of biofilms are characterized. The influence of external factors on the changes of cyanobacterial biofilms is shown. The features of biofilms localization on the granite surface in an urban environment and in natural outcrops are studied. Differences in the biofilms metabolites composition at the granite quarries and monuments of St. Petersburg are shown. The behavior of chemical elements during the bioweathering of granite is estimated. The role of biofilms in the accumulation of chemical elements on the surface of granite is established. The dynamics of chemical elements leaching from granite may depend on the type of biofilm developing on granite.

Keywords: granite, weathering, biofouling, biogeochemical process, leaching, biodeterioration, microorganisms, environment, model experiments, mobile forms of elements

1. Introduction

Granite is one of the most widespread types of a stone in architecture of northern Russian cities such as Saint Petersburg, Vyborg, Priozersk, Primorsk as well as Finish cities such as Helsinki, Lappeenranta, Kotka, Hamina, Kuopio (Finland). The destruction of granite in the northern cities is a result of interrelated physical, chemical, and biological processes [1]. Biogenic weathering is connected with the impact on the rock surface by microorganisms (bacteria, microfungi, and microalgae) as well as lichens and mosses. They form lithobiotic communities which have a noticeable effect on the state of the stone.

The study of this problem seems to be an interdisciplinary task, the solution of which is possible only on the basis of an integrated scientific approach and the use of modern research methods. Organisms of lithobiotic communities are able to actively influence on the mineral substance chemically and physically. They catalyze the destruction of rocks, contributing to the extraction of minerals from them. Microbial activity in combination with atmospheric pollution is one of the features of urban ecosystems that determine the rate of weathering of granite and other types of stone.

 IntechOpen

Most microorganisms on stone surface exist in the form of biofilms, which are composed of microbial cells and metabolites. Primary biofilms on granite most often consist of cyanobacteria and green algae. Aerophilic green algae are less resistant to adverse conditions than cyanobacteria and need more moisture. Green biofilm usually can be indicator of increased periodic or constant moisture of a stone [2, 3]. As organic matter accumulates on the surface of the stone, the participation of heterotrophic bacteria and fungi in the microbial community increases [4]. The close cooperation in microbial communities contributes to the successful growth and development of biofilms on stony substrates including granite. Biofilms can penetrate into cracks and pores. As a result, the absorption and retention of water in the rock mass increases, the intensity of diffusion and evaporation of water changes, and the processes of dissolution of the stone take place. The growth of biofilms causes pressure on the structural elements of the rock, acts on individual crystals and grains of stone.

Biochemical activity of microbial communities has a strong influence on mineral substance due to producing chemically active compounds such as polysaccharides, lipopolysaccharides, proteins, glycoproteins, lipids, glycolipids, fatty acids, and enzymes [5, 6]. Biomineral interaction leads to the leaching, formation of secondary minerals, primary soil formation, and thus, prepares the conditions for the further biological colonization of the stone.

The state of the stone surface has a particular importance for the biological colonization. A rough (uneven) surface is colonized much better than a smooth one [7]. Rough surface provides more opportunities for attachment and development of microorganisms (local humidity, microcracking, delay of various contaminants that serve as sources of nutrition for microorganisms, etc.). The biosusceptibility of natural stone may vary depending on the duration and conditions of its exposure in the open air [8].

Thus, natural stone together with biofouling is a peculiar and very complex lithobiotic system, the development of which depends on the properties of the stone, the composition of biological community, and environmental conditions. The aim of our investigation is the analysis of granite biological colonization peculiarities in different environment as well as the estimation of granite changes under the biofouling influence.

2. Materials and methods

2.1 Materials

The objects of research were selected in urban environment as well as in natural outcrops. Peter and Paul Fortress and monuments of the Museum Necropolises were studied in Saint Petersburg. Vyborg castle, fountain, tunnels, and outcrops in the Monrepos park were observed in Vyborg.

Granite outcrops were examined in the natural park Ristijärvi and on the Owl Mountain (Karelia). Also, four old quarries in the south part of Finland were examined: quarry I – (N 60° 34.207′ E 027° 43.835′); quarry II (N 60° 31.855′ E 027° 39.698′); quarry III (N 60° 32.101′ E 027° 39.823′); quarry IV (N 60° 44.413′ E 028° 00.564′). Granite mining at these quarries has long been discontinued. Currently, they undergo a process of natural overgrowth and are ideal model for studying of natural stone biofouling in low anthropogenic influence. More than 500 samples of destroyed rapakivi granite were investigated from 2013 to 2019. Rapakivi granite, as a rule, had its own unique image: large egg-shaped clusters of feldspar with a diameter of 3–6 cm, surrounded by an edge of greenish-gray plagioclase, placed in a fine-grained matrix of feldspars, quartz, and biotite.

2.2 Study of microorganisms

Primary attention was paid to the structure of granite, the presence of cracks, holes, and other surface irregularities, which can serve as a shelter for microorganisms. Traditional cultural methods of mycology and microbiology have been applied for isolation and identification of microorganisms in biofilms on the surface of the granite [9]. Also, metagenomic analysis was used to determine a wide range of microorganisms in biofilms. The work was carried out in the resource center of Saint Petersburg State University "Development of cellular and molecular technologies." Diversity of bacteria in biofilms on granite was carried out on the basis of the 16S rRNA genes analysis. Metagenomic study of fungal diversity in biofilms on granite was carried out with primers for site amplification ITS1-5.8S–ITS2. For the identification of cyanobacteria, direct microscopy of the samples was used. Cumulative cultures were also obtained in distilled water and in the Gromov 6 medium (period cultivation from week to month). Verification of species in accordance with the current nomenclature was carried out using the electronic database AlgaeBase (http://www.algaebase.org/).

2.3 Biochemical analysis

For analysis of small organic molecules in several types of biofilms samples were extracted with 15 mL methanol vigorously mixed and centrifuged (10 min, 400 × g) at room temperature. The supernatant was transferred to a new vial and dryed by a rotary evaporator at 40°C.

The dried extracts were soluble in pyridine (30 μL) and BSTFA (N,O-bis—3-methyl-silyl-3-F-acetamide) (30 μL), incubated at 100°C for 15 minutes. The derivatized samples were analyzed by gas chromatography-mass spectrometry (GC-MS) by Agilent MSD 597, column HP-5MS, 30m × 0.25 mm. Chromatography was carried out with linear temperature programming from 70 to 320° at a speed of 4°C/min. Data were collected using Agilent ChemStation software. Mass spectrometric information was processed and interpreted using AMDIS program (http://www.amdis.net/index.html), standard NIST2005 library, and the library of standard compounds of BIN RAS. Quantitative interpretation of chromatograms was carried out with hydrocarbon using UniChrom program http://www.unichrom.com/unichrome.shtml.

2.4 Scanning electron microscopy

Scanning electron microscopy was used in order to study peculiarities of localization of microorganisms in the surface layer of the stone and to characterize the relationship between lithobiotic organisms during colonization of the granite. Samples of the damaged stone (0.5–1.0 cm × 0.5–1.0 cm) were initially examined under binocular loupe. The criterion of selection for SEM analysis was the presence of structures of microorganisms on the stone surface as well as transformation of the granite surface. The material was examined (after fixation) under the scanning electron microscope in the range of magnification from 100× to 10000×. SEM studies were performed on electron microscope ABT-55 (Japan) and TM 3000 (HITACHI, Japan, 2010) with an attachment of an energy-dispersive microanalysis OXFORD in SPbU Resource Center "Microscopy and Microanalysis."

2.5 Geochemical study

The determination of elemental composition in fresh granite and various types of crusts was carried out using inductively coupled plasma (ICP MS, Agilent 7700) in the chemical laboratory of the All-Russian Geological Institute.

For the experiment on the dynamics of granite bioleaching, we took three types of samples from the surface of granite rapakivi from the Monrepos Park (Vyborg): surface layer of granite without biofilms, granite with black (lichens + fungi + cyanobacteria), and with gray (lichens + alga) biofilms. There are no local sources of pollutions in this area. Previously, the samples were powdered. Samples part (2 g) were diluted with 10 mL of bidistilled water each (in a ratio of 1:5) and mounted on a vibration panel for constant mixing of the sample and placed in a thermostat. The experiment lasted about a month. During this period, the temperature in the thermostat was 25^0C at normal pressure. Aliquots of the solution were taken from the upper part of the flasks in the following time intervals: 1, 3, 6 hours from the beginning of the experiment; then after 1, 3, 8, 11, 14, 18, 22, 28, 32 days. The solution was analyzed with the following parameters: pH, particle size (HORIBA LA-950 nano-sizer), and composition of elements (ICP-MS, Agilent 7700).

3. Results and discussion

3.1 Main types of granite biofilms (outward, biodiversity, and metabolism)

There are different types of granite destruction in St. Petersburg, Vyborg, and quarries in Finland: fissuring, granular disintegration, flaking, exfoliation, loss of color, crusts, biofilms of different composition, ovoid weathering, and macrofouling. Primary biological colonization usually connected with the formation of pigmented biofilms. The color and structure of biofilms usually depends on the dominance of certain groups of microorganisms (cyanobacteria, algae, microscopic fungi, and lichens).

Cyanobacteria typically prevailed in primary biofilms, especially in natural outcrops of granite. They formed the basis of lithobiotic communities in most of the studied habitats. Both mono-species and multi-species communities dominated by cyanobacteria were noted. The dominance of specific species often determined the morphology of the whole biofilm. So, on granite in quarry I, a rich biofilm with a dominance of *Stigonema ocellatum* Thuret ex Bornet & Flahault (Dillw.) was formed in the place of natural water seepage (**Figure 1**). This species forms the interwoven filaments, which are clearly visible in the SEM image. The upper part of the biofilm has a greenish-olive color and is represented by *Stigonema ocellatum*. In the lower part of the biofilm, a change in color to brown-red can be seen due to the change of the dominant species by *Gloeocapsopsis magma* (Brébisson) Komarék et Anagnostidis. Other representatives of cyanobacteria also appear in the lower part of the biofilm: *Lyngbya* sp., *Leptolyngbya foveolarum* (Rabenhorst ex Gomont) Anagnostidis et Komarek, *Synechocystis salina* Wislouch.

Lighting also plays an important role in the formation of biofilms on the granite surface. Thus, it was shown by comparative studies of the species composition of cyanobacteria in the Vyborg granite tunnels (with scarce of light) and open areas of granite near tunnels. Under natural lighting, six species of cyanobacteria were identified (for 1 sample): *Gloeocapsopsis magma*, *Nostoc commune* Vaucher ex Bornet et Flahault f. Commune, *Calothrix parietina* (Nägeli) Thuret ex Bornet et Flahault f. *parietina*, *Scytonema hofmannii* C. Agardh ex Bornet & Flahault, *Aphanocapsa* sp. 1, *Aphanocapsa* cf. *fusco-lutea*. In the same time, no more than three species of

Figure 1.
Stigonema ocellatum *dominated biofilm on the surface of granite rapakivi in the quarry I.*

cyanobacteria were detected in each of the studied samples which were collected inside the tunnels. Moreover, the diversity of biofilms types inside tunnels was much lower than in the open air. In total, only three types of biofilms with the dominance of cyanobacteria were found in the tunnels (**Figure 2**).

a. Dark green to black biofilm is represented by the dominant species *Chroococcus* sp. 1 with the participation of *Aphanocapsa* sp.;

b. Green algae predominate in the green biofilm; cyanobacteria *Chroococcus* sp. 1 and *Aphanocapsa* sp. also found;

c. White deposits represent a mineral layer and contain neither cyanobacteria nor microalgae;

d. Cyanobacteria *Gloeocapsopsis magma* dominates in the reddish-brown biofilm.

It is interesting to note that only five taxa were found in Monrepos Park (Vyborg). This is due to the super dominance of certain species of cyanobacteria in biofilms on the surface of granite (**Figures 3–5**).

In total, 78 cyanobacteria taxa belonging to 5 orders, 18 families, and 29 genera were identified in the studied habitats. Quarry IV was the richest in the number of species (**Figure 6**).

The largest number of families (4), genera (8), and species (29) was noted for the order Synechococcales, followed by the order Chroococcales (25 species). The most diverse is the genus *Leptolyngbya* (**Figure 7**). It occurs most often. Cosmopolitan species predominated among the identified cyanobacteria.

The most common in the studied areas are *Calothrix parietina* Thur. ex born. & Flah., *Gloeocapsopsis magma* (Brébisson) Komarék et Anagnostidis, *Leptolyngbya foveolarum* (Rabenhorst ex Gomont) Anagnostidis et Komarek, *Gloeocapsa atrata* Kützing, nom. illeg. (**Figure 8**).

Figure 2.
Different types of biofilms and deposits in the Vyborg tunnels.

Figure 3.
Biofilm formed by the cyanobacteria Microcoleus vaginatus *Gomont ex Gomont on the granite wall of the Vyborg castle (Vyborg).*

Organotrophic bacteria were also characterized by significant diversity at various granite sites in the city of Vyborg, including Monrepos Park. Their number reached 10^7 cells per 1 gram of material. A similar picture was observed in St. Petersburg (on the granite monuments of Museum Necropolises). The results of

Figure 4.
Biofilm formed by the cyanobacteria Lyngbya martensiana *Meneghini ex Gomont on a granite block in Monrepos Park (Vyborg).*

Figure 5.
Biofilm formed by the cyanobacteria from the genera Lyngbya *and* Synechococcus *on a granite wall in Monrepos Park (Vyborg).*

metagenomic analysis show that two main bacterial phyla dominate in biofilms on the granite rapakivi in city environment: Bacteroidetes and Proteobacteria. The Bacteroidetes phyla were characterized by a large presence in black biofilms. A significant part of the lithobiotic communities in all samples of granite was represented by actinomycetes. Acidobacteria were also isolated in a significant amount from black biofilms (**Table 1**).

In the heterotrophic block of biofilms on the surface of granite, a significant diversity of micromycetes was noted. In total, 64 species of micromycetes were isolated and identified (47 – St. Petersburg, 42 – granite outcrops, and 25 – common species). The domination of dark-colored fungi in biofilms on the granite

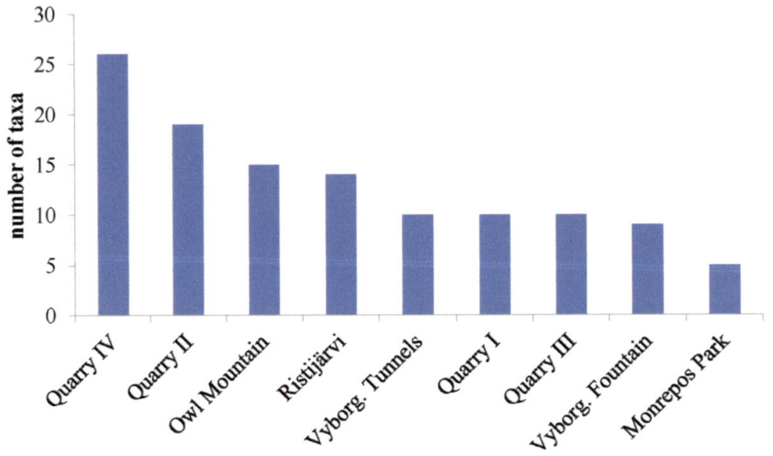

Figure 6.
The number of cyanobacterial taxa revealed on the granite in study areas.

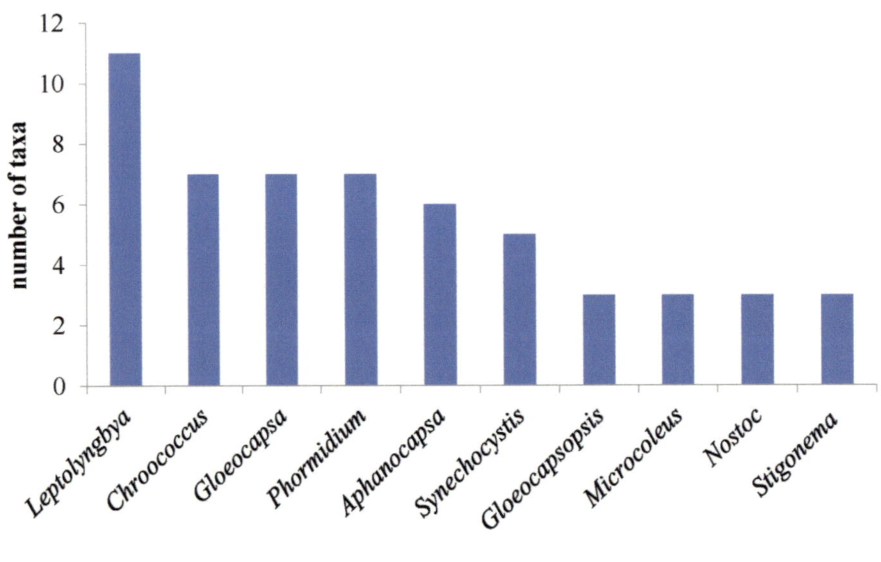

Figure 7.
Number of species in the richest genera.

surface was typical for the urban environment. It is interesting to note that some microfungi were superdominants in biofilms on granite in an urban environment (*Cladosporium, Alternaria, Aureobasidium*, and also black yeast-like fungi). Species of ascomycetes prevailed in the taxonomic relationship, which was shown using metagenomic analysis. Microcolonies and hyphae of microscopic fungi were typical for damaged granite surface (**Figures 9–11**). According to scanning electron microscopy study (SEM-analysis), microcolonies can be considered as the dominant form of the fungal existence on granite. Small compact clusters and chains of thick-walled cells (short hyphae) were noted in the granite surface in natural outcrops as well as in urban environment. Fungi are able to penetrate through microcracks into the rock substrate while causing weakening of the surface layer of granite. Fungal microcolonies were formed usually on feldspar and mica. Long hyphae were usually connected with the microrelief of K-feldspar. They were more typical for granite rapakivi in Saint Petersburg (**Figure 11**).

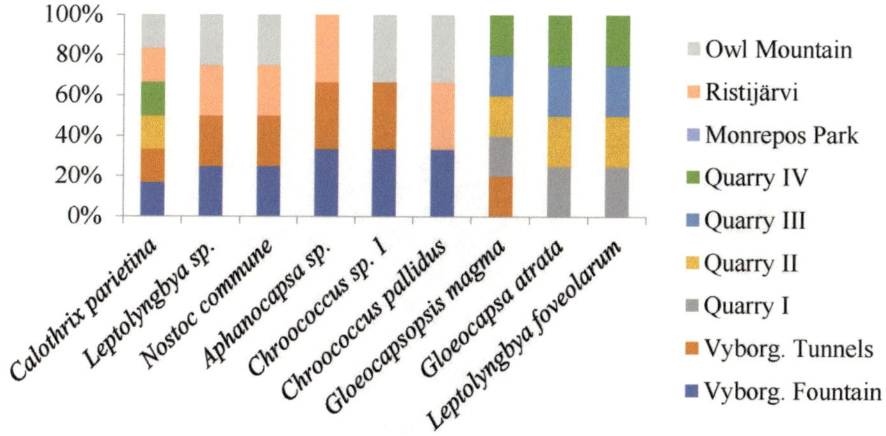

Figure 8.
The presence of common taxa of cyanobacteria in biofilms on granites in the studied places.

Bacteria phyla	Green biofilm	Black biofilm
Acidobacteria	0.2	6.6
Actinobacteria	15.2	7.5
Bacteroidetes	13.3	40.5
Firmicutes	2.7	0.0
Proteobacteria	48.7	33.4

Table 1.
The dominant groups of organotrophic bacteria in biofilms on granite at the monuments of the Museum Necropolises in St. Petersburg (share according to the results of metagenomic analysis).

As a result of biochemical studies, more than 200 different compounds were found in biofilms samples from granite quarries. Among them were identified: mono, di, and trisaccharides, aliphatic carboxylic acids, amino acids, sugar alcohols, phenolic compounds, diterpenes, sterols, ethanolamine, phosphate, glycerol-3-P, and urea. In samples of biofilms taken in an urban environment only about 100 different low molecular weight organic compounds were identified. In general, the biofilm samples from granite in urban environment had a significantly lower molecular diversity of metabolites than the samples taken in the quarries in Finland. At the same time, the quantitative content of some compounds, primarily sugar alcohols, was significantly higher in biofilms in the urban environment. Most likely, the revealed differences are associated with the species composition of microorganisms in biofilms.

The general patterns of the distribution of small organic molecules depending on the type of biofilms were similar for samples taken in the quarries and in Museum Necropolises (Saint Petersburg). In biofilms with a predominance of algae and cyanobacteria, the amount of mono- and disaccharides, amino acids and organic acids in free form was significantly higher in comparison with other types of biofilms. In samples dominated by fungi, the amount of free-form organic acids was lower and concentration of polyols was higher compared to algae.

Sugar alcohols and phenolic compounds predominated in the fouling formed by lichens. In samples of primary soil with a moss cover, the greatest variety of low molecular weight metabolites was observed; however, their quantitative content was lower than in other samples. The data obtained show the possibility

Figure 9.
Fungal microcolonies on the border of mica (Quarry I).

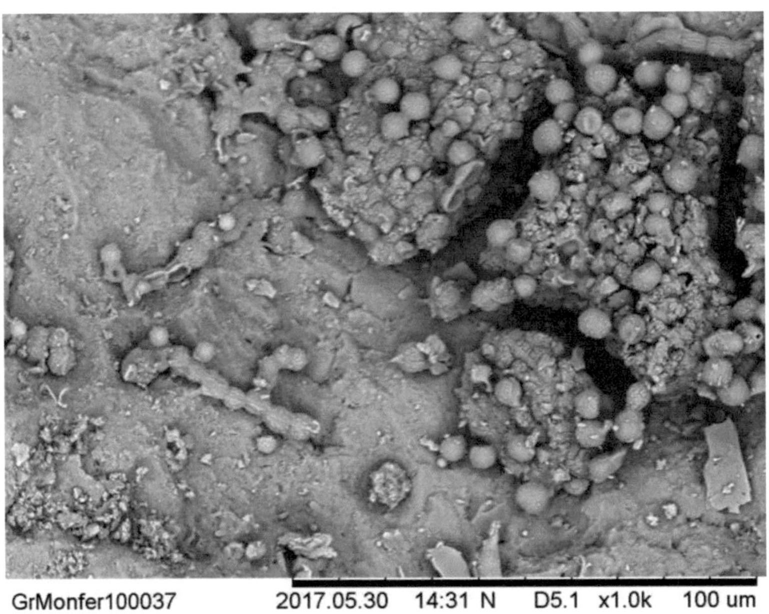

Figure 10.
Fungal microcolonies and short hyphae located on the K-feldspar (Quarry I).

of applying the metabolomic approach to the study of lithobiotic communities in different environment.

3.2 Geochemical peculiarities of granite bioweathering

To assess the effect of biofouling on the behavior of chemical elements during granite weathering, samples of granite rapakivi were taken in natural outcrops

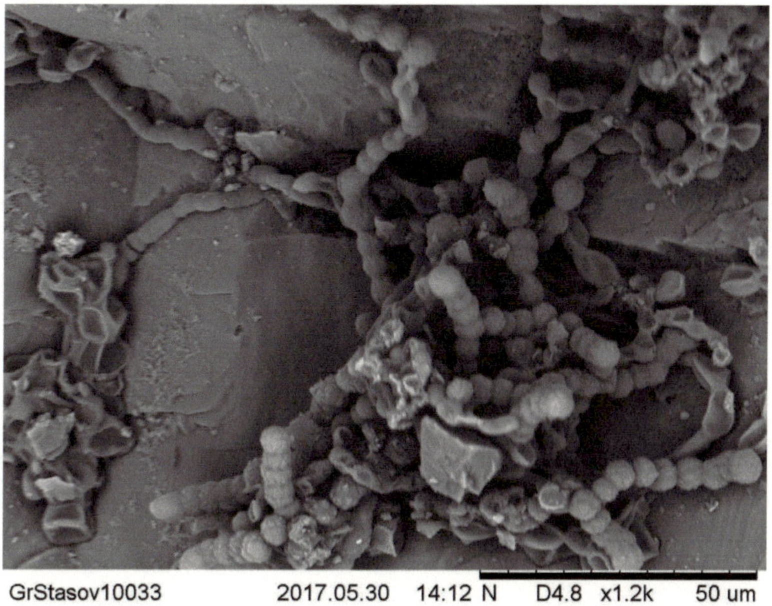

Figure 11.
Fungal hyphae in the granite surface of Stasov monument (Museum Necropolis of Saint Petersburg).

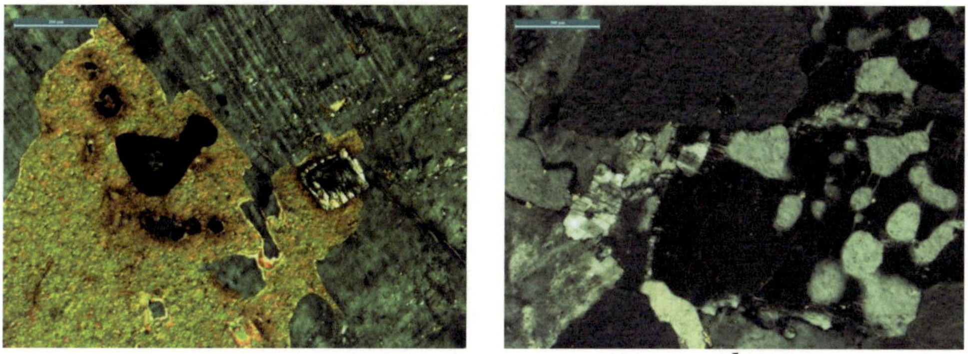

a *b*

Figure 12.
Thin skin of rapakivi granite: (a) plagioclase, microcline and biotite with pyrite and zircon; (b) quartz, plagioclase and microcline. XPL (a), PPL (b).

(granite wall) in Monrepos Park (Vyborg neighborhood) where the influence of the urban environment on natural ecosystem is insignificant. This type of granite is commonly called Wiborgite. Wiborgite is a porphyritic, coarse-grained granite with a typical rapakivi texture composed of round 1–3 cm potassium feldspar ovoids with a plagioclase mantle. The color of this rock can be brown, brownish red, red or green. The essential minerals are potassium feldspar, quartz, plagioclase, biotite, and hornblende (**Figure 12a** and **b**, **Table 2**).

Three types of samples were taken for comparative study: fresh granite, crust without biofilm (3 mm) and crust with biofilm (3 mm). The results of the analysis are presented in **Tables 3** and **4**.

It is shown that the content of almost all petrogenic oxides (except SiO_2), decreases in the crust without a biofilm (**Table 3**). This fact can be explained by the destruction of the granite structure and leaching of the most mobile chemical elements and particles of minerals under the influence of rain and wind. The crust is

Mineral	Mass %	Mineral	Mass %	Mineral	Mass %
Quartz	24–42%	Muscovite	0–0.1%	Ilmenite	0–0.5%
K-feldspar	28–42%	Allanite	0–0.2%	Rutile	0–0.1%
Albite	7–13.7%	Tourmaline	0–0.2%	Apatite	0–0.3%
Andesine	3–27%	Zircon	0.1–0.2%	Pyrite	0.0
Amphibole	0.2–11%	Kaolinite	0–0.1%	Calcite	0.0
Chlorite	0.0–0.2%	Thorite	0–0.1%	Bastnasite	0–0.9%
Biotite	2.9–7.5%	Magnetite	0–0.1%	Fluorite	0.2–1.9%

Table 2.
Mineral composition of rapakivi granite (Wiborgite).

Samples	SiO_2	Al_2O_3	Fe_2O_3	K_2O	Na_2O	CaO	MgO	TiO_2	P_2O_5	MnO	LOI
Granite (n = 7)	70.78	13.51	3.38	5.57	3.39	2.32	0.22	0.33	0.07	0.04	0.38
Granite crust (n = 7)	72.19	13.05	3.19	5.29	3.11	2.06	0.20	0.34	0.08	0.04	0.46
Granite crust with biofilm (n = 7)	69.51	14.50	3.54	5.41	3.34	2.29	0.23	0.36	0.06	0.03	0.76

Table 3.
Content of pertogenic oxides in fresh granite and two types of crust (mass%).

Elements	Granite (n = 7)	Crust (n = 7)	Crust with biofilm (n = 7)	CC_1 = crust / granite	CC_2 = crust with biofilm/granite
Ba	119	126	121	1.06	1.02
Sr	12.8	8.1	15.8	0.63	1.18
Li	38.3	34	40.2	0.89	1.05
Sc	4.88	4.36	5.1	0.89	1.05
U	7.37	2.37	12.4	0.32	1.68
Se	3.19	1.89	6.45	0.59	2.02
Mo	1.66	0.22	2.83	0.13	1.70
Cd	0.24	0.11	0.25	0.46	1.08
Sb	0.09	0.05	0.09	0.56	1.00
Ni	12.6	14.6	15.5	1.16	1.23
Co	3.3	3.09	3.79	0.94	1.15
Cu	6.28	6.1	7.98	0.97	1.27
Zn	65.3	72.2	79.1	1.11	1.21
As	11.7	10.3	11.1	0.88	0.95

Table 4.
Content of trace elements in fresh granite and two types of crust (ppm) and coefficient concentration (CC).

relatively enriched with the most stable mineral quartz. The organic matter content LOI (loss on ignition) increases slightly in comparison with fresh granite. In the crust with biofilm the situation is different. Particles of weathered granite can be accumulated in a biofilm. This probably explains the fact that the content of almost all basic elements in the crust with biofilm is close to the composition of unaltered granite. The organic matter content in the crust with biofilm is naturally the highest in comparison with other variants.

A similar situation is observed in the behavior of trace elements. It is shown using the concentration coefficient (CC) calculated as the ratio of the content of the element in the crust to its content in not weathered granite. In the weathered crust, in comparison with fresh granite, the removal of most chemical elements is observed (**Table 4**). The concentration coefficient in this case is less than 1. At the same time, trace elements (Se, Mo, U, Cu, Ni, Zn, and Sr) are accumulated in the crust with the biofilm (concentration coefficient is more than 1).

It is well known that the main environment of migration of chemical elements in the nature is water. Migration of elements in the liquid phase occurs in the form of ions, molecules, and colloidal particles. The chemical composition of water in the hypergenesis zone is formed primarily due to the dissolution of solid phases interacting with water. Granite biofouling may affect this process. For the experiment on the dynamics of granite bioleaching, we took three types of samples from the surface of granite rapakivi from the Monrepos Park (Vyborg): surface layer of granite without biofilms, with black (lichens + fungi + cyanobacteria), and with gray (lichens + alga) biofilms.

As a result, it was shown that the particle size changes over time that reflects the periods of their dissolution and coagulation. On the first day no changes are observed. Further until the 22nd day changes in particles size are observed and then alignment occurs (about 380 nm in size). The curves for the studied variants differ markedly. Largest particle size during the experiment is observed for granite with black biofilm a compared to granite with gray biofilm (**Figure 13**).

A comparison of the graphs of pH changes (**Figure 14**) shows that at the beginning of the experiment, the pH of solutions for the granite without biofilms and granite with biofilms is different. Amplitude of the pH values changes varies from 6.3 to 7.6 and does not connect with the changes in particles size. Correlation analysis confirmed the absence of any linear dependence of the change in the size of nanoparticle in solution on the pH of the solutions.

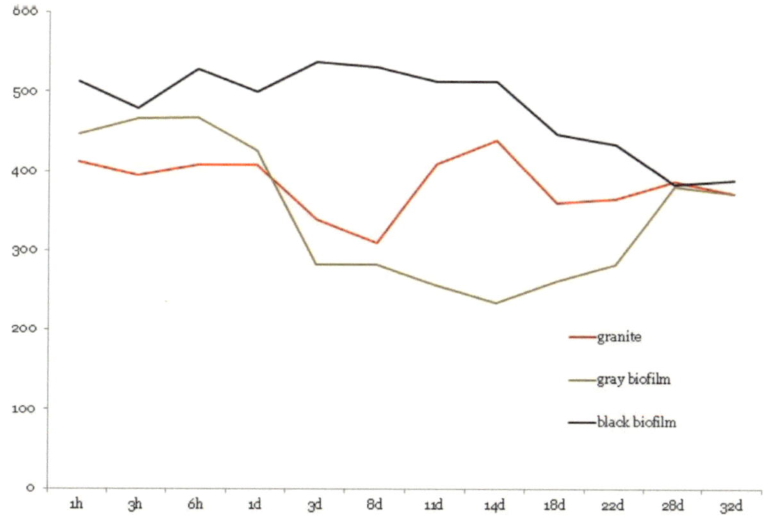

Figure 13.
Particles size changes in time for granite, granite with black, and gray biofilms (nm). h – hours; d – days.

The results show the periodic variation of the acid-alkaline properties of the solutions. As a whole, the variant with the black biofilm are characterized by a more alkaline medium; the variant with gray biofilm has a relatively more acidic medium.

In selected aliquots of solutions, the content of chemical elements was determined by the ICP MS method. The highest concentrations of elements in the solutions were observed for K, Na, Mg, and Ca (an example for calcium is shown in **Figure 15**). This indicates a fairly rapid leaching of these elements from the minerals of the rock, where they are in water-soluble form. Lower contents are typical for a group of elements: Al, Fe, Ba, and Li. Hundreds of mg per liter were found for: Mn, Rb, Sr, and Cs. Thousands of mg were found for the following elements: Sc, V, Ni, Pb, Cu, Zn, Mo, U, Th, Y, La, and Ce. An increased concentration of various groups of elements is observed on the 8th day (K, Na, Ni, As, Cd, and Mo) that can be associated with an increase in the pH of the solution.

The experimental results demonstrate the different behavior of chemical elements in the absence and presence of biofilms on granite. There is also a different

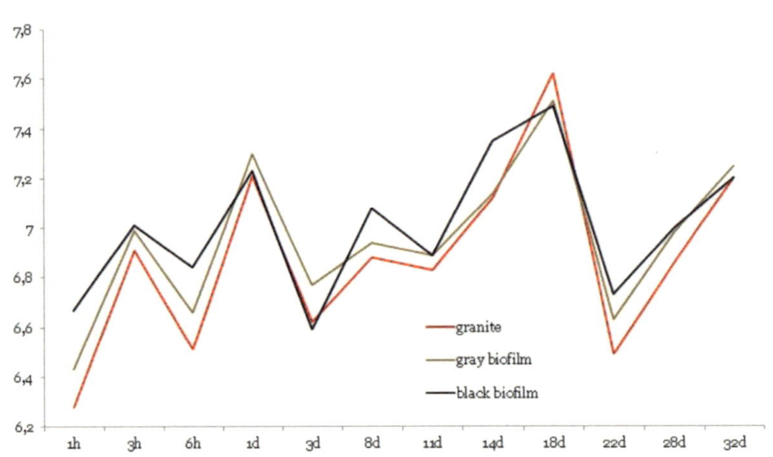

Figure 14.
pH values changes of solutions (granite, granite with black, and gray biofilms). h – hours; d – days.

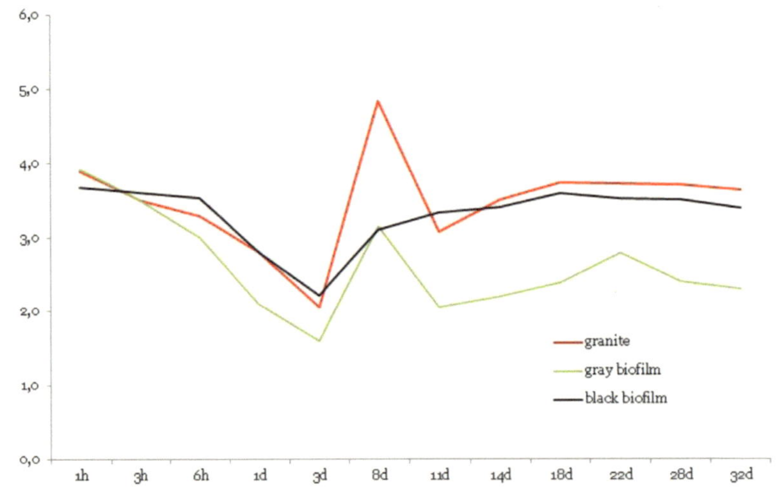

Figure 15.
Dynamics of the calcium content changes in solutions (mg/L) during the dissolution of granite, granite with gray and black biofilms. h – hours; d – days.

behavior of chemical elements in variants with different types of biofilms. The dissolution of granite with a black biofilm is the least intense, which is especially noticeable on the example of Na, Ca, and Mo. The content of these elements in granite with black biofilm practically does not change in solutions over time. Since fungi dominate in the black biofilm, it can be assumed that the migration of elements into the solution may be limited due to the immobilization of elements by fungal biomass. Due to metabolic processes (the release of organic acids and the binding of metals by specific proteins) as well as the physicochemical properties of the cell wall, fungi can efficiently bind metals and significantly reduce their mobility in solution [6].

4. Conclusion

Biogenic weathering of granite is connected with the impact on the rock surface by microorganisms of lithobiotic communities (bacteria, microfungi, microalgae, lichens, and mosses). The biological colonization of granite is a multifactorial process. It depends on the composition of the microbiota, the state of the stone, as well as external conditions. The ecological aspect of the problem is determined by the difference between granite biofouling in the anthropogenic (urban) and natural environment. The biofilms on granite are characterized by a wide diversity of cyanobacteria, micromycetes, and organotrophic bacteria. The species composition often determines the features of the appearance of a biofilm, the features of its development on granite, as well as the biochemical composition and degree of impact on granite. Behavior of chemical elements during the bioweathering of granite depends on the type of biofilm in which some elements can be accumulated. This problem seems as an interdisciplinary task and requires the collaboration of biologists and geologists.

Acknowledgements

The work was carried out in the resource centers of Saint Petersburg State University: "Chemical Analysis and Materials Research Center," "Development of cellular and molecular technologies," and "Microscopy and Microanalysis."

Researches are supported by the European Union, Russia, and Finland (KS 1528 project).

Conflict of interest

The authors declare no conflict of interest.

Author details

Dmitry Yu. Vlasov[1,3*], Elena G. Panova[1], Marina S. Zelenskaya[1], Oksana A. Rodina[1], Alexey D. Vlasov[2] and Katerina V. Sazanova[2,3]

1 Saint Petersburg State University, Saint Petersburg, Russia

2 The Archive of the Russian Academy of Sciences, Saint Petersburg, Russia

3 Komarov Botanical Institute of the Russian Academy of Sciences, Saint Petersburg, Russia

*Address all correspondence to: dmitry.vlasov@mail.ru

References

[1] Toreno G, Isola D, Meloni P, Carcangiu G, Selbmann L, Onofri S, et al. Biological colonization on stone monuments: A new low impact cleaning method. Journal of Cultural Heritage. 2018;**30**:100-109. DOI: 10.1016/j.culher.2017.09.004

[2] Grbić ML, Vukojević J, Simić GS, Krizmanić J, Stupar M. Biofilm forming cyanobacteria, algae and fungi on two historic monuments in Belgrade, Serbia. Archives of Biological Sciences. 2010;**62**(3):625-631. DOI: 10.2298/ABS1003625L

[3] Ozturk A, Karaca Z, Unsal T. The activity of oxygenic photosynthetic microbial consortia on different granites. Ekoloji. 2014;**23**(90):90-96. DOI: 10.5053/ekoloji.2014.9011

[4] Gorbushina AA. Life on the rocks. Environmental Microbiology. 2007;**9**(7):1613-1631. DOI: 10.1111/j.1462-2920.2007.01301.x

[5] Dakal TC, Cameotra SS. Microbially induced deterioration of architectural heritages: Routes and mechanisms involved. Environmental Sciences Europe. 2012;**24**(1):1-12. DOI: 10.1186/2190-4715-24-36

[6] Gadd MG. Fungi, rocks and minerals. Elements. 2017;**13**:171-176. DOI: 10.2113/gselements.13.3.171

[7] Prieto B, Silva B. Estimation of potential bioreceptivity of granitic rocks from their instrinsic properties. International Biodeterioration and Biodegradation. 2005;**56**:206-215. DOI: 10.1016/j.ibiod.2005.08.001

[8] Miller AZ, Sanmartín P, Pereira-Pardo L, Dionísio A, Saiz-Jimenez C, Macedo MF, et al. Bioreceptivity of building stones: A review. The Science of the Total Environment. 2012;**426**:1-12. DOI: 10.1016/j.scitotenv.2012.03.026

[9] Vlasov DY, Panova EG, Zelenskaya MS, Vlasov AD, Sazanova KV, Rodina OA, et al. Biofilms on Granite Rapakivi in Natural Outcrops and Urban Environment: Biodiversity, Metabolism and Interaction with Substrate. In: Processes and Phenomena on the Boundart Between Biogenic and Abiogenic Nature. Springer; 2020. pp. 535-559. DOI: 10.1007/978-3-030-21614-6_29

Section 2

Geochemistry of Isotopes and Exploration Geochemistry

Chapter 7

Geochemistry of Radioactive Isotopes

Salih Muhammad Awadh

Abstract

The chapter targeted the geochemistry of radioactive isotopes dealing with multidisciplinary topics and focusing on geochronology and tracer studies. The most common subjects are presented to include the basic principles of radioactive isotopes. The radioactive decay, the parent nuclide, the SI unit of radioactive decay as well as the historical discovery of radioactivity, the neutrons and protons in atomic nuclei, alpha and beta particles, gamma rays, electromagnetic radiation, decay and mode of decay, chain of decay, decay rates, decay timing, principle of dating, radiometric dating, isotope systems, the Rb/Sr System, the U, Th, Pb System, the age of the earth, Sm-Nd dating, K-Ar dating, ^{14}Carbon dating, the geochron, all those were included overall.

Keywords: geochemistry, isotopes, radioactive isotope, parent nuclide, dating

1. Introduction

The process in which an unstable atomic nucleus loses energy by emitting radiation in the form of particles or electromagnetic waves known as radioactive decay that causes the energy loss from the parent nuclide converting it to daughter nuclide [1]. This chapter has been authorized based mainly on published reference focusing on some basic properties and principles of radiation and how to use this phenomenon for the estimation the absolute geological age depending on the isotope half-life and provides brief summary of only a very few examples of dating applications. Geochronology and tracer studies are two principle applications of geochemistry of radiogenic isotope. Geochronology goes to estimate the absolute time based on the radioactive rate decay from the beginning of decay to its daughter by knowing how much nuclides have decayed. Tracer application relies on the variation in ratio of the radiogenic daughter isotope to other isotopes of the element. The purpose of authoring this chapter is to help those who are interested in this field and to provide what is useful and brief in a simplified way away from the complexity.

2. Radioactive decay and natural radioactive isotopes

The radioactive decay (a phenomenon of natural and artificial) means loss of energy that results in an atom named the parent nuclide converting it to an atom of a different type, called the daughter nuclide. The ^{14}C is a parent, emits radiation and transforms to a ^{14}N representing a daughter [2]. Accordingly, it is easy to

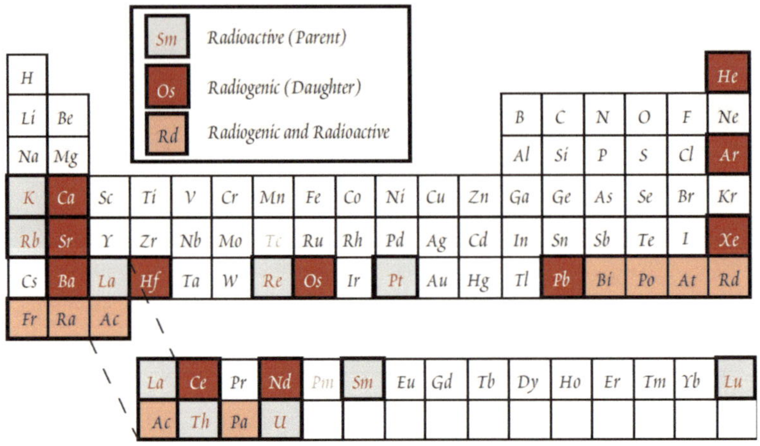

Figure 1.
Periodic table showing the elements of natural radioactive isotopes and their daughters.

understand that the radioactivity decay is that process by which an unstable atomic nucleus loses energy by emitting radiation in the form of particles or electromagnetic waves. Radioactive elements and their radiogenic daughters as well as the radiogenic and radioactive are illustrated in **Figure 1**.

3. Unit of radioactive decay and activity measurements

The becquerel (symbol Bq) is typically used as a SI unit of radioactive decay and it is defined as one decay/second. The Bq is just a tiny measure of activity; a small part of tera-becquerel (TBq) or giga-becquerel (GBq) that is commonly used. The curie (Ci) is an another unit of radioactivity that was basically defined as the activity of 1 g of pure radium ^{226}Ra. Currently, The Bq is ordinary equal to number of disintegrations per second; where Ci is equal to 3.7 × 1010 disintegrations per second. Low activities are also measured in disintegrations per minute (dpm) [2]. The name of the unit "becquerel" is originated and belonging to the Henri Becquere l, a French scientist, who discovered radiation while he working on phosphorescent materials in 1896. Later, many contributions by Becquerel, Marie Curie, Pierre Curie, Ernest Rutherford and others discovered that radioactivity was significantly more complicated [2].

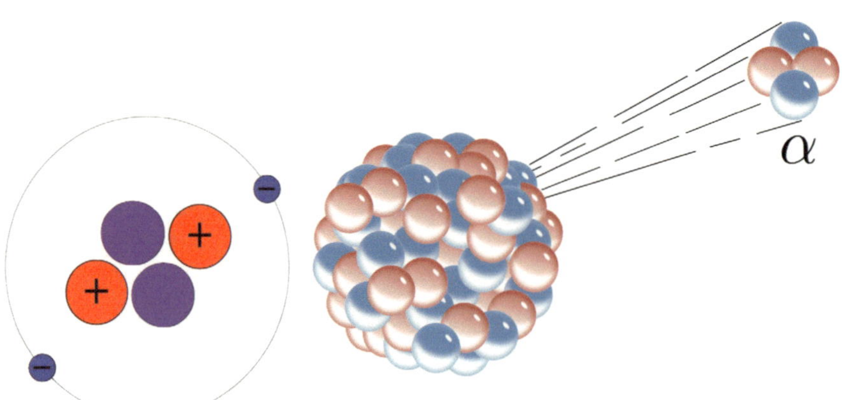

Figure 2.
Alpha particle represents the helium atom nuclei.

4. Alpha particles

They are particles (α) emitted during the radioactive decay from the nucleus consisting of two protons and two neutrons tightly bound together (**Figure 2**). Such this decay is known as alpha-decay. All chemical elements above Pb, in the Periodic Table have at least one isotope which decays by emitting alpha particles. This process is relatively rare due it requires high energy to release two neutrons and two protons out of a nucleus. The alpha particle is expressed as an identical to a helium nucleus.

5. Beta particles

They are also known as beta ray or beta radiation, symbolized by β. Beta has high-energy, high-speed electron or positron emitted during decay process of a nuclei and give β− and β+, which yield electrons and positrons respectively (**Figure 3**).

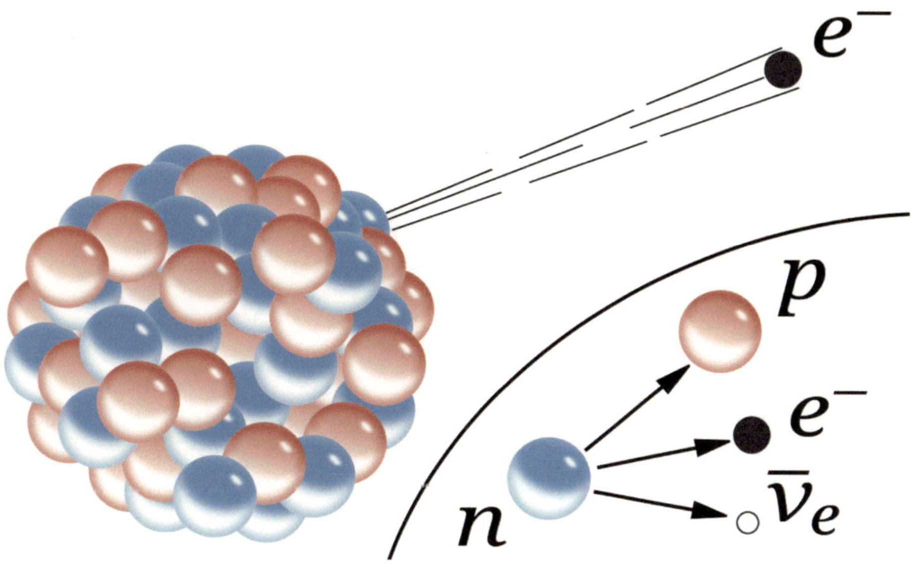

Figure 3.
Bata radiation.

6. Gamma ray

Gamma ray is also named gamma radiation symbolized with γ which is an electromagnetic radiation (**Figure 4**) emitting from the radioactive decay of atomic nuclei [3]. This type of radiation is very common.

7. Modes of decay

Different decay reactions of the radionuclides, the mass number A and atomic number Z of nucleus defined as A, Z are presented in **Table 1**. The column

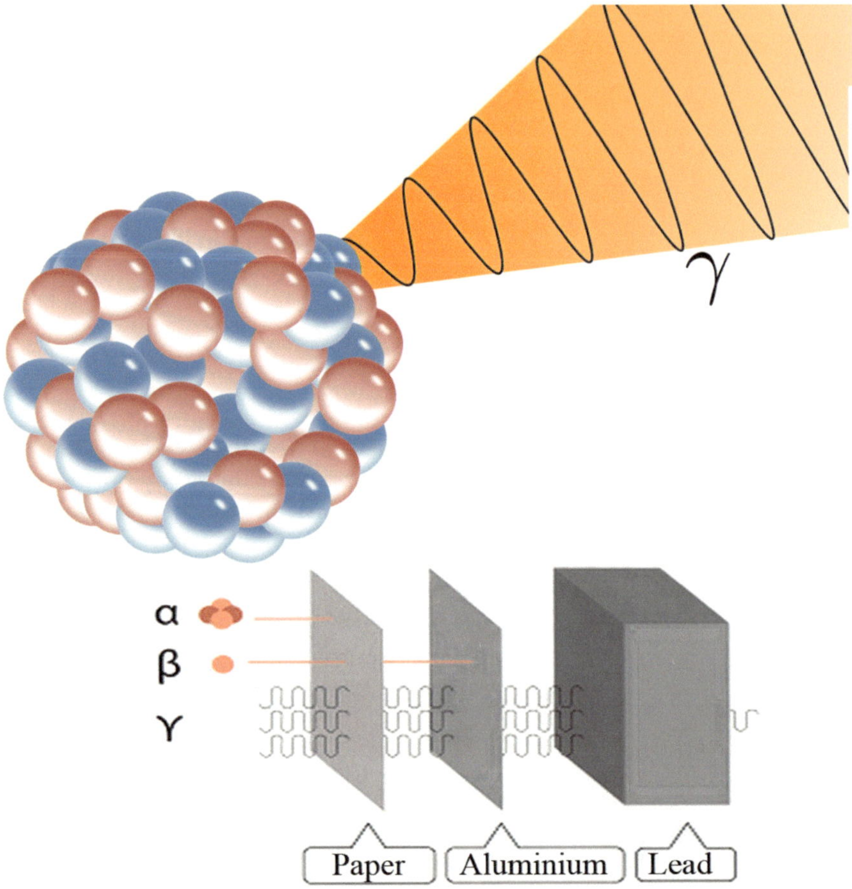

Figure 4.
Gamma radiation.

(daughter nucleus) represents the difference between the produced nucleus and the parent. Thus, (A–2, Z) means that the mass number is two less than before, but the atomic number is the same as before.

8. Decay chain and uranium isotopes

The daughter nuclide is a result of the radioactive decay of a certain radioactive element. Daughter is stable or may also be radioactive, so the chain still continues to decay. The resulting second and/or third daughter nuclide may be radioactive leading to sequential radiation, so the process known as decay chain. Uranium is very heavy element has 92 atomic number (**Figure 5**).

Three isotopes are most common of uranium; these are with their relative abundance and half-life (t1/2):

- ^{238}U has relative abundance 99.2739–99.2752% and half-life 4.4683 × 109 years.
- ^{235}U has relative abundance 0.7198–0.7202% and half-life 703.8 million years.
- ^{234}U has relative abundance 0.0050–0.0059% and half-life 245,500 years.

Mode of decay	Participating particles	Daughter nucleus
Decays with emission of nucleons		
Alpha decay	An alpha particle (A = 4, Z = 2) emitted from nucleus	(A−4, Z−2)
Proton emission	A proton ejected from nucleus	(A−1, Z−1)
Neutron emission	A neutron ejected from nucleus	(A−1, Z)
Double proton emission	Two protons ejected from nucleus simultaneously	(A−2, Z−2)
Spontaneous fission	Nucleus disintegrates into two or more smaller nuclei and other particles	—
Cluster decay	Nucleus emits a specific type of smaller nucleus (A1, Z1) smaller than, or larger than, an alpha particle	(A−A1, Z−Z1) + (A1, Z1)
Different modes of beta decay		
Beta-negative decay	A nucleus emits an electron and an antineutrino	(A, Z+1)
Positron emission, also beta-positive decay	A nucleus emits a positron and a neutrino	(A, Z−1)
Electron capture	A nucleus captures an orbiting electron and emits a neutrino - The daughter nucleus is left in an excited and unstable state	(A, Z−1)
Double beta decay	A nucleus emits two electrons and two antineutrinos	(A, Z+2)
Double electron capture	A nucleus absorbs two orbital electrons and emits two neutrinos - The daughter nucleus is left in an excited and unstable state	(A, Z−2)
Electron capture with positron emission	A nucleus absorbs one orbital electron, emits one positron and two neutrinos	(A, Z−2)
Double positron emission	A nucleus emits two positrons and two neutrinos	(A, Z−2)
Transitions between states of the same nucleus		
Gamma decay	Excited nucleus releases a high-energy photon (gamma ray)	(A, Z)
Internal conversion	Excited nucleus transfers energy to an orbital electron and it is ejected from the atom	(A, Z)

Table 1.
Detailed radioactive decay reaction after [1].

Half-life (t) means the time required for a given amount of radionuclide to lose 50% of its activity, and can be expressed as the exponential relationship (**Figure 6**) represents the interval of time required for one-half of the atomic nuclei of a radioactive sample to decay and spontaneously changes into other element by emitting particles and energy.

The best example for the radioactive decay can be illustrating by the uranium decay chain (**Figure 7**) [4]. The natural decay chain of ^{238}U which eventually decays to ^{210}Po emitting alpha with a half-life of 140 days to produce finally a stable isotope which is ^{206}Pb.

$$^{238}U \rightarrow {}^{234}Th \rightarrow {}^{234m}Pa \rightarrow {}^{234}U \rightarrow {}^{230}Th \rightarrow {}^{226}Ra \rightarrow {}^{222}Rn \rightarrow {}^{218}Po \rightarrow {}^{214}Pb \rightarrow {}^{214}Bi \rightarrow {}^{214}Po \rightarrow {}^{210}Pb \rightarrow {}^{210}Bi \rightarrow {}^{210}Po \rightarrow {}^{206}Pb.$$

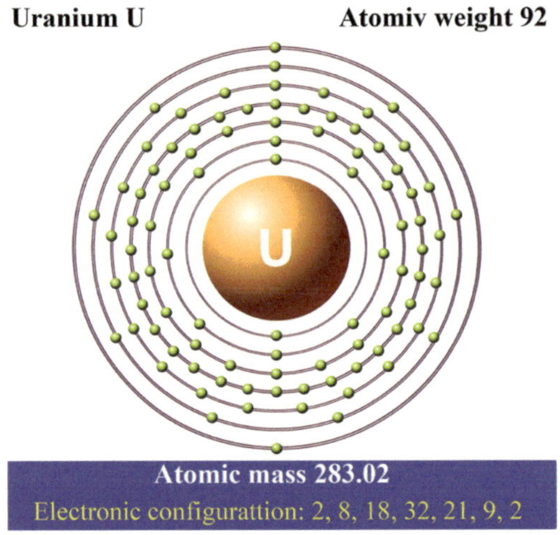

Figure 5.
Configuration of the uranium atom showing the atomic number and mass number.

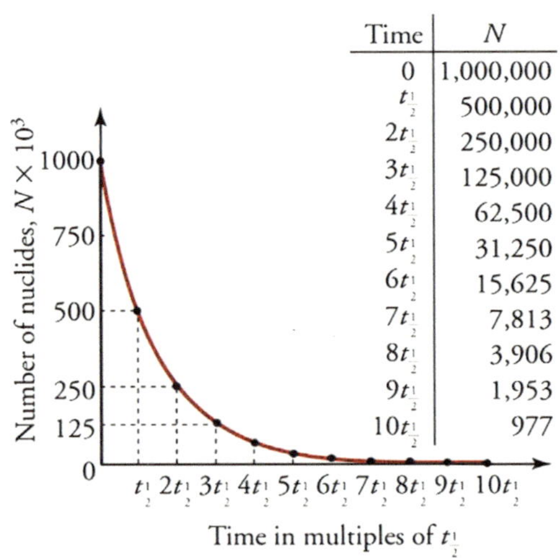

Figure 6.
The exponential relationship of half-life with radioactive decay.

9. Isotope systems in geological dating

Many isotope systems are mostly given as examples for dating geologic materials (**Table 2**).

It can be done to use this information to date rocks, for example; usually, the amount (N) of an isotope present today, and the amount of a daughter element produced by decay (D*), see Eqs. (1) and (2).

$$D^* = N_0 - N \tag{1}$$

$$N = N_0 \, e^{-\lambda t} \tag{2}$$

λ, the decay constant

$$D^* = Ne\lambda t - N = N(e\lambda t - 1) \tag{3}$$

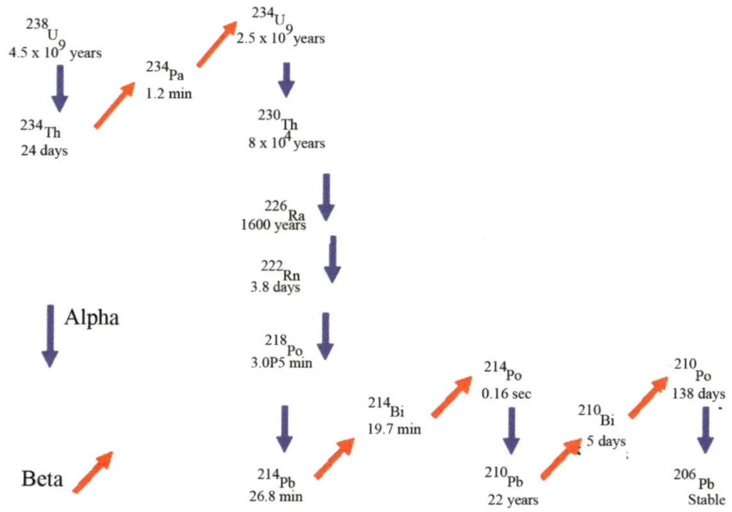

Figure 7.
Natural radioactive decay series of ^{238}U.

Parent	Daughter	Half-life	Range available	Lithology type
^{238}U	^{206}Pb	4.47 b.y	>10 m.y	Igneous & sometimes metamorphic rocks and minerals
^{235}U	^{207}Pb	707 m.y		
^{232}Th	^{208}Pb	14 b.y		
^{40}K	^{40}Ar & ^{40}Ca	1.28 b.y	>10,000 y	
^{87}Rb	^{87}Sr	48 b.y	>10 m.y	
^{147}Sm	^{143}Nd	106 b.y		
^{14}C	^{14}N	5730 y	100 – 70,000 y	Organic Material
b.y = billion years; m.y = million years; y = year				

Table 2.
Some radioactive elements with their daughters and dating application.

Consequently, it is possible to calculate the age if you have the number of daughter atoms produced by decay (D*), while the number of parent atoms (N) is known now with should pay attention for the number of daughter atoms that may have been present prior to the start of our clock.

9.1 The Rb/Sr system

We can simplify our isochron equation somewhat by noting that if x is small [7],

$$e^x = 1 + x + \frac{x^2}{2!} + \frac{x^3}{3!} + \frac{x^4}{4!} + \ldots = 1 + x \quad (4)$$

so that (eλt − 1) = λt, when λt is small.

$^{87}_{37}Rb \rightarrow ^{87}_{38}Sr$ by β decay. The neutron emits an electron to become a proton.

For this decay reaction, $\lambda = 1.42 \times 10^{-11}/y$, $t_{1/2} = 4.8 \times 10^{10}$ y, at present, 27.85% of natural Rb is ^{87}Rb.

If we use this system to plug into Eq. (3), then

$$^{87}Sr^{*} = {}^{87}Rb\ (e\lambda t - 1) \tag{5}$$

but,

$$^{87}Sr_t = {}^{87}Sr_0 + {}^{87}Sr^{*} \tag{6}$$

or

$$^{87}Sr^{*} = {}^{87}Sr_t - {}^{87}Sr_0 \tag{7}$$

Plugging this into Eq. (5)

$$^{87}Sr_t = {}^{87}Sr_0 + {}^{87}Rb\ (e\lambda t - 1) \tag{8}$$

We still do not know $^{87}Sr_0$, the amount of ^{87}Sr daughter element initially present. To account for this, we first note that there is an isotope of Sr, ^{86}Sr, that is:

1. non-radiogenic (not produced by another radioactive decay process),

2. non-radioactive (does not decay to anything else).

Thus, ^{86}Sr is a stable isotope, and the amount of ^{86}Sr does not change through time. If we divide Eq. (8) through by the amount of ^{86}Sr, then we get

$$\left(\frac{^{87}Sr}{^{86}Sr}\right)_t = \left(\frac{^{87}Sr}{^{86}Sr}\right)_0 + \left(\frac{^{87}Rb}{^{86}Sr}\right)_t (e^{\lambda t} - 1) \tag{9}$$

This is known as the isochron equation. In case of Sr was isotopically homogeneous, the time (t equal 0). For instance, $^{87}Sr/^{86}Sr$ was the same in the igneous mineral at the time of crystallization. Typically, rock – forming minerals more may be have a different amount of ^{87}Rb [5], and accordingly, those minerals are ordinary have a different $^{87}Rb/^{86}Sr$ at the crystallization time. During the natural cooling, the ^{87}Rb in each mineral will decay to ^{87}Sr, and each mineral will have a different ^{87}Rb and ^{87}Sr over time [6].

And simplify to:

$$\left(\frac{^{87}Sr}{^{86}Sr}\right)_t = \left(\frac{^{87}Sr}{^{86}Sr}\right)_0 + \left(\frac{^{87}Rb}{^{86}Sr}\right)_t \lambda t \tag{10}$$

Then time (t) can be computed as:

$$t = \frac{\left(\frac{^{87}Sr}{^{86}Sr}\right)_t - \left(\frac{^{87}Sr}{^{86}Sr}\right)_0}{\left(\frac{^{87}Rb}{^{86}Sr}\right)_t \lambda} \tag{11}$$

The initial ratio ($^{87}Sr/^{86}Sr$)0, is useful to use as a geochemical tracer because Rb distributed unequally through the Earth over time [7]. The amount of Rb in the earth mantle is typically low (<0.1 ppm). The mantle thus has a low $^{87}Rb/^{86}Sr$ ratio and would not change its $^{87}Sr/^{86}Sr$ ratio very much with time, whilst the earth crust has higher amounts of Rb (>20 ppm) and therefore start out with a relatively high $^{87}Rb/^{86}Sr$ ratio. Over time, this results in crustal rocks having a much higher $^{87}Sr/^{86}Sr$

ratio than mantle rocks. Thus, it will be expected if the mantle has a $^{87}Sr/^{86}Sr$ of say 0.7025, melting of the mantle would produce a magma with a $^{87}Sr/^{86}Sr$ ratio of 0.7025, and all rocks derived from that mantle would have an initial $^{87}Sr/^{86}Sr$ ratio of 0.7025.

On the other hand, if the crust with $^{87}Sr/^{86}Sr$ of 0.710 melts, then the resulting magma would have $^{87}Sr/^{86}Sr$ of 0.710 and rocks derived from that magma would have an initial $^{87}Sr/^{86}Sr$ ratio of 0.710. So, the rock derived from the mantle or crust can determine its initial Sr isotopic ratio accordingly.

9.2 The U, Th, Pb system

Many Pb isotopes are produced from U and Th isotopes. ^{238}U and ^{235}U and ^{232}Th can produce Pb isotopes during their radioactive decay that can be described as follows:

$$^{238}U \rightarrow 8\,^4He + {}^{206}Pb \text{ by } \alpha \text{ decay} \tag{12}$$

$$\lambda 238 = 1.551 \times 10^{-10}/y, t_{1/2} = 4.47 \times 10^9 \text{ y} \tag{13}$$

$$^{235}U \rightarrow 7\,^4He + {}^{207}Pb \tag{14}$$

$$\lambda 235 = 9.849 \times 10^{-10}/y, t_{1/2} = 0.707 \times 10^9 \text{ y} \tag{15}$$

$$\text{the present ratio of } \frac{^{235}U}{^{238}U} = \frac{1}{137.8} \tag{16}$$

$$^{232}Th \rightarrow 6\,^4He + {}^{208}Pb \tag{17}$$

$$\lambda 232 = 4.948 \times 10^{-11}/y, t_{1/2} = 1.4 \times 10^{10} \text{ y} \tag{18}$$

^{232}Th does not used in dating.

^{204}Pb is a stable non-radiogenic isotope of Pb, the two isochron equations and get two independent dates from the U-Pb system can be written as:

$$\left(\frac{^{206}Pb}{^{204}Pb}\right)_t = \left(\frac{^{206}Pb}{^{204}Pb}\right)_0 + \left(\frac{^{238}U}{^{204}Pb}\right)_0 \left(e^{\lambda_{238}t} - 1\right) \tag{19}$$

and

$$\left(\frac{^{207}Pb}{^{204}Pb}\right)_t = \left(\frac{^{207}Pb}{^{204}Pb}\right)_0 + \left(\frac{^{235}U}{^{204}Pb}\right)_0 \left(e^{\lambda_{235}t} - 1\right) \tag{20}$$

If these two independent dates are concordant, a concordia diagram will show the values of Pb isotopes that would give concordant dates and can plug in t and solve for the be calculated Eqs. (21) and (22) as follows:

$$\left(\frac{^{206}Pb^*}{^{238}U}\right) = \frac{\left(\frac{^{206}Pb}{^{204}Pb}\right)_t - \left(\frac{^{206}Pb}{^{204}Pb}\right)_0}{\left(\frac{^{238}U}{^{204}Pb}\right)_t} = \left(e^{\lambda_{238}t} - 1\right) \tag{21}$$

and

$$\left(\frac{^{207}Pb^*}{^{235}U}\right) = \frac{\left(\frac{^{207}Pb}{^{204}Pb}\right)_t - \left(\frac{^{207}Pb}{^{204}Pb}\right)_0}{\left(\frac{^{235}U}{^{204}Pb}\right)_t} = \left(e^{\lambda_{235}t} - 1\right) \tag{22}$$

The Concordia is particularly useful in dating of zircon, that usually contains a lot U and less amounts of Pb, so we expect it has large amounts of radiogenic Pb that can be produced. Apatite and sphene are the two minerals that are commonly can be used in radiometric dating as well. Zircon from the crystallization time to the present represents a closed system in case no loss or gain of uranium or lead. The age of the zircon can be determined from its position on the Concordia after plotting the $^{206}Pb^*/^{238}U$ and $^{207}Pb^*/^{235}U$ ratios on the Concordia diagram. The discordant dates fall out of the Concordia curve.

The both ends of the Discordia intersect are represented by t_0, the older and t^*, the younger. Many reasons lead to Pb leakage. Metamorphism for example, could heat the crystal to the point where Pb will become mobile. Another possible reason cause U leakage, where the discordia is represented by the two points that would give two ages $-t^*$ representing the possible metamorphic event and t0 representing the initial crystallization age of the zircon.

The Pb-Pb isochrons can be normally concluded from combining the two isochron Eqs. (19) and (20).

$$\frac{\left(\frac{^{207}Pb}{^{204}Pb}\right)_t - \left(\frac{^{207}Pb}{^{204}Pb}\right)_0}{\left(\frac{^{206}Pb}{^{204}Pb}\right)_t - \left(\frac{^{206}Pb}{^{204}Pb}\right)_0} = \left(\frac{^{235}U}{^{238}U}\right)\left[\frac{(e^{\lambda_{235}t} - 1)}{(e^{\lambda_{238}t} - 1)}\right] \tag{23}$$

$$\frac{^{235}U}{^{238}U} = \frac{1}{137.8} \tag{24}$$

Then, and by assuming that the ^{206}Pb and ^{207}Pb dates are the same, then Eq. (23) is the equation have a slope.

$$m = \frac{1}{137.88}\left[\frac{(e^{\lambda_{238}t} - 1)}{(e^{\lambda_{235}t} - 1)}\right] \tag{25}$$

that passes through the point.

$$\left(\frac{^{207}Pb}{^{204}Pb}\right)_{0,} \left(\frac{^{206}Pb}{^{204}Pb}\right)_0 \tag{26}$$

10. The age of the earth and dating

The oldest rock found in Canada, with an age of 3.962 b.y ±3 m.y. This provide only a minimum age of the Earth. The age of the earth can be computed based on the chemical concept that is the $^{235}U/^{238}U$ ratio may have been 1.0 when the elements formed. So, from:

$$N = N_0 e^{-\lambda t} \tag{27}$$

the

$$\frac{^{235}U}{^{238}U} = \frac{^{235}U_0 e^{-\lambda_{235}t}}{^{238}U_0 e^{-\lambda_{238}t}} \tag{28}$$

Finally, t is about 6 b.y.

This age represents the maximum age of the Earth. From the Pb-Pb isochron Eq. (23) and based on meteorites that may have been formed at the same time the solar system in which basically the Earth formed as well. The thing to be needed to date meteorites is knowing the initial ratios of the Pb isotopes. Two major types of meteorites are recognized; Fe- meteorites and stony (or chondritic) meteorites. The Fe-meteorites contain troilite (FeS) that has no U. Since the mineral troilite contains no U, all of the Pb present in the troilite is the Pb originally present, and none of it has been produced by U decay. Thus, the troilite in the Fe-meteorites will provide with the initial ratios of $^{206}Pb/^{204}Pb$ and $^{207}Pb/^{204}Pb$.

The Pb ratios can be determined in other meteorites and check if they fall on the isochron of Pb-Pb that passes through the initial ratios determined from troilite in Fe-meteorites. The slope of this isochron (Geochron) estimated the earth age is of $4.55 \pm 0.07 \times 10^9$ yr. Consequently, the best estimation of the age of the Earth is 4.55 billion years.

10.1 Sm-Nd dating

^{147}Sm decays to ^{143}Nd by alpha decay with half-life of 106 ± 2 b.y. [8], ^{147}Sm, ^{148}Sm, ^{149}Sm, and ^{144}Nd are radioactive, three nuclides accordingly generated ^{144}Nd, ^{145}Nd, and ^{140}Ce [9].

$$^{147}\text{Sm} \rightarrow \,^{143}\text{Nd} \tag{29}$$

$$\lambda = 6.54 \times 10^{-12}/\text{yr}, t_{1/2} = 1.06 \times 10^{11}\ \text{y} \tag{30}$$

The isochron equation is described based on whether the ^{144}Nd is stable and non-radiogenic as:

$$\left(\frac{^{143}\text{Nd}}{^{144}\text{Nd}}\right)_t = \left(\frac{^{143}\text{Nd}}{^{144}\text{Nd}}\right)_0 + \left(\frac{^{147}\text{Sm}}{^{144}\text{Nd}}\right)_t \left(e^{\lambda t} - 1\right) \tag{31}$$

The age of the rock can be estimated later from the isochron equation that basically can be drawn by the determination of the $^{143}Nd/^{144}Nd$ and $^{147}Sm/^{144}Nd$ ratios for several minerals.

10.2 K-Ar dating

In the nature, ^{40}K makes up 0.119% of natural K, as it is a radioactive element, its decay can be presented as follows [10]:

$$^{40}\text{K} \rightarrow \,^{40}\text{Ca by } \beta \text{ decay} \tag{32}$$

The equation above is not used, because 40Ca can be present as both radiogenic and non-radiogenic Ca [10].

$$^{40}\text{K} \rightarrow \,^{40}\text{Ar by electron capture} \tag{33}$$

For the combined process,

$$\lambda = 5.305 \times 10^{-10}/\text{y}, t_{1/2} = 1.31 \times 10^{9}\ \text{y} \tag{34}$$

and for the Ar branch of the decay scheme

$$\lambda e = 0.585 \times 10^{-10}/\text{y} \tag{35}$$

Argon is a gas easily can escape from a magma or liquid, therefore, the percentage of initial ^{40}Ar is expressed as zero; during the rapid cooling of magma, quanitity of the Ar may be trapped. The date consequently obtained will be older than the date at which the magma erupted.

The dating equation used for K-Ar is:

$$^{40}Ar = \frac{\lambda_e}{\lambda}\,^{40}K\left(e^{\lambda t} - 1\right) \quad (36)$$

where $\frac{\lambda_e}{\lambda} = 0.11$ and refers to fraction of ^{40}K that decays to ^{40}Ar.

Many points need to have attention when use K-Ar dating, the use of minerals like sanidine or biotite is better to use whole rocks because minerals not contain excess Ar. Other thing, some atmospheric argon originated from volcanic eruptions could be absorbed onto the sample surface, ^{40}Ar should be corrected for accordingly. Additionally, most minerals lose Ar during metamorphism due to high temperature, so the date will represent the metamorphic event (**Table 3**).

10.3 ^{14}C dating

Radiocarbon dating is different than the other methods of dating because it cannot be used to directly date rocks, but can only be used to date organic material produced by once living organisms [12]. Radiocarbon (^{14}C) has a short half-life (5730 y), it is therefore only used to date materials younger than about 70,000 years. The ratio of ^{14}C to ^{14}N in the Earth's atmosphere is constant and the organism have the same ratio of ^{14}C to ^{14}N as the atmosphere. When an organism dies, the ^{14}C decays back to ^{14}N, with a half-life of 5730 years. Measuring the amount of ^{14}C in this dead material thus enables the determination of the time elapsed since the organism died. Bones, teeth, charcoal, fossilized wood, and shells are materials can be used for dating.

10.4 Re-Os dating

Rhenium has stable ^{185}Re and the radioactive ^{187}Re. The latter is the most abundant (62.6%) and decays to ^{187}Os based on beta decay, typically with a half-life of 41.6 × 109 y [13]. Osmium has seven isotopes; only two are the product of natural decay of radioactive isotopes: ^{186}Os is produced from ^{190}Pt by α-decay (half-life 4.7 × 1011 y, [14]) and ^{187}Os by β-decay of ^{187}Re. The two radiogenic isotopes ^{187}Os (~2%) and ^{186}Os (~1.6%) are typically normalized to the stable ^{188}Os (13.24%). Rhenium-osmium (Re-Os), an applicable method was first applied to meteorites [15]; it provides a chronometer for directly dating both of sulfides and oxides ore minerals.

Decay	Decay factor	value
$^{40}K \rightarrow {}^{40}Ca$ by β-	λβ-	$4.962 \times 10^{-10}\ a^{-1}$
$^{40}K \rightarrow {}^{40}Ar$ by electron capture and γ	λe	$0.572 \times 10^{-10}\ a^{-1}$
$^{40}K \rightarrow {}^{40}Ar$ by electron capture	λ' e	$0.0088 \times 10^{-10}\ a^{-1}$
combined value	λ = λβ- + λec + λ' ec	$5.543 \times 10^{-10}\ a^{-1}$
	present day $^{40}K/K$	0.0001167

Table 3.
Decay constants for K-Ar and Ar-Ar dating [11].

11. Discussion and conclusions

This chapter deals with the various types of radiation emitted by radioactive nuclides with principles of radionuclide decay and its radiations. Here an overview of some of the many dating radioactive techniques that play as significant role in our day-to-day lives. The dating techniques developed for defining reliable ages of geologic events other geochronological studies are recording of the isotope concentrations. All radiometric clocks depend on a radioactive "parent" isotope that decays to a daughter stable isotope of another element at a constant rate on geologic timescales. This process may take single step, or it may involve many stages of decay products before reaching the final stable daughter isotope. The half-life of the initial quantity of parent isotope to decay must be on the same order of magnitude as the time span to be measured. The Concordia–Discordia model has been developed for the U and Pb isotopes. The ^{235}U transforms to ^{207}Pb through a chain of radioactive nuclides, releasing (235–207)/4 = 7α-particles with the constant λ5 = 9.8·10–10. The ^{238}U turns to ^{206}Pb releasing (238–206)/4 = 8α-particles with the constant λ8 = 1.55 · 10–10 [16]. Currently, the ratio of $^{238}U/^{235}U$ (137.88) is growing. Both isotopes of uranium are closely connected to each other in kinetic processes due to the value $\sqrt{}238/235$ = 1.0063, which is close to 1. The Discordant values can be obtained from the development of the Concordia–Discordia model as an open system with losses of radiogenic lead in accessory minerals such as zircon, monazite, apatite etc. [17]. A very wide time range, not only the 12 b.y. of the Universe age and the 4.5 b.y. of the Earth's age can be explore, but also the details related paleontology through the history of the Earth and recent events of the last millennia [12]. The K-Ar dating technique developed soon after the discovery of ^{40}K and provided an important dating tool beside U-Pb and U-He dating methods. The half-life (1250 m.y) made this method is most popular for dating geological events [18]. The K-Ar dating are based on the decay of a ^{40}K to an isotope of ^{40}Ar by a branching process; 10.48% of ^{40}K decays to ^{40}Ar by β + decay, and 89.52% decays to 40Ca by β- to the ground state [10]. The age measured by K-Ar techniques reflects the time since radiogenic argon produced by decay of ^{40}K, became trapped in the mineral or rock. The radiogenic noble gas daughter nuclides provide many methods for determining not only the chronology of events but also thermal histories combined with U-Pb and Rb-Sr dating techniques. This technique uses to conclude the cooling history based on use mineral closure or field estimates [19]. It can be applied for dating young volcanic eruptions and for low-temperature phases such as clay minerals like illite. In addition, they can be used for exploring. Despite the Rhenium–osmium (Re-Os), an applicable method was first applied to meteorites [15]; it also provides a chronometer for directly dating both of sulfides and oxides ore minerals. Recently, this technique is developed and become capable to estimate dating via dealing with very low content of Re and Os. The relative abundance of Osmium is the earth's core and extraterrestrial material with a very leaser amount (ppt) in the mantle and it can be stored in sulfide and oxide minerals in the crust. It 8is best method for dating the age of gold in auriferous pyrites, it also used for dating marine shale containing coal.

It is difficult to obtain good precision and accuracy for radiocarbon due to its abundancy in the environment and it is possible to contaminate from material of a different age. Consequently, the methods for radiocarbon measuring are well tested, reproduced and carefully controlled under specific lab conditions. Recently, the radiocarbon methods have been developed, over the last 30 years to cover most of the materials suitable for radiocarbon measurement. The AMS system at Oxford was built with very high precision and accuracy for radiocarbon dating by High Voltage Engineering Europa BV. For high precious in situ age dating of Pb-U, Hf and U-Th isotope ratios in very small minerals like zircons, it is recommended to use the Thermo Scientific Neptune XT MC-ICP-MS or Thermo Scientific Element 2 and Thermo Scientific Element XR High-Resolution ICP-MS, combined with a laser ablation system. The Thermo Scientific Triton XT Multicollector Thermal Ionization Mass Spectrometer (TIMS), provides the ultimate precision for U-Pb geochronology, while The Triton XT TIMS is an equipment with high-quality age dating for Rb-Sr, Sm-Nd and Re-Os.

To be useful, result must be accurate, so uncertainty must always be taken into account. The geochronological result is influenced by uncertainty. So, if it does not be known well, result is scientifically meaningless. The uncertainty is coming from error in sampling, laboratory procedure; adaption of methods to the problem in question [12]. Overall, the source of uncertainty obtained from:

a. collecting samples

b. Parent decay methods

c. Long half-life parent daughter methods

d. Age calculations

11.1 Analytical methods

The radiogenic isotopes are typically separated from the no-radiogenic isotopes using spectrometer whenever they used as dating tools or tracers. Sample is ionized normally in thermal-ionization mass spectrometer (TIMS). Recently, the induced coupled plasma (ICP) is technique used for the chemical purification before mass spectrometry method. The laser ablation is also used for analyzing mineral with high concentration of radiogenic elements. However, in samples of whole-rocks, in which the concentration of radiogenic isotope is mostly low, it is necessary to pre-concentrate after dissolution and chemical extraction. The silicate geological samples are routinely dissolved in hot method using either concentrated HF acid or $HCLO_4$ acid at atmospheric pressure; so, the most rock-forming minerals are dissolved, but the resistant minerals like zircon must be dissolved under pressure in a bomb at 220°C. The bomb liner and beaker are made of poly-fluorinated ethylene [20]. The formation of fluoride is the most common problem that may encounter after dissolution in HF as an insoluble in HCl acid. Consequently, the refluxing with HNO_3 is needed [21]. The additional adding of HNO_3 before completely evaporation of HF leads to promote process [22]. If at some stage, complete digestion is not achieved, decant off the solution is recommended and return to undissolved fraction at the previous stage for a second acid attack [23]. Thereafter, the solution rich in isotope is split for isotope-dilution analysis and for accurate-isotope ratio analysis.

11.2 Dating of igneous rocks

11.2.1 Sr-model ages

The Rb-Sr is used widely as a method provides a great information of igneous rock dating. The naturel process begins from the decay of ^{87}Rb occurred in minerals to ^{87}Sr, so the number of ^{87}Sr daughters produced informs us about t years ago:

$$^{87}Sr = {}^{87}Sr_1 + {}^{87}Rb\left(e^{\lambda t} - 1\right) \tag{37}$$

Where the $^{87}Sr_1$ is the number of the initial ^{87}Sr atoms. The given nuclides are of so difficult to measure their absolute abundance, so, it is suitable to find isotope ratio by dividing by ^{86}Sr which is not radiogenic and accordingly still constant with time as follows:

$$\left(\frac{^{87}Sr}{^{86}Sr}\right)_p = \left(\frac{^{87}Sr}{^{86}Sr}\right)_I + \frac{^{87}Rb}{^{86}Sr}\left(e^{\lambda t} - 1\right) \tag{38}$$

Currently, strontium isotope ratio (p) can be measured by mass spectrometry, and the $^{87}Rb/^{86}Sr$ is calculated from Rb/Sr weigh ratio. From the initial ratio ($^{87}Rb/^{86}Sr$)I estimated, time (t) can be computed as:

$$t = \frac{1}{\lambda}\ln\left\{1 + \frac{^{86}Sr}{^{87}Rb}\left[\left(\frac{^{87}Sr}{^{86}Sr}\right)_p - \left(\frac{^{86}Sr}{^{87}Sr}\right)_I\right]\right\} \tag{39}$$

Over geological time, Rb-rich minerals "like lepidolite" develop ratio (0.712) of $^{87}Rb/^{86}Sr$ and may use in chronological studies without error. The Rb-Sr method was extended to include other mineral such as mica (biotite and muscovite) as well as potash feldspar that have lower Rb/Sr ratios. The discordant dates are suggested based on the initial ration (0.712) when the real initial ratio was higher. This expressed as a problem and was overcome by the isochron diagram designed by [24] who developed a new way for treating Rb-Sr data based on the principle of linear equation:

$$y = c + xm \tag{40}$$

In this way, $^{87}Sr/^{86}Sr$ (y) is plotted versus $^{87}Rb/^{86}Sr$ (x), then the intercept c is the initial $^{87}Sr/^{86}Sr$ ratio (**Figure 8**).

This figure presents a suite of co-magmatic minerals of same age and initial of $^{87}Sr/^{86}Sr$ ratio, forming a line called isochron. From slop of isochron, $m = e^{\lambda t} - 1$, the mineral age can be determined. The mineral with very low Pb may yield the

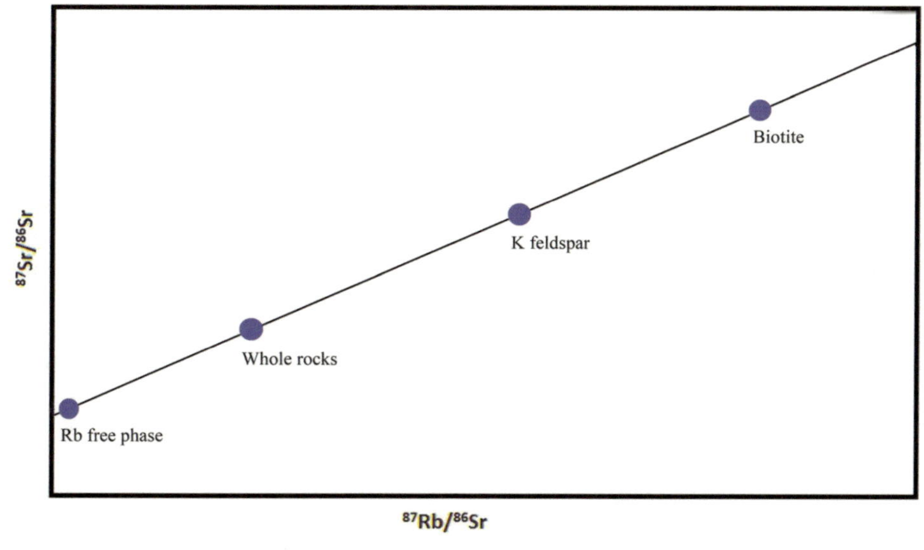

Figure 8.
Schematic Rb-Sr isochron diagram for a suit of co-magmatic igneous minerals [25].

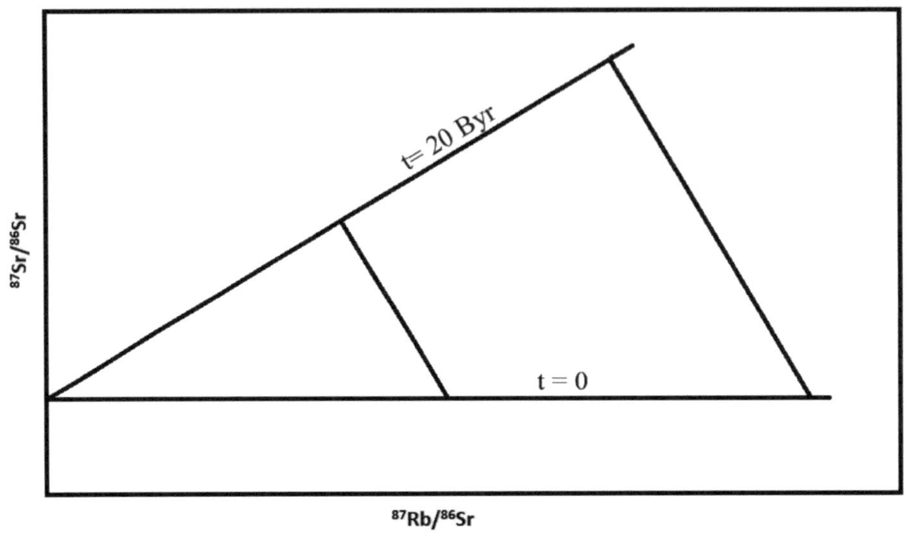

Figure 9.
Rb-Sr isochron diagram on axis of equal magnitude showing production of ^{87}Sr as ^{87}Rb is consumed in two hypothetical samples [25].

initial ratio directly on y-axis. On **Figure 9**, the $^{87}Sr/^{86}Sr$ increase with decrease $^{87}Rb/^{86}Sr$ due to decay Rb over time. The slope of ischron is increased accordingly with time. Practically, the *y*-axis is expanded much to displays rocks of geological time clearly, and the growth lines become vertical accordingly.

11.3 Erupted isochro

The original isotopic composition of mantle is basically inherited in the primary basic magma. The alkali ocean-island basalts were investigated for the Rb-Sr system [26]. The results of fourteen wide range type of ocean- island basalt samples plotted on an isochron displayed a proportional correlation to a slop age of about 2B.y representing the time of mantle isolation [25]. This age is known as a mantle isochrones that is also extended to continental igneous rocks [27]. The ratio of $^{87}Sr/^{86}Sr$ should correct back to initial ratio at the time of magmatism before plotting versus $^{87}Rb/^{86}Sr$, so these are termed as pesud-oisochrons. Data of plotting 30 samples from both volcanic and plutonic continental igneous rock suits formed a roughly linear array. The pesud-oisochrons generates from two lines representing crustal contamination of mantle-derived basaltic magma, Scientist replaced that by timing the mantle differentiation events that established mantle domains of different ratio of Rb/Sr in subcontinental lithosphere. This suggestion does not provide reliable results to ascribe age significantly to erupted isochrones. However, the only isotope-isotope mantle isochrones is reliable and can be interpreted and used significantly as tool for dating the age of mantle differentiation events.

Author details

Salih Muhammad Awadh
Department of Geology, College of Science, University of Baghdad, Baghdad, Iraq

*Address all correspondence to: salihauad2000@yahoo.com

References

[1] L'Annunziata MF. Radioactivity Introduction and History. Oxford: Elsevier; 2007. p. 610

[2] L'Annunziata MF. Radioactivity, Introduction and History. The Netherland: Library of Congress Cataloging in Publication Data; 2007. p. 610

[3] Cardarelli F. Encyclopaedia of Scientific Units, Weights and Measures. springer-Germany: Springer-Verlag London Ltd; 2003. ISBN: 978-1-4471-1122-1

[4] Mitchell N, Pérez-Sánchez D, Thorne MC. A review of the behaviour of U-238 series radionuclides in soils and plants. Journal of Radiological Protection. 2013;**33**:R17-R48

[5] Nebel O, Scherer E, Mezger K. Evaluation of the ^{87}Rb decay constant by age comparison against the U–Pb system. Earth and Planetary Science Letters. 2011;**301**:1-8

[6] Jenkin GRT, Ellam RM, Rogers G, Stuart FM. An investigation of closure temperature of the biotite Rb–Sr system: The importance of cation exchange. Geochimica et Cosmochimica Acta. 2001;**65**:1141-1160

[7] White WM. Isotope Geochemistry. New York, USA: Wiley-Blackwell; 2014. p. 496

[8] Gupta MC, McFarlane RD. The natural alpha radioactivity of samarium. Journal of Inorganic and Nuclear Chemistry. 1970;**32**:3425-3432

[9] Lugmair GW, Marti K, Scheinin NB. Incomplete mixing of products from r-, p-, and sprocess nucleosynthesis: Sm-Nd systematics in Allende inclusion EKI-4-1. Lunar Planet. 1978;**IX**:672-674

[10] Beckinsale RD, Gale NH. A reappraisal of the decay constants and branching ratio of ^{40}K. Earth and Planetary Science Letters. 1969;**6**:289-294

[11] Steiger RJ, Jäger E. Subcommission on geochronology: Convention on the use of decay constants in geo- and cosmocchronology. Earth and Planetary Science Letters. 1977;**36**:359-362

[12] Allegre CJ. Isotope Geology. London: Cambridge University Press; 2008. p. 511

[13] Smoliar MI, Walker RJ, Morgan JW. Re-Os ages of group IIA, IIIA, IVA, and IVB iron meteorites. Science. 1996;**271**: 1099-1102

[14] Begemann F, Ludwig KR, Lugmair GW, Min K, Nyquist LE, Patchett PJ, et al. Call for an improved set of decay constants for geochronological use. Geochimica et Cosmochimica Acta. 2001;**65**:111-121

[15] Shirey SB, Walker RJ. The Re-Os isotope system in cosmochemistry and hightemperature geochemistry. Annual Review of Earth and Planetary Sciences. 1998;**26**:423-500

[16] Jaffey AH, Flynn KF, Glendenin LF. Precision measurement of half-lives and specific activities of ^{235}U and ^{238}U. Physics Review. 1971;**C4**:1889-1906

[17] Rasskazov SV, Brandt SB, Brandt IS. Radiogenic Isotopes in Geologic Processes. Springer Dordrecht Heidelberg London New York, Library of Congress Control Number: 2009938100. New York; 2010

[18] McDougall I, Harrison TM. Geochronology and Thermochronology by the ^{40}Ar/^{39}Ar Method. New York: Oxford University Press; 1999. p. 212

[19] Grove M, Harrison TM. ^{40}Ar diffusion in Fe-rich biotite. American Mineralogist. 1996;**81**:940-951

[20] Krogh TE. A low contamination method for hydrothermal decomposition of zircon and extraction of uranium and Pb for isotopic age determination. Geochimica et Cosmochimica Acta. 1973;**37**:484-494

[21] Parrish RR. An improved micro-capsule for zircon dissolution in U-Pb geochemistry. Chemical Geology: Isotope Geoscience Section. 1987;**66**: 99-102

[22] Croudace IW. A possible error source of silicate wet chemistry caused by insouluble fluoride. Chemical Geology. 1980;**31**:153-155

[23] Patchett PI, Tatsumoto M. A routine high precision method for Lu-Hf isotope geochemistry and chronology. Contributions to Mineralogy and Petrology. 1980;**75**:263-267

[24] Nicolyasen IO. Graphic interpretation of discordant age measurements of metamorphic rocks. Annals of the New York Academy of Sciences. 1961;**91**:189-206

[25] Dickin AP. Radiogenic Isotope Geology. Cambridge University Press; 2005. p. 492

[26] Sun SS, Hanson GN. Evaluation of the mantle: Geochemical evidence from alkali basalt. Geology. 1975;**3**:297-302

[27] Brooks C, Harts SR, Hofmann A, James DE. Rb-Sr mantle isochrons from oceanic regions. Earth and Planetary Science Letters. 1967;**32**:52-62

Chapter 8

Exploration for Fe-Mn Oxides Using Geochemical Signatures in Soil: A Case Study of Part of Northwestern Nigeria

Olufemi Sijuade Bamigboye

Abstract

Part of northwestern Nigeria was investigated with the aim of delineating concealed mineralization using geochemical signatures in soils. To achieve this, 30 selected soil samples were analysed geochemically. The result of the elemental analysis was subjected to Principal Component Analysis (PCA) and isograde plotting, while selected elements were correlated. From the geochemical result, most of the analysed elements have anomalous value in the southern part of the area, while the least values are in the southwestern. From the PCA analysis, six factor groups were distinct. The factor groups were interpreted geochemical to fingerprint mineralization in the area. The result of correlation analysis shows that Fe is negatively correlated with most of the correlated elements. The study concluded that the central part of the study area is mineralized with both manganite and goethite. In addition, manganese mineralization is indicated by elemental association: Zn+As+Be+Bi+Co+Nb+Ni+CsP+Al+Ca+Cd+Li+K, while iron mineralization is indicated with Zr+Th+Pd+Mo+V+Sn+Cr+Ce+InSc+P+Pb association.

Keywords: manganite, goethite, Kaojc, exploration geochemistry, soil survey

1. Introduction

Traditional exploration geochemistry usually employs earth materials such as rock, stream sediment and soil to detect unusual concentration of elements that may serve as pathfinder for concealed ore body [1–3]. The distribution of pathfinder elements in these media especially stream sediment and soil is governed by weathering and hydromorphic conditions of the area under investigation in addition to the mobility of such elements. Despite their dispersion, trace and rare earth elements still retain their bedrock characteristics. These elements are adsorbed unto surfaces of weathered products and possess the ability to remain on the surfaces of these weathered particles for long periods unless they are mobilized by decomposition processes such as redox conditions [4]. Bowen [5] believes that the residence time of these elements in temperate soils is not the same. For example, the residence time of Pb is between 740 and 5900 years, Zn has a residence time of 70–510 years, 13–1100 years is the residence time for Cd, and 310–1500 years is the residence time for Cu, while the residence time in tropical soils is 40 years due to the shorter rate of leaching. Soil geochemical surveys have proven to be highly effective with long and impressive history of discovering blind deposits [6].

Akilolu [7] reports the gossan manganite occurrences in Buya and Derena area, northwestern Nigeria. He groups these ores into three which are Buya, Derena north and Derena south clusters. In his paper, he argues that the Buya gossans are purer than other clusters. Bamigboye et al. [8] report goethite mineralization in addition to the manganite mineralizations reported by Akilolu [7] in Kaoje, northwestern Nigeria. Furthermore, an exposure of brown laminated manganite ore, south of Barkin-Ruwa, was also reported to earlier reported manganite clusters in these areas. This exposure is relatively unique in terms of its brown colouration as against the black colour of those in Buya and Derena clusters. The report of Akilolu [7] and Bamigboye et al. [8] did, however, not give account of the distribution of trace and rare earth elements in the soil in Kaoje and its environs, neither is vivid account given on the possibility of discovering concealed bodies of these ores or others within Kaoje and its environs. Other work related to this include the work of Adekoya [9], Mucke [10], Okorie et al. [11, 12] and Fillie [13, 14], among others. This work is therefore aimed at identifying the mineralized zones in Kaoje and its environs using geochemical signatures in soils around Kaoje and its environs, northwestern Nigeria. The study area is bounded by Long. 3°55′ and 4°10′E and Lat. 11°00′ and 11°15′N covering an approximate area of 676 km^2.

2. Geology of the study area

The geology of the area can grossly be divided into the basement complex rock and sedimentary rocks. The basement complex rocks are mainly banded gneiss, migmatite, granite gneiss, quartz-mica schist and granite (**Figure 1**).

The banded gneiss occurs as elongated rock body within the schist and is seen about 2 km south-west of Buya and midway between Idowa and Derena. The banded gneisses are rich in quartz with pervasive jointing, filled by quartz and quartzofeldspathic vein in most cases. Migmatite trends in SE-NW direction mainly. The granite gneisses have a sharp contact with the mica schist. Mineralogically, quartz, biotite and sodic feldspar predominate, but crystals of tourmaline are seen in some localities. The quartz-mica schist is extensively weathered and covers about one-fifth of the area studied. It occurs as a low-lying exposure and stream-cut exposure. Some of the schists occur as xenolith with a large pegmatitic body close to Barkin-Ruwa, along Barkin-Ruwa-Buya road.

The granitic bodies occur towards the southern part of the study area. The granitic rocks occur as boulders and cobbles of rocks and are closely associated with weathered schist in the southernmost part of the area. The pegmatitic rocks in this area are seen associated with the schist and migmatite mainly. Some of the pegmatites are also seen at the contact between the schist and the sedimentary rock in the northwestern area. Mineralogically, the pegmatites are made up of quartz, biotite, k-feldspar tourmaline (in the southeastern part) and chalcedony (in the northwestern part).

Sedimentary rocks in this area occupy about 60% of the area. This rock includes the Biongbe, Morongba and Koremi hills. These hills are essentially clastic sedimentary rocks that include the sandstone and silty sandstone. In the northeastern part of Kaoje, sedimentary rocks occur as remnant of gully eroded sedimentary rocks. These rocks are finely laminated with characteristic ichnofossil. Other characteristic of these rocks are their semi-consolidated nature that is typical of Taloka Formation in the Sokoto Basin. The rocks in this area include silty sandstones that are finely laminated, while some are massive. Other sedimentary rock types seen associated with the sandstone are goethite, kaolinitic mudstone, finely laminated

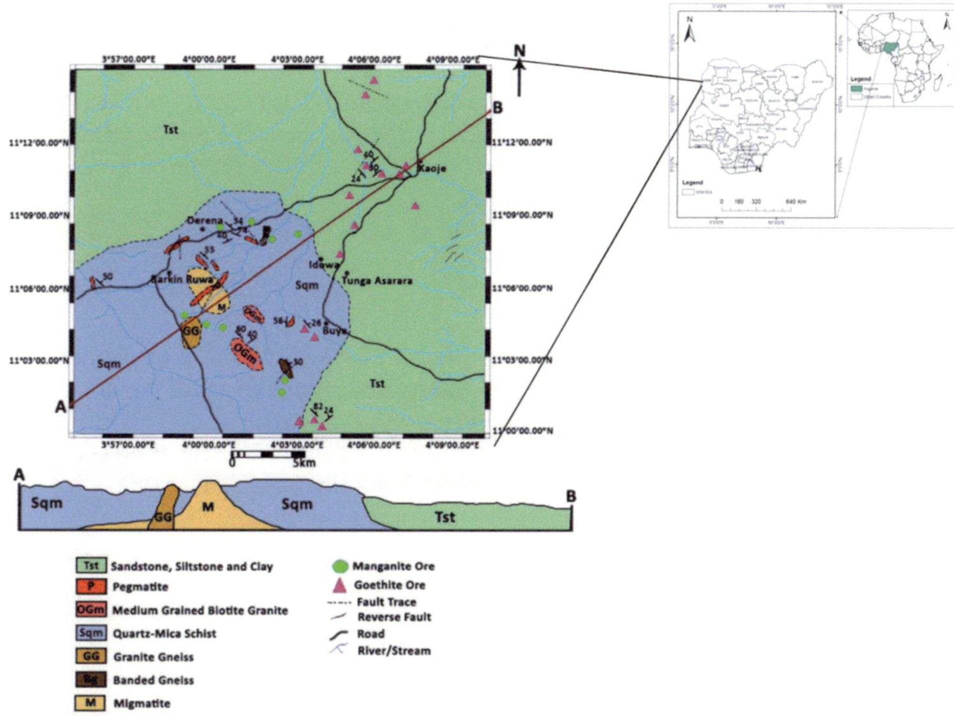

Figure 1.
Geological map of Kaoje and its environs with insect of map of Nigeria and Africa.

siltstones and kaolinitic purple claystone. All these are exposed along Kaoje-Idowa road and towards the edge of the area studied in the northeastern part.

The iron ore occur as vein, boulders and sedimentary capping in the northeast and southeast mainly. It shows conspicuous alternating bands of iron and chert-rich layers in most of the location, while few are massive iron ore with little or no cherty contents. Bamigboye et al. [8] describe these irons to be of hydrothermal origin but now remobilized. Manganite gossans occur as boudins in an almost NE-SW direction from Buya through Derena to the eastern part of Barkin-Ruwa. These ores occur as black laminated exposures mainly around the Buya-Derena axis and as brown folded brown laminated body south of Barkin-Ruwa. Mn is associated with quartz veins. From the work of Bamigboye et al. [8], the iron ores are made up of feroxhyte, gibbsite, goethite, maghemite, magnetite and haematite essentially with subordinate manganite, while the manganites are composed of manganite, gibbsite, magnetite, bayerite, akaganeite and nordstrandite among others.

3. Methodology

The methods adopted in this work are grossly divided into two. These are fieldwork and laboratory work. The fieldwork was carried out between 2011 and 2012. During this period, rock and soil samples were taken. The soil samples were taken from a grid of 1 km × 1 km from a depth range of 30–50 cm corresponding to the B-horizon as recommended by Vareikiene and Lehtonen [15]. Thirty soil samples were prepared and subjected to geochemical analysis at ACME Analytical laboratory, Vancouver, Canada. In the laboratory, 0.5 g of the soil samples were digested with a modified aqua regia solution of equal parts concentrated HCl, HNO_3 and DI H_2O for 1 h in a heating block. The resulting solution was thereafter analysed by ICP-ES and ICP-MS for trace and rare earth elemental concentration. The selected elements that were considered related to iron and manganite mineralization from the geochemical data was correlated. Others were subjected to principal component statistical analysis (PCSA) using the SPSS software, while isograde plotting of the data was done using Surfer 12.

4. Results and interpretation

The result of the geochemical analysis where the concentration of 53 elements was determined is shown in **Table 1**. From the result, sample 83a and 86 had the highest concentration of most of the elements, while sample 64 and 222 had the lowest concentration of most of the elements. S had a constant value of <0.02 except in sample 83a where the value is 0.02%. This same sample had 0.05 ppm concentration of W as against <0.05 in the remaining samples. This is equally similar to the slight higher concentration of 0.02 ppm concentration of Se as against <0.1 in other samples.

The good correlation coefficient of Mn with Ni and Mg and the weak correlation especially between Mn and Cu are related to Mn-rich end member (**Table 2**). Similarly, good correlation coefficients between Fe and Ti and P coupled with the correlation between Co and Ti and the weak-to-low correlation between Ti and Ca, Ca and P, Ca and Co, and Ca and Fe are all related to Fe-rich end member. Likewise, the strong correlation between Al and K ($r = 0.845$) is related to aluminosilicate minerals, while the fairly good correlation coefficients between Mg and Mn and Cr and Ca are related to mantle-derived ultramafic materials.

The result of the PCA shows a total of six factor groups (**Table 3** and **Figure 2**). These are factors 1–6. Factor 1 with Mg, Ni, Cs, K, Te, Li, Ti, Rb, Bi, Al, Zn, Cu, Tl, U, Cr, Co, Na and Be accounts for a total of 43.59% before rotation and 31.86% after rotation. This is factor 2 with Zr, Hf, Th, Ga, In, Sc, Sn, Fe, V, Pd, Mo, Sb and As, and Ce accounts for 20.701% before rotation and 23.09% after rotation. Factor 3 has Se, Nb, Ca, Sr and P. This factor group accounts for 9.95% before rotation and 14.697% after rotation. Factor 4 with Ba, Pb, Y, Mn, La and Cd gave an account of 8.518% before rotation and 10.078% after rotation, while factor 5 with the following elements: Hg, Ag and Au account for 3.412% before rotation and 2.758% after rotation. Factor 6 does not have the highest loading of any element. Despite this, it still accounts for 2.658% before rotation and 3.106% after rotation.

Factor 1 of the principal component analysis of the soil was related to the weathering of the hydrothermal sulphide ore and silicate-rich banded iron formation [16, 17], while factor 2 was the product of weathered chert-rich magnetite and ores that are rich in terrigenous materials. The inclusion of elements, like Sn and Mo, indicates weathering of granitic rocks like pegmatite that had protore of some sulphide minerals [18]. The presence of Pd, Ce and Sb in this factor group attested to the presence of these minerals [16].

Factor 3 was related to the rocks and minerals such as phosphate REE-Nb deposits in an evaporite environment, while factor 4 was interpreted as the product of weathered manganese ore. Factor 5 was related to the weathered precious metals that were originally formed from hydrothermal sulphide ore [18]. Factor 6 had the second loading value of Mn, Sb and Co. This factor though not having the highest loading factor of any of the elements showed the weathering of Co-bearing manganese ore [19] and incompleteness or leakages in the system. These leakages typified by factor 6 were pronounced by the mobility of some of the elements. Factors 1–5

Sample no.	Cd (ppm)	Sb (ppm)	Bi (ppm)	V (ppm)	Ca (%)	P (%)	La (ppm)	Cr (ppm)	Mg (%)	Ba (ppm)	Ti (%)	B (ppm)	Al (%)	Na (%)	K (%)	W (ppm)	Sc (ppm)	Tl (ppm)	S (%)	Hg (ppb)
10	<0.01	0.04	0.06	17	0.02	0.006	9.4	9.0	0.02	18.7	0.009	<20	0.43	<0.001	0.03	<0.05	2.3	0.05	<0.02	11
18	0.02	0.05	0.05	22	0.15	0.009	15.8	15.1	0.03	42.4	0.019	<20	0.41	<0.001	0.03	<0.05	2.4	0.04	<0.02	30
33	<0.01	0.04	0.05	14	0.02	0.004	8.9	11.3	0.02	14.5	0.021	<20	0.23	<0.001	0.01	<0.05	1.6	0.03	<0.02	6
44	<0.01	<0.02	0.11	20	0.03	0.006	12.4	12.5	0.07	40.3	0.018	<20	0.45	0.002	0.08	<0.05	2.1	0.11	<0.02	9
60	<0.01	0.06	0.07	45	0.08	0.011	13.4	20.8	0.03	50.8	0.008	<20	0.49	<0.001	0.03	<0.05	2.3	0.07	<0.02	14
64	<0.01	<0.02	0.02	10	0.03	0.004	7.1	6.0	0.02	42.4	0.004	<20	0.14	<0.001	0.02	<0.05	1.0	0.03	<0.02	8
73	<0.01	0.02	0.04	13	0.03	0.005	7.3	8.1	0.02	33.4	0.008	<20	0.25	<0.001	0.02	<0.05	1.3	0.04	<0.02	7
76	0.01	0.03	0.05	33	0.01	0.010	10.9	21.3	0.02	15.1	0.014	<20	0.69	<0.001	0.01	<0.05	5.2	0.04	<0.02	14
78	<0.01	0.03	0.05	21	0.06	0.008	16.1	14.2	0.02	19.1	0.012	<20	0.47	<0.001	0.02	<0.05	2.7	0.03	<0.02	19
83a	0.02	<0.02	0.15	19	0.74	0.020	24.4	10.7	0.24	135.9	0.068	<20	1.55	0.003	0.42	0.05	2.0	0.33	0.02	17
86	0.01	<0.02	0.22	59	0.13	0.011	24.2	44.8	0.50	144.3	0.096	<20	1.98	0.003	0.51	<0.05	6.9	0.29	<0.02	11
112	<0.01	0.05	0.05	26	0.05	0.008	14.8	17.5	0.02	26.5	0.023	<20	0.38	<0.001	0.02	<0.05	3.2	0.04	<0.02	23
122	<0.01	0.07	0.11	73	<0.01	0.010	21.0	24.7	0.01	9.1	0.024	<20	1.07	<0.001	0.02	<0.05	7.8	0.07	<0.02	23
128	<0.01	0.04	0.05	45	0.07	0.011	14.1	21.2	0.02	45.0	0.015	<20	0.55	<0.001	0.03	<0.05	3.4	0.03	<0.02	15
135	<0.01	0.05	0.09	39	0.02	0.008	15.4	19.1	0.03	14.6	0.022	<20	0.74	<0.001	0.03	<0.05	5.4	0.06	<0.02	15
151	0.01	0.04	0.06	25	0.14	0.009	17.4	11.2	0.04	86.3	0.009	<20	0.42	<0.001	0.05	<0.05	2.4	0.06	<0.02	28
187	0.02	0.04	0.03	18	0.05	0.006	12.5	14.3	0.02	30.8	0.019	<20	0.25	<0.001	0.02	<0.05	2.1	0.03	<0.02	34
190	<0.01	0.04	0.05	23	0.03	0.007	12.2	17.0	0.02	18.7	0.027	<20	0.44	<0.001	0.02	<0.05	3.0	0.04	<0.02	28
194	<0.01	0.05	0.08	43	<0.01	0.011	11.0	19.6	0.01	7.4	0.013	<20	0.87	<0.001	0.02	<0.05	5.3	0.05	<0.02	16
199	<0.01	0.05	0.06	24	0.11	0.011	17.8	18.8	0.03	17.4	0.024	<20	0.46	<0.001	0.04	<0.05	3.0	0.05	<0.02	29
214	<0.01	<0.02	0.06	17	0.26	0.007	20.0	9.5	0.11	154.3	0.009	<20	0.53	0.001	0.08	<0.05	3.1	0.10	<0.02	22
217a	<0.01	<0.02	0.04	14	0.09	0.006	14.4	10.1	0.04	40.0	0.005	<20	0.38	<0.001	0.03	<0.05	1.7	0.05	<0.02	21

Sample no.	Cd (ppm)	Sb (ppm)	Bi (ppm)	V (ppm)	Ca (%)	P (%)	La (ppm)	Cr (ppm)	Mg (%)	Ba (ppm)	Ti (%)	B (ppm)	Al (%)	Na (%)	K (%)	W (ppm)	Sc (ppm)	Tl (ppm)	S (%)	Hg (ppb)
219	0.02	0.05	0.07	40	0.15	0.010	21.6	13.9	0.06	259.8	0.012	<20	0.46	0.003	0.03	<0.05	2.9	0.16	<0.02	13
222	<0.01	<0.02	0.02	6	0.04	0.003	8.5	5.8	0.02	28.8	0.005	<20	0.17	0.002	0.02	<0.05	1.5	0.03	<0.02	8
225	<0.01	<0.02	0.08	24	0.11	0.007	17.3	15.2	0.08	59.2	0.004	<20	0.76	<0.001	0.07	<0.05	3.7	0.13	<0.02	13
244	<0.01	0.05	0.07	36	0.04	0.011	14.9	14.9	0.03	27.4	0.025	<20	0.56	0.001	0.03	<0.05	3.5	0.06	<0.02	20
258	0.02	0.03	0.06	18	0.11	0.010	24.5	10.8	0.07	190.6	0.008	<20	0.43	0.001	0.06	<0.05	2.6	0.06	<0.02	27
261	0.02	0.04	0.07	29	0.09	0.009	22.9	11.6	0.04	68.9	0.010	<20	0.47	<0.001	0.04	<0.05	2.9	0.06	<0.02	25
266	0.02	0.05	0.07	30	0.08	0.013	30.6	15.6	0.04	43.8	0.021	<20	0.48	<0.001	0.04	<0.05	3.7	0.07	<0.02	29
286	0.01	0.04	0.09	42	0.06	0.009	18.3	12.6	0.03	43.9	0.011	<20	0.82	<0.001	0.04	<0.05	4.7	0.08	<0.02	21

Sample no.	Se (ppm)	Te (ppm)	Ga (ppm)	Cs (ppm)	Ge (ppm)	Hf (ppm)	Nb (ppm)	Rb (ppm)	Sn (ppm)	Ta (ppm)	Zr (ppm)	Y (ppm)	Ce (ppm)	In (ppm)	In (ppm)	Re (ppb)	Be (ppm)	Li (ppm)	Pd (ppb)	Pt (ppb)
10	<0.1	<0.02	2.6	0.48	<0.1	<0.02	0.15	8.7	0.5	<0.05	1.4	8.11	33.8	0.03	0.03	<1	0.3	1.2	<10	<2
18	<0.1	<0.02	3.1	0.36	<0.1	0.09	0.65	7.4	0.7	<0.05	3.2	9.57	33.1	0.02	0.02	<1	0.2	1.2	<10	<2
33	<0.1	<0.02	2.1	0.28	<0.1	0.07	0.37	4.7	0.6	<0.05	2.8	5.35	21.8	<0.02	<0.02	<1	0.1	0.7	<10	<2
44	<0.1	<0.02	2.7	1.16	<0.1	<0.02	0.18	17.9	0.5	<0.05	0.9	5.62	25.8	0.02	0.02	<1	0.4	3.4	<10	<2
60	<0.1	<0.02	3.1	0.39	<0.1	0.04	0.26	8.9	0.5	<0.05	2.3	8.98	33.2	<0.02	<0.02	<1	0.5	1.1	<10	<2
64	<0.1	<0.02	0.8	0.22	<0.1	<0.02	0.10	3.9	0.2	<0.05	0.6	3.85	12.7	<0.02	<0.02	<1	0.1	0.6	<10	<2
73	<0.1	<0.02	1.7	0.26	<0.1	<0.02	0.17	5.1	0.3	<0.05	1.1	4.63	21.8	<0.02	<0.02	<1	0.3	0.6	<10	<2
76	<0.1	<0.02	5.5	0.44	<0.1	0.17	0.35	2.8	0.9	<0.05	8.7	9.11	35.2	0.04	0.04	<1	0.2	1.0	<10	<2
78	<0.1	<0.02	3.0	0.42	<0.1	0.03	0.33	5.4	0.6	<0.05	1.9	10.48	55.0	<0.02	<0.02	<1	0.2	1.1	<10	<2
83a	0.2	<0.02	7.2	3.66	<0.1	0.11	2.63	57.1	1.1	<0.05	3.8	4.52	49.0	<0.02	<0.02	<1	1.4	28.8	<10	<2
86	<0.1	0.05	7.5	4.66	<0.1	0.03	0.49	55.1	1.0	<0.05	1.7	10.84	60.8	0.03	0.03	<1	1.1	32.2	<10	<2
112	<0.1	<0.02	3.5	0.41	<0.1	0.04	0.50	7.1	0.8	<0.05	3.3	9.72	41.0	0.03	0.03	<1	0.3	0.9	<10	<2

Sample no.	Se (ppm)	Te (ppm)	Ga (ppm)	Cs (ppm)	Ge (ppm)	Hf (ppm)	Nb (ppm)	Rb (ppm)	Sn (ppm)	Ta (ppm)	Zr (ppm)	Y (ppm)	Ce (ppm)	In (ppm)	In (ppm)	Re (ppb)	Be (ppm)	Li (ppm)	Pd (ppb)	Pt (ppb)
122	<0.1	<0.02	9.5	0.81	<0.1	0.36	0.35	5.5	1.6	<0.05	19.6	10.67	79.6	0.05	0.05	<1	0.2	1.4	14	<2
128	<0.1	<0.02	4.5	0.41	<0.1	0.05	0.45	6.0	0.9	<0.05	2.4	11.38	38.4	<0.02	<0.02	<1	0.3	1.1	<10	<2
135	<0.1	<0.02	7.0	0.75	<0.1	0.22	0.25	10.8	1.3	<0.05	11.2	9.41	85.1	0.04	0.04	<1	0.3	1.7	<10	<2
151	<0.1	<0.02	2.3	0.39	<0.1	0.06	0.26	9.9	0.4	<0.05	2.4	13.53	47.9	0.02	0.02	<1	0.4	1.2	<10	<2
187	<0.1	<0.02	2.0	0.29	<0.1	0.05	0.40	5.4	0.7	<0.05	2.9	9.95	26.0	<0.02	<0.02	<1	0.2	0.7	<10	<2
190	<0.1	<0.02	3.9	0.42	<0.1	0.04	0.39	6.5	0.8	<0.05	3.9	8.94	40.4	0.03	0.03	<1	0.2	1.2	<10	<2
194	<0.1	<0.02	6.8	0.55	<0.1	0.19	0.43	5.0	1.4	<0.05	10.5	5.06	48.3	0.04	0.04	<1	0.2	1.5	<10	<2
199	<0.1	<0.02	4.1	0.41	<0.1	0.08	0.54	7.9	1.0	<0.05	3.4	10.10	45.8	<0.02	<0.02	<1	0.3	1.1	<10	<2
214	<0.1	<0.02	2.3	0.46	<0.1	0.04	0.25	14.1	0.7	<0.05	2.1	15.96	43.3	<0.02	<0.02	<1	1.0	3.1	<10	<2
217a	<0.1	0.02	1.6	0.32	<0.1	<0.02	0.13	8.0	0.3	<0.05	1.1	8.44	31.5	<0.02	<0.02	<1	0.3	1.1	<10	<2
219	<0.1	<0.02	3.1	0.50	<0.1	0.07	0.40	8.2	0.6	<0.05	3.6	17.46	78.6	0.02	0.02	<1	0.6	1.2	<10	<2
222	<0.1	<0.02	0.9	0.26	<0.1	<0.02	0.09	4.0	0.2	<0.05	0.4	4.74	17.6	<0.02	<0.02	<1	0.4	0.6	<10	<2
225	<0.1	<0.02	4.1	0.82	<0.1	0.04	0.18	14.3	0.6	<0.05	2.0	9.58	39.4	<0.02	<0.02	<1	0.7	3.1	<10	<2
244	<0.1	0.03	4.5	0.56	<0.1	0.07	0.48	9.8	1.1	<0.05	4.4	10.24	54.8	<0.02	<0.02	<1	0.4	1.5	<10	<2
258	<0.1	<0.02	1.9	0.44	<0.1	0.03	0.26	11.2	0.4	<0.05	1.1	24.93	48.4	<0.02	<0.02	<1	0.7	1.3	<10	<2
261	<0.1	<0.02	3.5	0.57	<0.1	0.06	0.25	8.9	0.8	<0.05	3.3	14.57	83.8	<0.02	<0.02	<1	0.4	1.4	<10	<2
266	<0.1	<0.02	3.7	0.66	<0.1	0.04	0.53	12.4	1.2	<0.05	3.1	22.52	58.9	<0.02	<0.02	<1	0.4	1.8	<10	<2
286	<0.1	<0.02	6.0	0.82	<0.1	0.15	0.35	10.8	1.0	<0.05	7.6	15.61	45.8	0.04	0.04	<1	0.4	2.1	<10	<2

Table 1.
Geochemical result of the analysis soil samples.

	Mo	Pb	Zn	Ni	Co	Mn	Fe	As	Th	Sr	Cd	Bi	V	Ca	P	Cr	Mg	Al	K	Sc	Cs	Nb	Rb	Sn	Zr	Ce	In	Be	Li	Pd
Mo	1																													
Pb	0.610	1																												
Zn	0.499	0.332	1																											
Ni	0.674	0.412	0.767	1																										
Co	0.523	0.602	0.525	0.752	1																									
Mn	0.374	0.593	0.439	0.615	0.924	1																								
Fe	0.930	0.667	0.482	0.521	0.388	0.267	1																							
As	0.827	0.506	0.693	0.605	0.352	0.240	0.854	1																						
Th	0.634	0.564	0.091	0.090	0.043	−0.050	0.755	0.537	1																					
Sr	0.031	0.284	0.758	0.310	0.256	0.317	0.100	0.341	−0.106	1																				
Cd	−0.094	0.427	0.202	0.052	0.311	0.407	−0.013	−0.034	−0.095	0.469	1																			
Bi	0.764	0.497	0.834	0.874	0.627	0.435	0.693	0.769	0.379	0.400	0.030	1																		
V	0.911	0.666	0.244	0.455	0.378	0.248	0.926	0.718	0.739	−0.127	−0.113	0.581	1																	
Ca	0.037	0.222	0.768	0.277	0.198	0.238	0.115	0.378	−0.074	0.987	0.425	0.403	−0.132	1																
P	0.502	0.576	0.676	0.373	0.354	0.321	0.603	0.664	0.382	0.653	0.398	0.558	0.454	0.659	1															
Cr	0.866	0.395	0.430	0.751	0.510	0.365	0.753	0.669	0.426	−0.093	−0.161	0.684	0.777	−0.098	0.369	1														
Mg	0.516	0.287	0.856	0.950	0.671	0.537	0.381	0.557	−0.066	0.502	0.086	0.840	0.258	0.480	0.380	0.598	1													
Al	0.774	0.478	0.849	0.814	0.482	0.290	0.763	0.828	0.449	0.452	0.001	0.930	0.616	0.470	0.643	0.711	0.808	1												
K	0.493	0.253	0.951	0.866	0.575	0.437	0.399	0.633	−0.033	0.653	0.146	0.849	0.211	0.655	0.526	0.499	0.957	0.845	1											
Sc	0.808	0.534	0.270	0.487	0.272	0.120	0.856	0.605	0.799	−0.164	−0.186	0.592	0.864	−0.164	0.348	0.766	0.309	0.680	0.242	1										
Cs	0.597	0.301	0.937	0.880	0.593	0.414	0.511	0.695	0.097	0.565	0.109	0.913	0.329	0.578	0.540	0.576	0.938	0.901	0.982	0.366	1									
Nb	0.213	0.215	0.755	0.243	0.173	0.163	0.312	0.532	0.121	0.839	0.412	0.438	0.024	0.881	0.776	0.057	0.392	0.532	0.616	−0.014	0.596	1								
Rb	0.449	0.271	0.964	0.805	0.565	0.427	0.377	0.613	−0.002	0.718	0.194	0.855	0.180	0.726	0.584	0.417	0.911	0.829	0.982	0.202	0.968	0.684	1							
Sn	0.676	0.505	0.324	0.264	0.189	0.090	0.810	0.643	0.860	0.055	0.046	0.496	0.719	0.090	0.602	0.554	0.134	0.587	0.194	0.798	0.310	0.362	0.221	1						
Zr	0.577	0.413	0.012	−0.033	−0.137	−0.201	0.744	0.472	0.913	−0.191	−0.124	0.239	0.695	−0.151	0.261	0.351	−0.172	0.358	−0.128	0.776	0.003	0.069	−0.123	0.786	1					
Ce	0.582	0.821	0.264	0.328	0.462	0.412	0.640	0.376	0.654	0.136	0.277	0.461	0.633	0.109	0.484	0.391	0.187	0.447	0.167	0.627	0.245	0.134	0.198	0.637	0.537	1				
In	0.571	0.242	0.021	0.103	−0.091	−0.244	0.644	0.355	0.793	−0.314	−0.326	0.296	0.630	−0.284	0.097	0.433	−0.036	0.415	−0.034	0.802	0.101	−0.067	−0.059	0.647	0.864	0.368	1			
Be	0.218	0.377	0.819	0.623	0.475	0.419	0.219	0.418	−0.068	0.831	0.263	0.657	0.076	0.794	0.548	0.189	0.770	0.652	0.810	0.092	0.746	0.576	0.849	0.072	−0.211	0.224	−0.207	1		

	Mo	Pb	Zn	Ni	Co	Mn	Fe	As	Th	Sr	Cd	Bi	V	Ca	P	Cr	Mg	Al	K	Sc	Cs	Nb	Rb	Sn	Zr	Ce	In	Be	Li	Pd
Li	0.512	0.247	0.956	0.838	0.550	0.402	0.433	0.660	0.005	0.651	0.141	0.844	0.227	0.665	0.548	0.502	0.934	0.861	0.994	0.259	0.985	0.661	0.977	0.239	−0.076	0.170	0.013	0.781	1	
Pd	0.515	0.330	0.013	0.002	−0.120	−0.144	0.618	0.483	0.588	−0.[illegible]47	−0.102	0.195	0.565	−0.118	0.076	0.245	−0.094	0.240	−0.072	0.542	0.014	−0.024	−0.089	0.448	0.738	0.348	0.548	−0.137	−0.049	1

Table 2.
Result of correlation analysis of selected elements.

	Component					
	1	**2**	**3**	**4**	**5**	**6**
Mg	0.959					
Ni	0.956					
Cs	0.921		0.343			
K	0.909		0.393			
Te	0.900					
Li	0.893		0.430			
Ti	0.881					
Rb	0.854		0.488			
Bi	0.850	0.376				
Al	0.812	0.481				
Zn	0.795		0.563			
Cu	0.770	0.354	0.319	0.315		
Tl	0.769		0.511			
U	0.734	0.351		0.379		
Cr	0.711	0.476				
Co	0.662			0.576		0.349
Na	0.632		0.358	0.358	−0.335	
Be	0.628		0.520	0.380		
Zr		0.969				
Hf		0.948				
Th		0.943				
Ga	0.420	0.864				
In		0.845				
Sc	0.388	0.841				
Sn		0.827				
Fe	0.428	0.826				
V	0.347	0.818				
Pd		0.735				
Mo	0.602	0.706				
Sb	−0.320	0.618			0.305	0.564
As	0.584	0.608	0.313			
Ce		0.603		0.542		
Se			0.942			
Nb	0.307		0.914			
Ca	0.311		0.903			
Sr	0.318		0.862			
P	0.350	0.367	0.629		0.361	
Ba	0.300			0.843		
Pb		0.518		0.754		

	Component					
	1	2	3	4	5	6
Y				0.737	0.569	
Mn	0.483			0.663		0.443
La	0.342			0.620	0.511	
Cd			0.455	0.521	0.381	
Hg					0.838	
Ag					−0.575	
Au					−0.500	

Table 3.
Rotated component matrix.

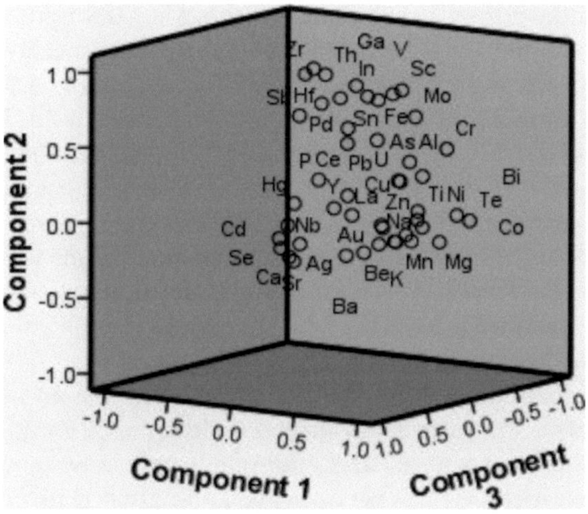

Figure 2.
3D component plot for soil factor analysis in rotated space.

were strongly related to mineralization. The presence of the country rock in the factor groups as accounted for by the principal component analysis decreases from factor 1 to factor 5. Factor 6 does not have the highest loading of any element but had the second highest loading of Mn, Sb and Co. This factor group was interpreted to imply the presence of minor or protore of Co-bearing manganite and also accounts for very strong leakages in the system during weathering of the rocks and ores in the area.

The depressions in the isograde plots were interpreted to be the zones of negative anomaly, while the positive anomalous zones were the peaks in the plots. The brown-coloured area defines the regional threshold, while the pinkish sky blue-coloured zones on the plots were interpreted as the local aureole. Deep blue zones were interpreted as the anomalous regions. The values of these zones were estimated from the corresponding values in the colour scale that is beside each plot.

From the plots, Ag had three positive anomalies that are defined by long. 4.13°E and lat. 11.125°N, long. 4.0°E and lat. 11.075°N and long. 3.35°E and 11.1°N, while its negative anomalies were defined by long. 4.09°E and lat. 11.07°N, long 4.05°E and lat. 11.07°N and long. 4.075°E and long. 11.09°N. Its regional threshold was ≤55 ppm, while the local aureole and anomaly had 70 ppm and >90 ppm, respectively. The positive anomalies of Au were defined by long. 11.225°E and lat. 4.0°N, while its negative anomaly was defined by long. 4.05°E and lat. 11.1°N. Its regional aureole had ≤4ppb, while its local aureole is 5.5 ppb. Its anomalous area had a concentration value of ≥6.5 ppb. Ba had its positive anomaly on the spot defined by long. 3.975°E and lat. 11.13°N. Its regional threshold had a concentration value of ≤150 ppm, while its local threshold and anomalous values were 230 ppm and ≥240 ppm, respectively.

Al had its positive anomaly defined by long. 4.05°E and lat. 11.09°N, while the negative anomaly was defined by long. 3.92°E and lat. 11.09°N. Its regional aureole, local aureole and anomalous value were ≤1.2 ppm, 1.5 ppm and >1.6 ppm, respectively. As (**Figure 3**) had its positive anomalous zone defined by long. 4.05°E and lat. 11.1°N and long. 4.15°E and lat. 11.12°N. Its negative anomalies were defined by long. 4.075°E and lat. 11.075°N and long. 4.0°E and lat. 11.1°N. As had regional aureole, local aureole and anomaly with concentration values of ≤0.8 ppm, 1.0 ppm and ≥1.2 ppm, respectively.

The positive anomalous zone of Be was defined by long. 4.06°E and lat. 11.12°N (**Figure 4**). Its negative anomaly was defined by long. 4.1°E and lat. 11.11°N. The concentration of Ba described as regional threshold is ≤0.8 ppm and its local and anomalous concentrations were 1.2 ppm and ≥1.3 ppm, respectively. The positive anomaly of Bi is on long. 4.04°E and lat. 11.075°N, while long. 3.93°E and lat. 11.1°N define the negative anomaly (**Figure 5**). The regional threshold, local threshold and anomaly values were ≤0.13 ppm, 0.18 ppm and ≥0.21 ppm, respectively. Ca has its positive anomaly defined by long. 4.04°E and lat. 11.075°N, while its negative anomaly is defined by long. 4.05°E and lat. 11.06°N. The regional threshold value of Ca is ≤0.45 ppm while its local aureole and anomalous values were 0.6 ppm and ≥0.65 ppm. Cd positive anomaly was on the spot defined by long. 4.075°E and lat. 11.1°N. Its regional threshold, local threshold and anomalous values were ≤0.013 ppm, 0.017 and ≥0.019 ppm, respectively.

The positive anomaly of Co was defined by long. 4.04°E and lat. 11.05°N (**Figure 6**). Its negative anomaly was defined by long. 3.97°E and lat. 11.16°N, while its regional threshold, local aureole and anomalous values were ≤10 ppm, 14 ppm and ≥15 ppm, respectively. Cs has no negative anomaly but its positive anomalous point is defined by long. 4.04°E and lat. 11.06°N (**Figure 7**). Its regional threshold

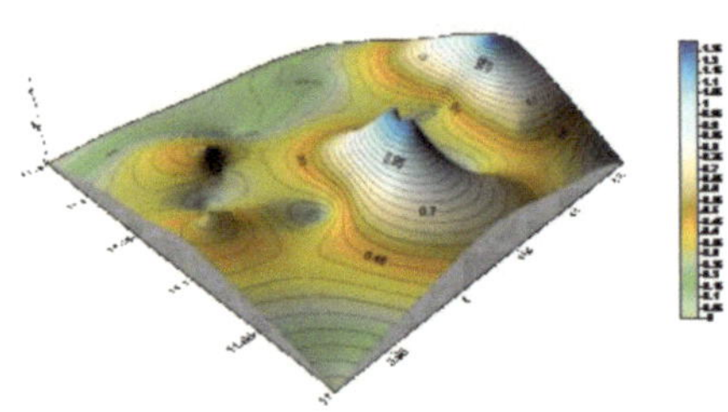

Figure 3.
3D isograde plot of As.

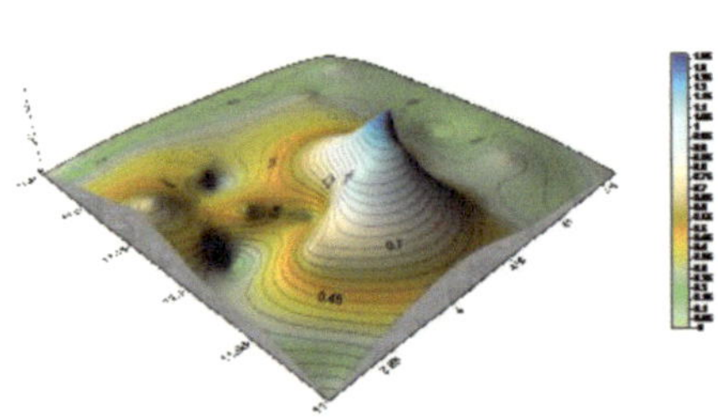

Figure 4.
3D isograde plot of Be.

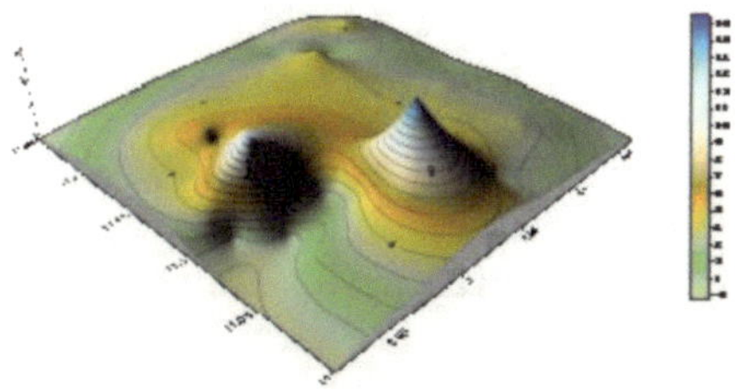

Figure 5.
3D isograde plot of Bi.

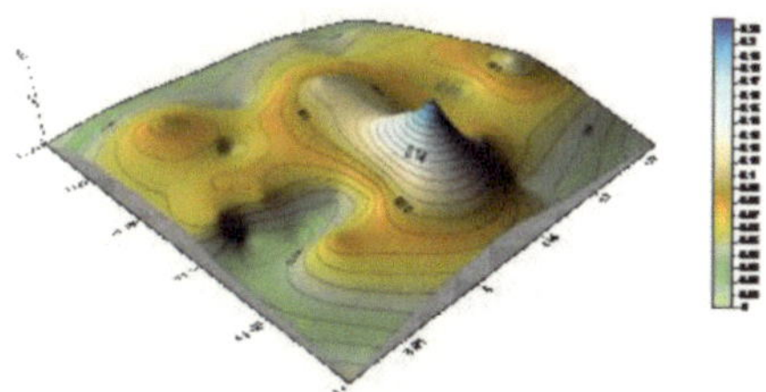

Figure 6.
3D isograde plot of Co.

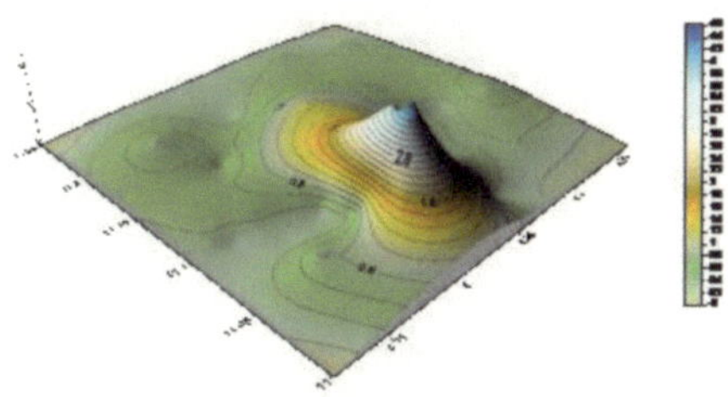

Figure 7.
3D isograde plot of Cs.

value was ≤2.8 ppm, while its local threshold and anomalous values were 4 ppm and ≤4.2 ppm, respectively. The positive anomaly of K was on the spot defined by long. 4.04°E and lat. 11.07°N. Its regional threshold value was ≤0.3 ppm, while its local aureole and anomaly had values 0.44 ppm and ≥0.48 ppm, respectively. Li had its positive anomaly defined by long. 4.04°E and lat. 11.07°N. Its regional threshold, local aureole and anomalous values were ≤8 ppm, 26 ppm and ≥30 ppm, respectively. Mg also had its positive anomaly defined by lat. 11.07°N but long. 4.035°E, while its negative anomaly was defined by long. 4.06°E and lat.11.025°N. Its regional threshold value, local aureole and anomaly had a concentration value of ≤0.012 ppm, 0.42 ppm and ≥0.48 ppm respectively. Mn had two positive anomalies that were defined by long. 3.36°E and lat. 11.12°N and long. 4.04°E and lat. 11.06°N (**Figure 8**). Its negative anomaly was defined by long. 4.07°E and lat. 11.05°E. Its regional threshold value, local aureole and anomalous values were ≤550 ppm, 800 ppm and ≥850 ppm, respectively.

The positive anomaly of Nb was on the spot defined by long. 4.04°E and lat.11.075°N (**Figure 9**). Its negative anomaly was defined by long. 4.05°E and lat. 11.075°N. Its regional threshold value was ≤0.6 ppm, while the local aureole was 2.2 ppm. Its anomaly had a value of ≥2.5 ppm. Ni had its positive anomaly on long.

4.03°E and lat. 11.075°N (**Figure 10**). Its regional threshold value and local and anomalous values were ≤6 ppm, 22 ppm and ≥25 ppm, respectively. P had its positive and negative anomalies defined by long. 4.04°E and lat. 11.1°N and long. 4.1°E and lat. 11.125°N, respectively. Its regional threshold, local aureole and anomalous values were ≤0.005 ppm, 0.017 ppm and ≥0.019 ppm, respectively. Rb has long. 4.05°E and lat. 11.1°N defining its positive anomaly. Its regional threshold, local aureole and anomalous values are ≤13 ppm, 47 ppm and ≥55 ppm, respectively. The positive anomaly of Sr was defined by long. 4.03°E and lat. 11.075°N, while its negative anomaly was defined by long. 4.07°E and lat. 11.05°N. Its regional threshold, local aureole and anomalous values were ≤12 ppm, 44 ppm and ≥48 ppm, respectively. Zn had its positive anomaly on the spot defined by long. 4.05°E and lat. 11.1°N (**Figure 11**). Its regional threshold, local aureole and anomalous values were ≤12 ppm, 42 ppm and ≥48 ppm, respectively. Ce had three positive anomalies defined by long. 3.97°E and lat. 11.15°N, long. 4.075°E and lat. 11.05°N and long. 4.16°E and lat. 11.1°N (**Figure 12**). Its regional threshold and local aureole had a concentration value of ≤55 ppm and 75 ppm, respectively, while the anomalous concentration is ≥80 ppm. Similarly, Cr had its positive anomaly defined by the same longitude and latitude that defined that of Co, but its negative anomaly was

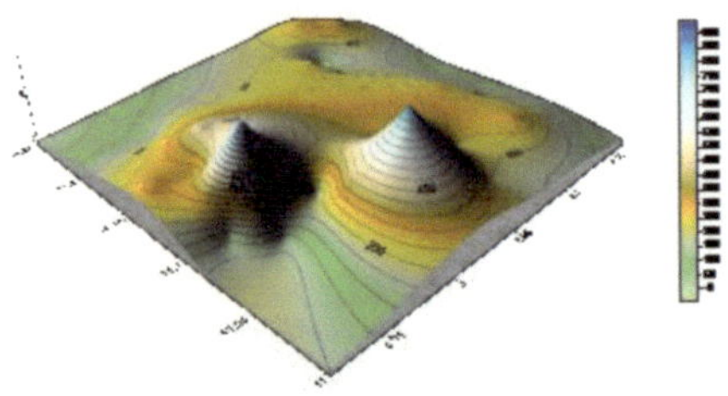

Figure 8.
3D isograde plot of Mn.

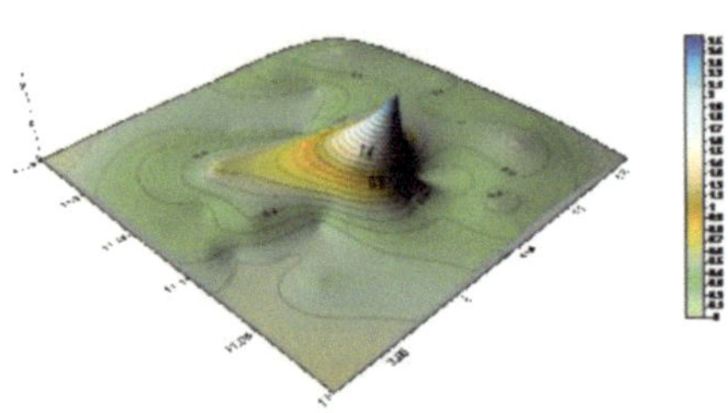

Figure 9.
3D isograde plot of Nb.

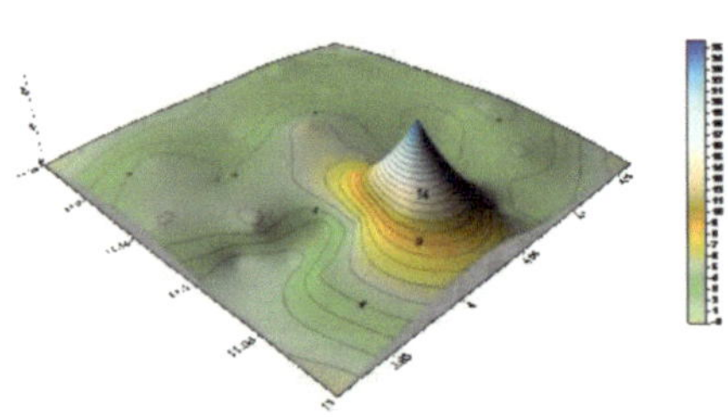

Figure 10.
3D isograde plot of Ni.

defined by long. 4.05°E and lat. 11.075°N (**Figure 13**). Its regional threshold value, local aureole and anomalous value were ≤28 ppm, 36 ppm and ≥40 ppm, respectively.

Fe had its highest concentration value on the spot defined by long. 4.16°E and lat. 11.1°N (**Figure 14**). Its regional threshold value was ≤2.4%, while its local aureole and anomalous values were 3.2% and ≥3.4%, respectively. The positive anomalies of Ga and Hf were defined by the same longitude and latitude that defined the anomalous point of Fe. The regional threshold, local aureole and anomalous value of Ga are ≤6 ppm, 8 ppm and ≥8.5 ppm, respectively, but the regional threshold value, local aureole and anomalous value of Hf were ≤0.22 ppm, 0.3 ppm and ≥0.32 ppm, respectively.

The positive anomaly of Hg was defined by long. 4.16°E and lat. 11.25°N, while its negative anomaly was defined by long. 4.1°E and lat. 11.1°N. Its regional threshold value was ≤22 ppm, while its local aureole and anomalous values were 26 ppm and ≥30 ppm, respectively. Like Fe, Ga and Hf, the positive anomaly of In was defined by long. 4.16°E and lat. 11.11°N (**Figure 15**). Its regional threshold value, local aureole and anomaly had values ≤0.03 ppm, 0.044 and ≥0.046 ppm, respectively. The positive anomaly of Mo was on the spot defined by long. 4.05°E and lat.

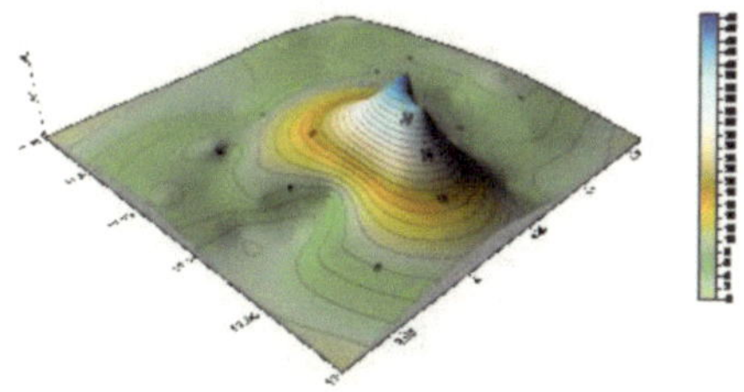

Figure 11.
3D isograde plot of Zn.

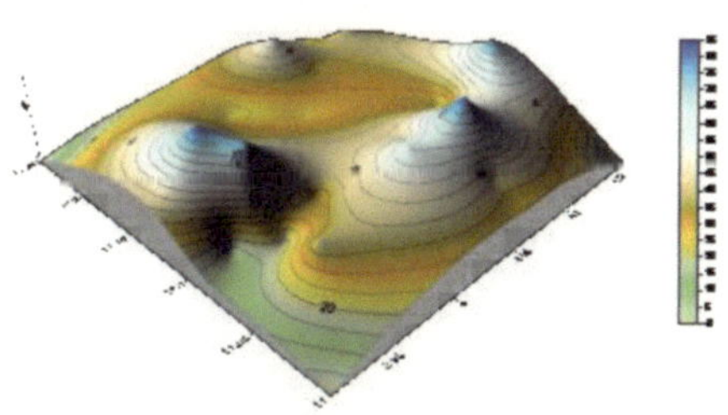

Figure 12.
3D isograde plot of Ce.

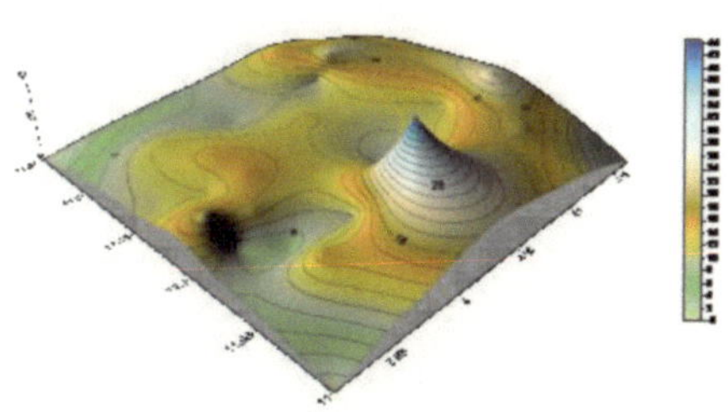

Figure 13.
3D isograde plot of Cr.

11.07°N (**Figure 16**). Its regional threshold had ≤0.25 ppm concentration, while its local aureole and anomalous value were 0.5 ppm and ≥0.9 ppm, respectively.

The positive anomaly of Pb is defined by long. 3.35°E and lat. 11.125°N. Its regional threshold, local threshold and anomalous values were ≤5 ppm, 17 ppm and ≥19 ppm, respectively. Pd anomaly was on the spot defined by long. 4.16°E and lat. 11.1°N (**Figure 17**). Its regional threshold value, local aureole and anomalous

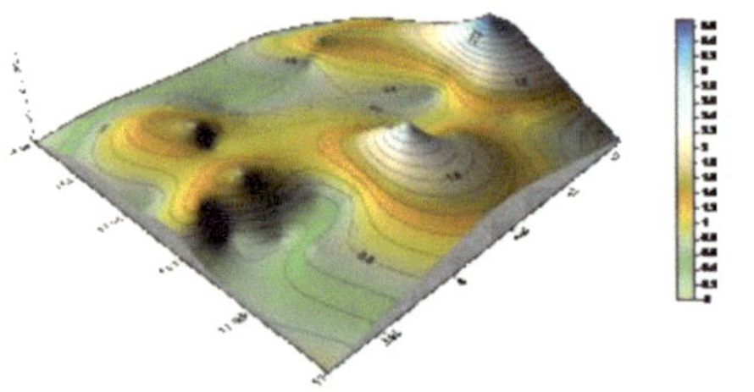

Figure 14.
3D isograde plot of Fe.

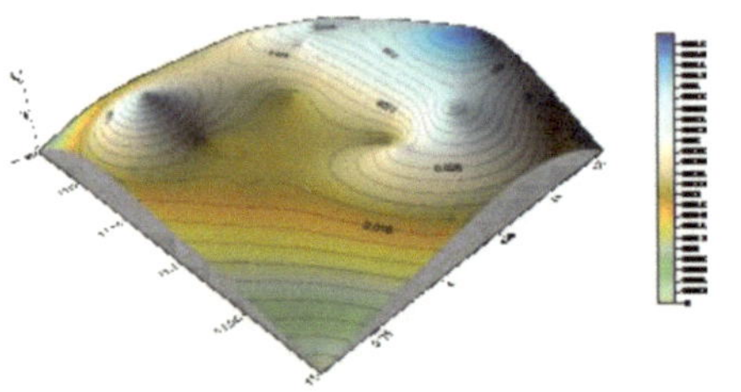

Figure 15.
3D isograde plot of In.

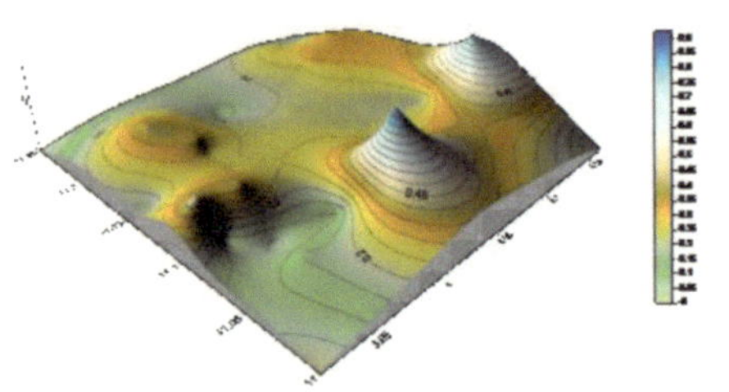

Figure 16.
3D isograde plot of Mo.

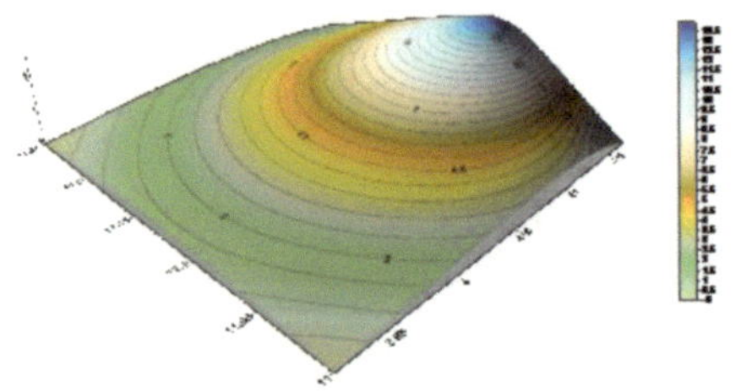

Figure 17.
3D isograde plot of Pd.

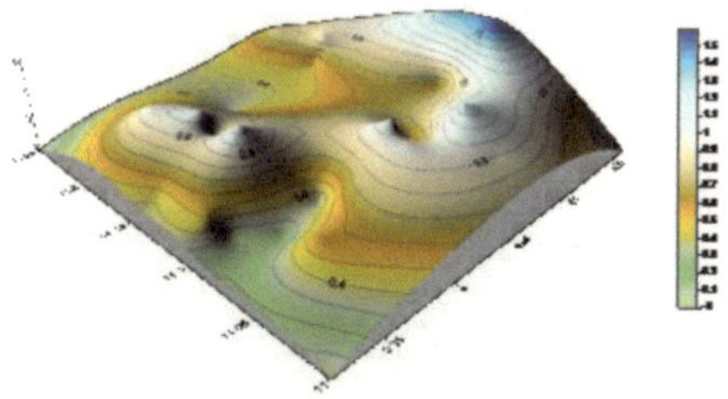

Figure 18.
3D isograde plot of Sn.

concentrations are ≤3 ppm, 12.5 ppm and ≥13.5 ppm. Sb, Sc and Sn (**Figure 18**) all have their positive anomalies defined by long. 4.16°E and lat. 11.1°N. The negative anomaly of Sc was defined by long. 3.95°E and lat. 11.1°N, while its regional threshold value, local aureole and anomalous values were ≤2 ppm, 6.5 ppm and ≥7.5 ppm, respectively. The regional threshold value, local aureole and anomalous values for Sb were ≤0.015 ppm, 0.06 ppm and ≥0.065 ppm. The negative anomaly of Sn was defined by long. 4.05°E and lat. 11.075°N. Its regional threshold value, local threshold and anomalous values were ≤0.4 ppm, 1.3 ppm and ≥1.5 ppm, respectively.

The anomalous value of Th is on the spot defined by long. 4.16°E and lat. 11.09°N (**Figure 19**). Its regional threshold value was ≤3 ppm, while its local aureole and anomalous values are 12 ppm and ≥13.5 ppm, respectively. Long. 4.05°E and lat. 11.05°N define Ti positive anomaly. Its regional threshold, local aureole and anomalous values were ≤0.025 ppm, 0.08 ppm and ≥0.095 ppm, respectively. Tl positive anomaly was defined by long. 4.04°E and lat. 11.1°N. Its regional threshold value, local aureole and anomalous value were ≤0.08 ppm, 0.28 ppm and ≥0.32 ppm, respectively. U had its positive and negative anomalies defined by long. 4.05°E and lat. 11.05°N and long. 4.1°E and lat. 11.1°N. Its regional threshold value, local aureole and anomaly were ≤0.6 ppm, 2.1 ppm and ≥2.4 ppm, respectively.

Long. 4.16°E and lat. 11.1°N defined the positive anomaly of V (**Figure 20**), while its regional threshold, local aureole and anomalous values were ≤18 ppm, 60 ppm and ≥70 ppm, respectively. Y had its positive and negative anomaly defined by long. 3.92°E and lat. 11.17°N and long. 4.03°E and lat. 11.075°N. Its regional threshold, local aureole and anomalous values are ≤6 ppm, 22 ppm and ≥24 ppm, respectively. Long. 4.16°E and lat. 11.11°N defined the Zr positive anomaly (**Figure 21**). Its regional threshold, local aureole and anomalous values were ≤5 ppm, 17 ppm and ≥19 ppm, respectively. La had its positive anomaly defined by long. 3.98°E and lat. 11.15°N. Its regional threshold, local aureole and anomalous values were ≤18 ppm, 26 ppm and ≥28 ppm, respectively. Na had its own positive anomalies defined by long. 3.96°E and lat. 11.12°N and long. 4.05°E and lat. 11.1°N, while its regional threshold value is ≤0.0008 ppm. Its local and anomalous values were 0.0026 ppm and ≥0.0028 ppm. Te had its positive anomaly defined by long. 4.06°E and lat.

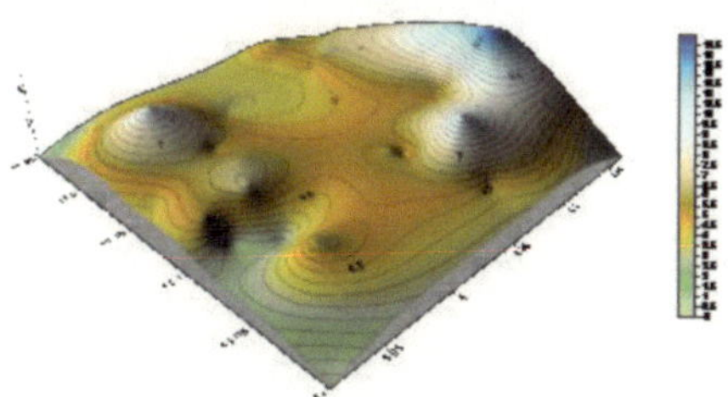

Figure 19.
3D isograde plot of Th.

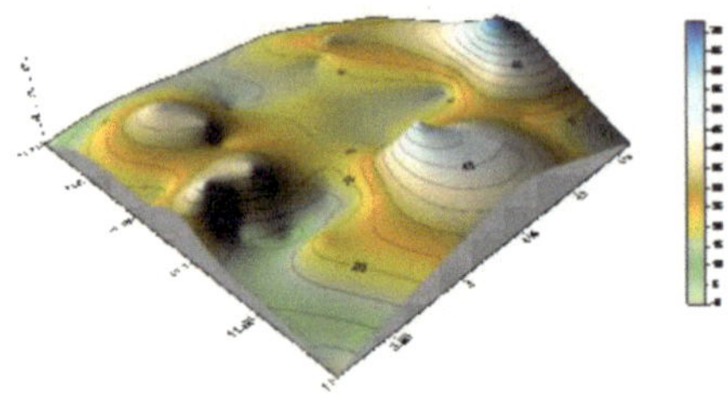

Figure 20.
3D isograde plot of V.

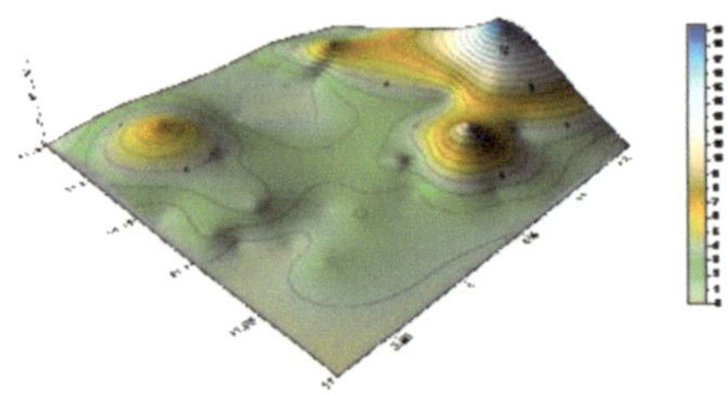

Figure 21.
3D isograde plot of Zr.

11.05°N. Its regional threshold value, local aureole and anomalous value were ≤0.012 ppm, 0.044 ppm and ≥0.048 ppm, respectively.

5. Discussion

Factor 1 of the soil factor analysis is predominated by the lithophile, chalcophile and siderophile elements. The distribution in this group is similar to the corresponding factor 1 in the ore factor analysis. In this factor group, most of the elements have mobilities which vary from low mobility to immobile [20]. Some of them, for example, Cu and Zn have varying mobilities between high to very low depending on the condition as against most of the elements that are immobile in the varying conditions. The association though related to weathered hydrothermal sulphide ore and sulphide-rich BIF is also aided by the mobility of the elements in fluctuating environmental conditions. The fluctuation in the condition is inferred from the inconsequential position occupied by the three types of elements that form the factor group. This scene is also repeated in factor 2 of the soil factor analysis. Most of the elements in this group are mainly transition metals. Their varying oxidation states coupled with the varying component of the materials that are coming together down the slope as the material is transported could account for the association in this group. At the beginning of this phase, the environmental condition favoured the lithophile element then followed by the condition that favoured the chalcophile. Midway into the phase, the condition became inconsistent before the condition that favoured the siderophile prevailed and capped by the chalcophile.

Factor 3 is typically of the lithophile elements with trapped single chalcophile element. The mobility of the chalcophile is similar to the mobility of the lithophile, thereby making it easy to be transported down the slope through the weathering path despite the differences in their chemical affinities and chemical properties. Factor 4 is also formed under relatively unstable environmental conditions that fluctuated between the condition that favoured the siderophile, chalcophile and lithophile elements. The conditions that favoured the lithophile existed over a relatively long period especially towards the end of the phase and thereafter changed to the condition that favoured the chalcophile elements and then capped by the condition favouring siderophile. Factor 5 with Hg, Ag and Au is related to the weathered precious metals that were originally formed from hydrothermal sulphide ore [16]. Factor 6 with no highest loading of any element but has Co, Sb and Mn with fair loading implies the presence of minute Co-bearing manganese ore in the area.

Ag has three positive anomalies. These are in the eastern and southwestern parts of the study area. The local aureoles are aligned in NW-SE and NE-SW towards the southern part of Kaoje and its environs. Closely associated with these aureoles are negative anomalies that trend parallel to the positive anomalies. Al has its anomaly at long. 4.05°E and lat. 11.09°N. There are other positive anomalies in the NW (long. 3.95°E and lat. 11.2°N) and at the eastern end (long. 3.92°E and lat. 11.2°N). The negative anomaly of Al is well-defined at the western part of the mapped area (long. 3.92°E, lat. 11.09°N). These anomalies trend in NW-SE. Another spot of negative anomaly is to the right of the positive anomaly that extends from the southern part of the area towards the central part. As has its positive anomaly in almost the same place with Al. Though there are three peaks described as anomaly having the highest values, two of these peaks are twinned towards the centre of the area, while the third one is at the eastern end of the study area. The negative anomaly that is closely associated with the twinned anomalies towards the centre is also similarly positioned like the negative anomaly in Al. At the western end of this twinned peaks are two of each positive and negative anomalies. While the positives trend in NE-SW, the negative anomalies trend perpendicular to them in NW-SE direction.

Au has relative different distribution patterns from Ag, Al and As. Its negative anomaly is positioned where the three elements (Ag, Al and As) have their positive anomalies. This relationship in the position of the negative anomaly of Au indicates its depletion in the minerals formed or path-found by the three elements. Other peaks of anomaly of Au are in the SE corner and eastern and western parts of the area. These peaks like the negative anomalies are aligned in almost east-west direction with the exception of the peak in the northern part. The three positive anomaly peaks in Ba trend essentially in NW-SE with its highest concentration point defined by lat. 11.13°N and long. 3.975°E. This point of highest concentration of Ba corresponds to local aureole of elements like As, Al, Ag and Au. There is no negative anomaly except sharp decrease to the background value within the area of study. These slopes flank the peaks and also separate the peaks from each other.

Like As, Be has its anomaly in the southern part extending towards the central part of the study area. There are positive anomalies that are similarly positioned in the western half of the area except that while the positive anomalies in this half trend in NE-SW for As, those in Be trend in NW-SE. In like manner, the negative anomalies in Be trend in NE-SW, but those in As trend in NW-SE. Be also has a twin peak of anomaly like As, but one of it is not well-defined. Their distribution pattern is not identical but has similar position of their anomalies. Bi has a distribution pattern that is closely similar to that of As in terms of their positive and negative anomalies. Its anomalous point is defined by lat. 11.075°N and long. 4.04°E. These peaks slope towards the north. There are few negative anomalies that are evenly distributed in the four quadrants of the area studied. Ca has a point of positive anomaly towards the centre of the area. This area is defined by long. 4.05°E and lat. 11.1°N. The negative anomaly is poorly defined in the NE and SE parts of Kaoje and its environs.

Cd has ridge of anomaly that trend in NW-SE. The other anomaly is almost perpendicular to the broad NW-SE trending anomaly. Ce has its aureole in the western and SE part essentially. The extreme NE and SW part are relatively depleted in Ce. The Ce anomaly defines an arc curving back at the southern part. Cr has its anomaly at lat. 4.05°E and long. 11.075°N. Its negative anomalies trend in almost E-W direction. It shares some point of anomaly with elements like As and Bi. With the exception of slightly elevated concentration at the western part of the area (lat. 3.965° E and long. 11.120°N) essentially, the distribution of Co is almost identical with that of Cr. They (Cr and Co) have their anomaly in the same spot, while the distribution pattern of Cs and K is identical and similar with that of Ca. Their anomaly point is defined by lat. 4.05°E and long. 11.06°E. This anomaly trends in N-S direction. The anomaly of Ca does not cover larger area; the anomaly of Cs, K, Li and Mg covers relatively larger area compared with that of Ca, Bi and Cr. Their distribution pattern is also identical sharing the same point of anomalous concentration. There are passive zones of depletion which fall generally under regional threshold. Fe has its highest concentration in the eastern part of the study area and an anomaly that is described as local aureole in the southern part of the study area. These two peaks of anomalies are also observed in the isograde plot of As. With little variations, its distribution pattern is closely related to the distribution pattern of Cr and As.

Ga and Hf have different distribution patterns, but their zones of anomaly are similarly positioned in the eastern part of the area. This part is similarly shared by Fe, As and Cr essentially. While the negative anomaly in Ga trends in NW-SE direction essentially, the negative anomaly in Hf is randomly oriented. Hg has its peaks trending in NW-SE and at the extreme NE. This aureole is closely associated with negative anomaly that trends also in NW-SE. The anomaly at the extreme NE end is similarly positioned like the one in Cd, while the aligned NW-SE aureoles are similar to those in Cd. In is enriched in the eastern half of the study area. There is an aureole in the NW part, while the remaining area is relatively depleted in In. this depleted zone coincide with the boundary between the sedimentary rocks and the basement rock exposures. It zones of highest concentration are in the extreme eastern part where Fe, Cd, etc. also have theirs. No negative anomaly is seen as the values of depression are relatively higher than the values in the extreme NW and SW part of the area.

The distribution pattern of La is similar to that of Hg. Its point of highest aureole is defined by lat. 11.15°N and long. 3.98°E. The aureoles in La form a poorly defined curve shape that trends mainly in NW-SE direction. The extreme SE part is also slightly enriched in La but not as high as the central to NW part of the study area; the depleted zones flank the aureoles. Mn plot is identical with that of Co; it shares the same spot of anomalous concentration with Co and one of Cr anomalous concentration area. The aureole trends in NW-SE and decreases from the SW half towards the extreme NE part. Though not identical, other elements like Cs, K, Li, Mg, Ca, Bi, Cr, etc. have their anomalous point in either point of the anomalous concentration of Mn and Co. Like Mn, Mo has its highest point of concentration in the southern part of the area. There is also a relatively high concentration in the eastern part where Fe has one of its aureoles. Other slightly enriched zone of Mo is the western part. The positive aureole of Mo trends almost perpendicular to the positive aureole of Mn. With the exception of two well-defined peaks of aureoles, other patterns of distribution of these elements are alike.

Na has two peaks of positive anomaly. These peaks are located in the western and central parts of the area study. The western aureole is related to the western aureole of Mn though the Mn aureole is not as broad as that of Na but is more steep sided than that of Na. The area study is slightly more enriched in Na except the extreme four corners of the area. Nb has similar distribution pattern with Ca. The study area is slightly more enriched in Nb relative to Ca, although they both have their positive aureole defined by the same latitude and longitude. The SW part of the area is more depleted in Nb, while the central part is enriched and trends perfectly in NW-SE direction. At the SE end of the positive anomaly, there is a zone of negative anomaly. This zone of Nb negative anomaly coincides with the zone of the most pronounced Au negative anomaly. Other elements with their positive anomaly or one of their positive anomalies defined by lat. 11.075°N and long. 4.05°E or closely related to that of Nb include Ca, Mg, K, Cs, Li, Be and As. Ni has its distribution pattern identical with that of Mg but the western part of the area is slightly depleted in Ni. The southern part of the study area is more enriched in Ni relative to the rest of the area. The aureole has >23 ppm concentration, while its local aureole has a concentration value of 21 ppm. Besides the fact that Li, Cs and K have primitive peaks closely associated with their aureoles, their distribution patterns and the distribution pattern of Ni are alike when their zones of anomaly are compared together. Ni also has its positive anomaly similarly positioned like the aureole of elements like Co, Bi and Al even though their distribution pattern is different. The positive aureole of Ni coincides with one of the positive aureoles in As while it is within the enriched zone of Be that trend NE-SW from the southern part. Au has its negative anomaly where Ni has its positive anomaly. This implies that Au depletion indicate the present of mineralization that is indicated by anomalous high concentration of Ni in this area.

Like Nb, P has its positive aureole defined by lat. 11.1°N and long. 4.05°E. The spot is shared or closely related to by the aureole of other elements like Li, Al, As, Be, Ca, Cs, K, etc. Its distribution pattern is similar to the distribution pattern of Mo, La, Hg, Ga, Fe and Ce essentially. Though with relatively lower concentration values and in rare cases slightly high concentration, the zones of positive enrichment and depletion of these elements are closely related. Cr has relatively high enrichment in all the areas where P also has its high enrichment but the zone of P depletion in the eastern part is not seen in Cr isograde plot. Pb has its highest concentration (>19 ppm) in the western part of the area defined by long. 3.35°E and lat. 11.125°N. Like P, La, Fe, Ce, etc., its distribution pattern forms a poorly defined U-shape. The zone of its highest positive anomaly has negative anomalies closely associated with it in the SW and NE parts. Other peaks of positive anomalies are in the southern and extreme eastern parts of the area. Its zone of highest concentration is related to the zone of highest concentration of elements like La, Ce, Cd and Ba. The peaks of positive anomaly in the western and southern parts relate very well with the two peaks of positive anomaly in Mn, while the peaks of positive anomaly of Pb in the southern and eastern parts are closely associated with the peaks of highest Fe concentration. Fe also has slightly elevated value in the western part where Pb has its highest concentration value. Other elements have their positive anomalies in one or more zone of Pb enrichment even though their distribution patterns are not the same.

Pd has its aureole in the western end of the study area. Its concentration decreases gradually towards the western end of the area. The zones of its highest concentration in the eastern part of the area coincide perfectly with one of the positive aureoles of Fe, Mo, In, Ga, Hf, etc. The Rb style of distribution is identical with the distribution pattern of Li, K and Cs. Its zone of highest concentration is wholly or partly shared besides the aforementioned elements by Al, As, Be, Bi, Ca, Cd, Co, Mg, Mn, Mo, Nb, Ni, P, etc. The peaks of its aureoles trend essentially in NE-SW direction. The negative anomalies are not well-defined but are closely associated with positive anomaly in the eastern part of the area.

Sb has its zones of highest concentration in the extreme eastern and western end of the area. Other zones of enrichment are closely associated with these two peaks defining a slightly different NE-SW trending direction. The distribution pattern is relatively unique, but its zone of highest concentration in the eastern end is related to the zones of highest concentration of Fe, Mo, In, Ga and Hf. The extreme NW and SW are the most depleted zones of Sb. Other areas have slightly elevated value relative to other elements.

With the exception of break in the ridge of positive anomaly of Sc in the western part of the area, the distribution pattern of Sc and Sn is identical. They both have a ridge of positive anomaly that rises in the NW going through the southern part and turning through the eastern part to end in the NE part of the area study. They have two peaks of positive aureoles. One of the aureoles is situated in the eastern part where elements like Fe, Mo, In, Ga and Hf have their or one of their positive anomalies. The second positive aureole of Sc is in the southern part of the area, while that of Sn shifts slightly to the SE part of the area. They have their negative anomalies trending essentially NW-SE in the western end and NE-SW in the eastern end. Sr has slight enrichment in the western part. These enrichments trend in NW-SE terminating in the central part. Its zone of highest concentration is in the centre. This centre where it has its highest concentration is defined by long. 4.05°E and lat. 11.075°N. Its zone of highest concentration is similarly shared by Rb, Al, As, Be, Bi, Ca, Cd, Co, Mg, Mn, Mo, Nb, Ni, P, etc. Its distribution pattern is identical with the distribution pattern of Ca.

The distribution pattern of Te is similar to the distribution pattern of Na. Its peak (positive anomaly) is in the central to SE part of the area. Its anomaly trends in NW-SE but generally decreases to the background value of $\leq$0.028 ppm. Th has its highest concentration in the eastern part. Other area with relatively high concentration of Th is in the SE and is closely associated with the zone of Th positive anomaly. The concentration of Th in the western half is also relatively high but not as high as the concentration in the eastern part. The western aureole trends in N-S direction with about three alternations in its concentration with about three alternations in its concentration. The extreme NW and SW corners of the area are relatively the most depleted zone for Th. The position of the positive anomaly of Th is the same with the position of positive anomaly of Fe, Sn, Sc, Sb, Pd, In, Hf and As. Mo and Ce has one of their positive anomaly in this eastern part, but the zone of highest concentration of Ag is just slightly off the zone of Th highest concentration to the west.

Ti, Tl and U have their positive anomalies defined by lat. 11.075°N and long. 4.04°E and lat. 11.08°N and long. 4.04°E and lat. 11.07°N and long. 4.04°E, respectively. Though with slight variation, these elements have their positive anomalies where elements like Sr, Rb, Ni, P, Nb, Li, K, Cs, Co, Cd, Ca, Bi, Be, As and Al have theirs or one of their positive anomalies. This zone is also though not exactly but similarly positioned like one of the Mn positive anomaly. Ti and Tl have their highest concentration within the central part of the area. Ti concentration is slightly high in the eastern and extreme NE part, but Tl has its own concentration decreasing gradually towards the extreme west, thereby making the western half slightly more enriched than the eastern half. Both Ti and Tl have sharp depression to the background value. These depressions are similarly oriented in poorly defined NWW-SEE in the western half of the area. Unlike Ti and Tl, U has its concentration in the central part which forms a ridge of positive anomaly extending from the central area to the NW of the area where it forms a twin peak. The peak to the east among these two peaks has slightly higher concentration than that in the western part. There are negative anomalies in the western half and eastern part of the area. The negative anomalies in the western half that are well-defined trend in NE-SW. The poorly defined negative anomaly trends with one of the well-defined in NW-SE. The U negative anomalies in the eastern half of the area trend mainly in NE-SW.

The distribution pattern of V is similar to the distribution pattern of Th, Sc, Sn, Pb, P, Mo, La, Hg, Ga, Fe, Cr and Ce. Its zone of highest concentration (>65 ppm) is in the extreme eastern part, while it is also enriched in the southern part. These two zones of highest concentration are similar to the zone of highest concentration of Fe, Sc and Mo. The enrichment in the western half is almost identical with that of Fe when the position of the zones of highest enrichment and depletion are compared. The central part of the area is the most depleted zone of V. Like some other elements, e.g. Hf, its distribution pattern has poorly defined U-shape. The zone of highest concentration of Y (>23 ppm) is in the NW part of the area. This area is flanked by negative anomaly in the SW and NE part. The zone of relative high concentration of the anomaly, usually ≥16 ppm, is restricted to the western half of the study area and trends in NW-SE direction. The extreme SW and SE corner is relatively depleted of Y (≤5 ppm), while other areas have elevated value that range between 5 and 16 ppm. The positive anomalous zone of Y is similarly enriched in Pb, La, Hg, Cd, Ce, Ba and Mn.

The distribution pattern of Zn is identical with the distribution pattern of Rb and almost identical with the distribution pattern of Li, K and Cs. Both Zn and Rb have two peaks at the crest of their anomalous zone in the central part of the study area. The peak in the northern part out of these two peaks is slightly more prominent than the one in the south. These two peaks are also present in the same position in the isograde plot of Tl despite the fact that its style of distribution is slightly different from that of Zn and Rb. Other similarities in the distribution pattern of Zn and Rb include the twin slightly enriched area in the NW part of the area and also the negative anomaly in their isograde plot. Other elements that have their positive anomaly where Zn has its own and with similar distribution pattern include Ti, Mg, Cs, etc. Cs and K have the two peaks, but their peak in the southern part is more prominent than the one in the northern part. The zone of Zn anomaly is shared by the anomaly of Al, As, Be, Bi, Ca, Cd, Co, Mn, Nb, P and Ni essentially. Zr highest concentration is ≥19 ppm. This concentration is in the eastern part of the area. With the exception of elevated value (≤9 ppm) in the NW part of the area, the enrichment of Zr is restricted to the eastern half of the area especially where Fe, Th and Pd have their highest concentration.

6. Conclusion

From integrated field observation, geochemical study, statistical analysis and isograde plottings, it was concluded that the area investigated is mineralized with goethite and manganite ores, while there are minor enrichments of tourmaline and silicate minerals like chalcedony. It was also concluded that the eastern and southern parts of the study area is mineralized with goethite, while the western and southern parts are mineralized with manganite. Furthermore, in exploring for iron-manganite mineralization in this area, manganite has as its pathfinder the following associations Zn + As + Be + Bi + Co + Nb + Ni + Cs ± P + Al + Ca + Cd + Rb + Li + K, while Zr + Th + Pd + Mo + V + Sn + Cr + Ce + In ± Sc + P + Pb associations are related to iron mineralization.

Author details

Olufemi Sijuade Bamigboye
Department of Geology and Mineral Sciences, Kwara State University, Malete, Nigeria

*Address all correspondence to: olufemi.bamigboye@kwasu.edu.ng

References

[1] Fabris AJ, Keeling JL, Fidler RW. Soil geochemistry as an exploration tool in areas of thick transported cover, Curnamona Province. MESA Journal. 2009;**54**:32-40

[2] Martin AP, Turnbull RE, Rattenbury MS, Cohen DR, Hoogewerff J, Rogers KM, et al. The regional geochemical baseline soil survey of southern New Zealand: Design and initial interpretation. Journal of Geochemical Exploration. 2016;**167**: 70-82

[3] Omorinoye OA, Adekeye JID. Soil geochemical survey of eruku and environs. Journal of Environment and Earth Science. 2013;**3**(7):105-115

[4] Filgueiras AV, Lavilla I, Bendicho C. Evaluation of distribution, mobility and binding behavior of heavy metals in surficial sediments of Louro River (Galicia, Spain) using chemometric analysis: A case study. Science of the Total Environment. 2004;**330**:115-129

[5] Bowen HJM. Environmental Chemistry of the Elements. London: Academic Press; 1979. pp. 333

[6] David RH. An Assessment of Soil Geochemical Methods for Detecting Copper-Gold Porphyry Mineralization through Quaternary Glaciofluvial Sediments at the WBX-MBX and 66 Zones, Mt. Geoscience BC Report. North-Central British Columbia: Milligan; 2010. p. 75

[7] Akilolu F. Exploration and Resource Estimation of the Manganese Deposit in Kaoje, Kebbi State. NGSA Report. 2007. p. 70

[8] Bamigboye OS, Adekeye JID, Kadioglu YK, Adedoyin AD, Omorinoye OA. Geochemistry and origin of Fe-Mn oxide mineralization in Kaoje-Derena and their environs, Northwestern Nigeria. Arabian Journal of Geosciences. 2018;**11**:570-587

[9] Adekoya JA. The geology and geochemistry of the Maru banded iron formation, Northwestern Nigeria. Journal of African Earth Sciences. 1998; **27**(2):241-257

[10] Mucke A. The Nigerian manganese-rich iron-formations and their host rocks from sedimentation to metamorphism. Journal of African Earth Sciences. 2005;**41**(5):407-436

[11] Okorie IM, Ayoade EO, Okunola OW, Okunlola OA. Manganese Mineralization in Parts of Northwestern Nigeria. In: Book of Abstract, 45th NMGS Annual International Conferences; Owerri, Nigeria. 2009

[12] Okorie IM, Ayoade EO, Okunola OW, Okunlola OA. Geological setting and occurrences of copper mineralization around Talata Mafara and Anka Areas, Nortwestern Nigeria. In: Book of Abstract, 45th NMGS Annual International Conferences; Owerri, Nigeria. 2009

[13] Fillie S. Genesis and paragenesis of the tetravalent Mn ores and Mn-Fe ores in secondary banded iron formations from Nigeria. In: Proceedings of the 3rd World Conference on Applied Sciences, Engineering and Technology, Kathmandu, Nepal. 2014. pp. 318-328

[14] Fillie S. Chemical mineralogy of supergene ores in banded iron formations from Northern and South-Central Nigeria. In: Proceedings of the 3rd World Conference on Applied Sciences, Engineering and Technology, Kathmandu, Nepal. 2014. pp. 329-339

[15] Vareikiene O, Lehtonene M. Heavy minerals study of soils: Techniques, their limitations and advantages. Geologija. 2004;**46**:1-7

[16] Levinson AA. Introduction to Exploration Geochemistry. Wilmete, USA: Applied Publishing Ltd; 1981. p. 613

[17] McQueen KG. Element fractionation and mineral hosts in the regolith. In: Book of Abstracts—LEME AMEC Minerals Exploration Seminar. 2004. pp. 19-23

[18] Dempster M, Cooper MR, Dunlop P, Scheib AJ. Using soil geochemistry to investigate gold and base metal distribution and dispersal in the glaciated North of Ireland. In: Young ME, editor. Unearthed: Impacts of the Tellus Surveys of the North of Ireland. Dublin: Royal Irish Academy; 2016. DOI: 10.3318/978-1-908996-88-6.ch7

[19] Tsikos H. Petrographic and geochemical constraints on the origin and post-depositional history of the hotazel iron-manganese deposits, kalahari manganese field, South Africa [Ph.D. thesis]. Rhodes University; 1999. p. 217

[20] Andrew-Jones DA. The Application of Geochemical Techniques to Mineral Exploration. Vol. 11(6). Golden/Colorado: Colorado School of Mines. Mineral Industries Bulletin; 1968. pp. 1-31

Section 3

Sedimentology

Chapter 9

Main Features of Sedimentogenesis and Ecogenesis in Late Paleozoic Sea Pools of Northeast Asia

Victor Gdal'evich Ganelin

Abstract

The main characteristics of the Late Paleozoic sediments and ecogenesis in the Late Paleozoic basins of Northeast Asia are considered. The authigenic bacterial nature of carbonate bodies, previously considered as detrital, was revealed. The assumption about the lithotrophic nature of the Late Paleozoic ecosystems of the region is substantiated, the basis of the trophic chains of which are prokaryotes. As it turned out, a feature of the Late Paleozoic benthic biota is associated not with climatic factors but with the trophic nature of the communities. Similar types of paleoecosystems in the same age layers of Gondwana are indicated.

Keywords: Northeast Asia, Gondwana, late Paleozoic basins, riftogenesis, black shale sedimentology, bacterial carbonate rocks, autolithotrophy, fluidolites, benthic biota

1. Introduction

Throughout the Late Paleozoic, there was a large marine basin at the northeast of Asia, with a total area covering about 2,500,000 square meters. Together with the southern part of Taimyr Peninsula, this territory represented a biogeographically and sedimentologically integral system of sedimentary basins, also known as "Taimyr-Kolyma" paleogeographic region, also including Transbaikalia and Northern Mongolia. From the mid-Early Permian, the Taimyr-Kolyma, East European, and West European paleogeographic regions are considered as parts of the Biarmian subregion, framing the modern Arctic [1] (**Figure 1A**). In its main outlines, this system of basins is apparently preserved during the Early Mesozoic.

The main features of sedimentary and biotic evolution of the Late Paleozoic basins of the northeast of Asia were connected with the destruction of the Middle Paleozoic continental shelf by establishing rift basins across it [2]. This process started in the mid-Early Carboniferous and, up to the end of the Early Permian, had been accompanied by intense mafic magmatism—the Omolon-Selenyakh volcanic belt of Lychagin and Lyckman [3] (**Figure 2**). By the mid-Middle Carboniferous time, these processes foreordained the general paleogeographic patterns of the Late Paleozoic basin system.

According to the features of sedimentation and the appearance of fossil assemblages, three paleogeographic areas can be distinguished within the northeast of Asia: Verkhoyansk-Okhotsk, Kolyma-Omolon, and Novosibirsk-Chukotka subregions [4, 5] (**Figure 1B**). Despite biota was very close in all of them, there are differences

IntechOpen

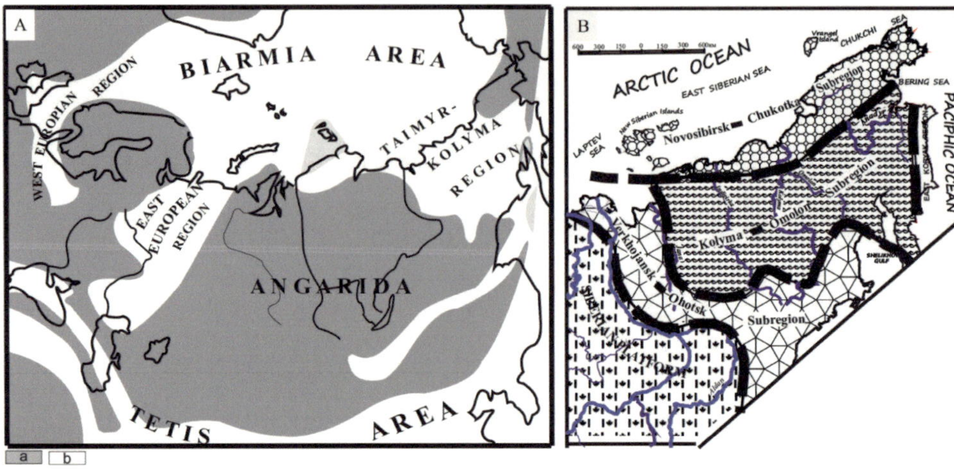

Figure 1.
Zoning of the late Paleozoic basins of northern Eurasia. (A) Zoning of perm basins of northern Eurasia (after Ganelin, modified by Kotlyar [1]). (a) land, (b) sea. (B) Zoning of the late Paleozoic basins of Northeast Asia (after Ganelin, Tschernjak [5]).

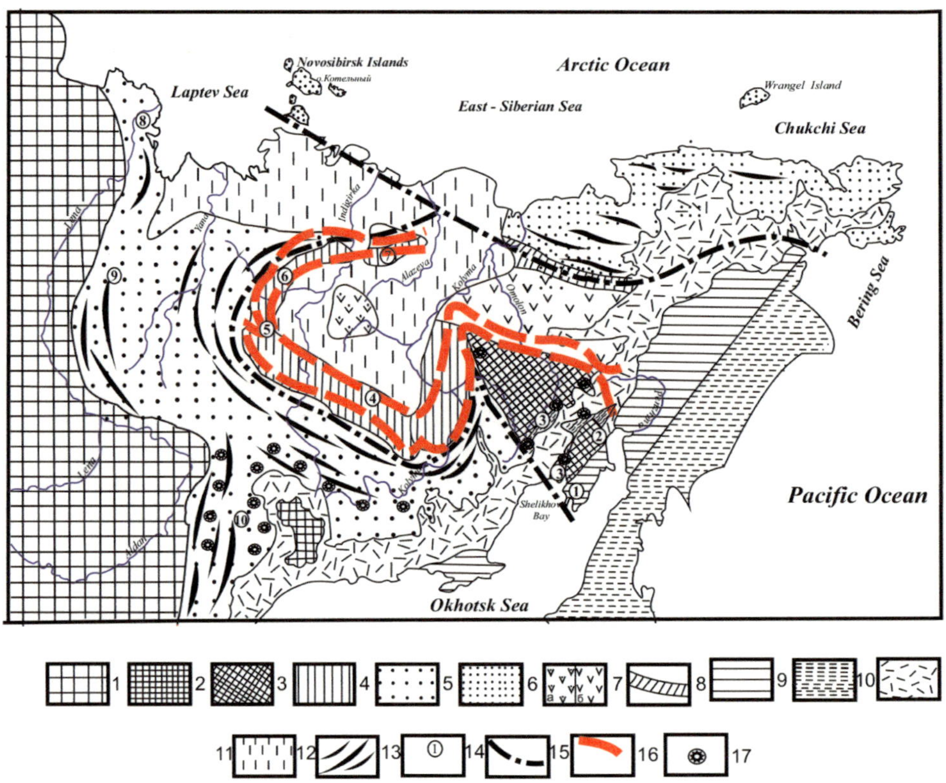

Figure 2.
Principal tectonic elements of Northeast Asia and the position of the late Paleozoic magmatic belt. 1 Siberian platform, (2 and 3 massifs) 2 Okhotsk, 3 Omolon, 4 uplifts, composed mainly of Paleozoic, (5–10 orogens) 5 Yana-Kolyma, 6 Novosibirsk-Chukotka, 7 Alazey-Oloy [(a) Alazey and (b) Oloy parts], 8 south Anyui, 9 Taiganos, 10 Koryak-Kamchatka, 11 Okhotsk-Chukotka volcanic belt, 12 Kolyma-Indigirka and Primorye lowlands covered by quaternary, 13 stretch of folds, 14 smaller tectonic structures, mentioned in the text (1 Taiganos, 2 Avekov block, 3 Gizhiga deflection, 4 Arga-Tass block, 5 Tas-Hayakhtakh uplift, 6 uplift of the Selenyakh and Polousny ridges, 7 Ulakhan-sis uplift, 8 Kharaulakh zone, 9 Orulgan zone, 10 South Verkhoyansk synclinorium), 15 borderline between Yana-Kolyma, Indigirka-Omolon, and Novosibirsk-Chukchi regions, 16 boundaries of the late Paleozoic Omolon-Selenyakh magmatic belt of Lychagin and Lyckman [3], and 17 presumed areas of fluid-explosive activity.

in sedimentation type and, also, in the composition of benthic assemblages at certain stratigraphic intervals, especially for the Novosibirsk-Chukotka subregion. Clearly aside these three was the East Koryak subregion, characterized by Tethyan fusulinid assemblages, which are unique for the territory under consideration. East Koryak subregion still remains poorly studied and is not considered in the present paper.

2. Paleogeographic and sedimentological structure

The Verkhoyansk-Okhotsk subregion was a passive margin of Angarida continent, with intense accumulation of material brought here by prograding deltaic systems. This resulted in the formation of thick clastic sequence—so-called Verkhoyansk complex Cheraskov [6]. Two particular areas are characterized by the most intensive accumulation—these are Yana-Indigirka area and Ayan-Yuryakh area. The Novosibirsk-Chukotka subregion corresponds to the marginal basins of another continent, the hypothetical Hyperborea (Arctida). Kolyma-Omolon subregion, separating two subregions mentioned above, looks to be an offshore area, containing shallow sediments of mid-ocean elevations and deep-sea formations framing them (**Figure 3**).

A remarkable characteristic is predominant black shale sedimentation. It is defined by the presence of carbon-rich, clayish, carbon-siliceous, and ash-siliceous deposits, sometimes manganese-bearing and mostly sulfide-bearing within the succession, and these types of sediments are regionally distributed. All these features were previously explained by the author as connected with the influence

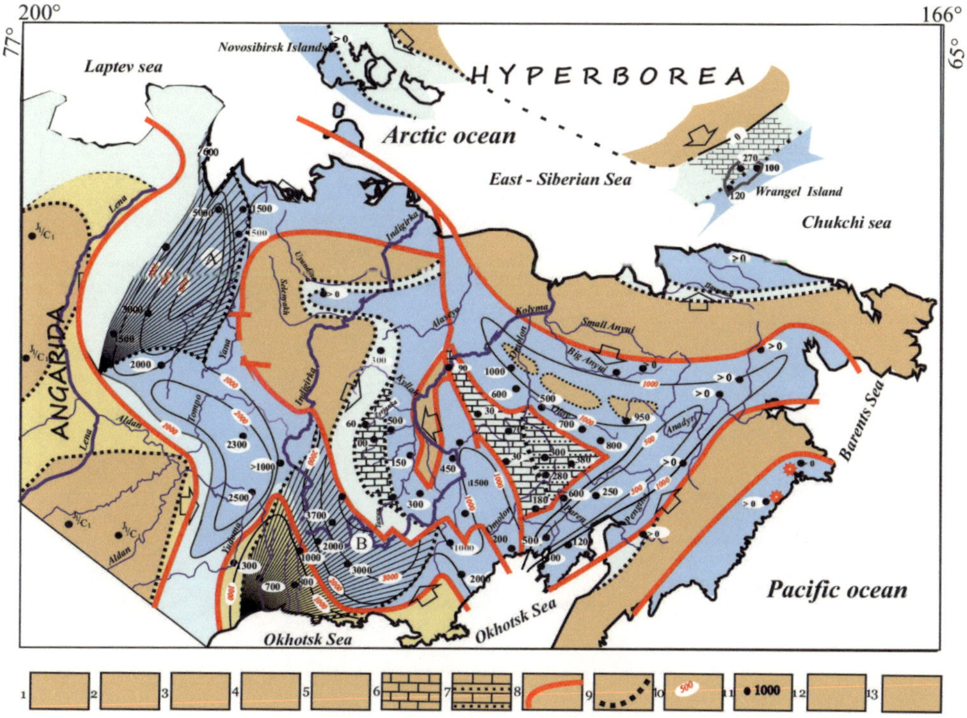

Figure 3.
Thickness and facies of the middle Permian in Northeast Asia: 1 presumable land (no deposit), 2 continental facies, 3 shallow-water facies, 4 offshore facies, 5 prodelta ((A) Yana-Indigirka, (B) Ayan-Yuryakh), 6 limestone facies, 7 limestone facies, intercalating with terrigenous rocks, 8 tectonic boundaries of structural zones, 9 facial boundaries, 10 isopachytes of the middle Permian, 11 total thickness of the middle Permian, 12 volcanic facies, and 13 clastic material transport direction.

of low-temperature hydrotherms, which caused the formation of the carbon-rich, substantially siliceous, sulfide-bearing sediments in the basin of euxinic type [7].

2.1 Carbonate rocks of the upper Paleozoic black shales

Considering the interpretation discussed above, particularly interesting are the carbonate rocks reported from the black shale successions. Those are represented both by isolated bioherms and biostromes inside the black shale series and by fully carbonate intervals within the succession, sometimes reaching 1000 m thick and replacing the black shales laterally (**Figure 4**). Such intervals are widely spread in the Kolyma-Omolon and Novosibirsk-Chukotka subregions. Within the shallow-water facies of mid-elevations (Omolon massif, Prikolymye), carbonate facies are characterized by the greatest variety of shelly benthic fauna, which makes it possible to consider them as centers of diversifications for the corresponding communities. The isotopic studies of these rocks, along with the analysis of their macro-, micro-, and ultramicrostructures, revealed the microbial origin of carbonates. Moreover, we have two different groups of these carbonate formations, which replace each other over time.

Carbonate formations of the first group are distributed from Middle Carboniferous to the first half of the Lower Permian. Biofacies of the Verkhoyansk type of fossil communities are associated with them. Carbonate rocks of the second group occupy the age interval from the second half of the Early Permian to the end of the Permian period. This second interval is characterized by the fossil assemblages of the so-called Kolyma-Omolon type [9].

Carbonate bodies of the first type are very rare. Findings of shell benthos, which belong to the Verkhoyansk type of communities, are also noted only locally. The carbonatolites of this age interval were studied by us in the upper reaches

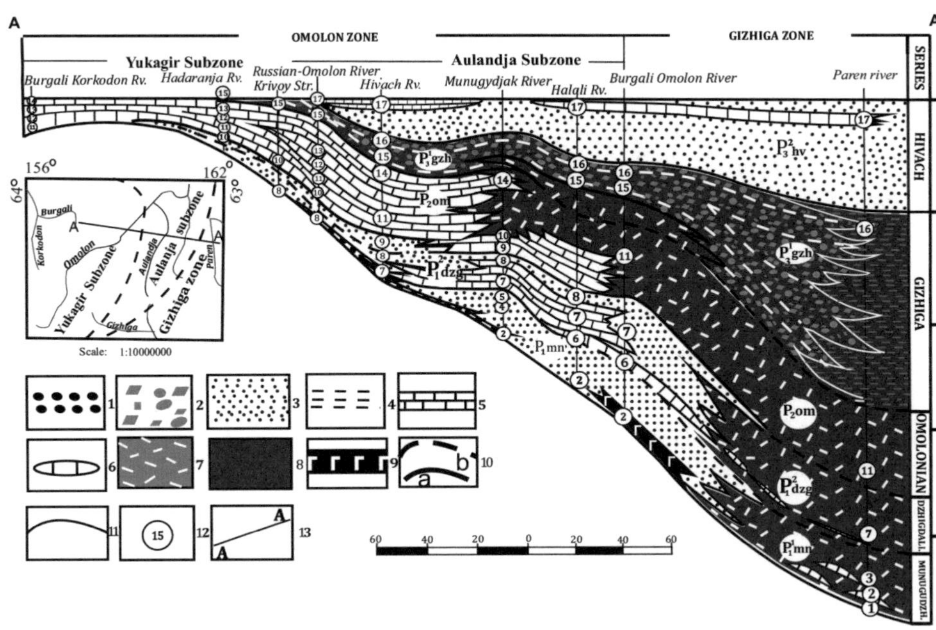

Figure 4.

Facies profile for the Permian deposits of the Omolon massif and Gizhiga zone. 1 conglomerates, 2 mixtites, 3 sandstones and siltstones, 4 flysch, 5 hydrogen sulfide limestones, 6 bioherms and biostromes, 7 fine pyroclastics, 8 carbon-rich black shales, 9 basalts, 10 boundaries of series, 11 boundary of formations, 12 stratigraphic zone numbers (Ganelin, Biakov [8]), 13 profile line of the map. $P_L{}^1mn$, lower Permian and Munugudzhak series; $P_1{}^2$ dzg, lower Permian Dzhigdali series; P_2 om, middle Permian, Omolon series, Kolyma series; $P_3{}^1$ gzh, upper Permian; Gizhiga regional stage; $P_3 2khv$, upper Permian, Khivach regional stage.

of the Paren' River (**Figure 4**). Here they are represented by bioherms, which have a lenticular, dome-shaped, and loaf-shaped form (**Figure 5A–D**). Their size reaches up to 7x0.7 m and they lie among carbon thin clastic tuffs. Tuffs contain an increased amount of organic carbon—up to 5%. A characteristic feature of these limestone bodies is their very thin layering, which, apparently, was formed by bacterial mats (**Figure 5E**). The rocks are saturated with finely dispersed pyrite,

Figure 5.
Carbonatolites of the middle carboniferous-early Permian. (A, B) bioherms in siliceous pyroclastic-rich rocks of the Olcha formation, middle carboniferous of the upper reaches of Paren' river, (C) bioherm in carbon-rich siliceous rocks of the Magiveem formation, early Permian, same location, (D) biostrome in carbon-rich siliceous rocks of the Magiveem formation, early Permian, (E) the same, showing thin bedding, (F) the same as (D), showing assemblage of brachiopods and bivalves.

which is often framboidal. Although benthic fossils are extremely rare in this interval (especially the Middle and Upper Carboniferous), some bodies are real "oases of life" (**Figure 5F**). Dense settlements of brachiopods, sometimes bivalve mollusks and, to a lesser extent, gastropod mollusks, conuliria, hyolites, as well as attached agglutinated foraminifers. Under scanning electron microscope (SEM), the microbial structure of the matrix is clearly visible (**Figure 6**). Isotopic data

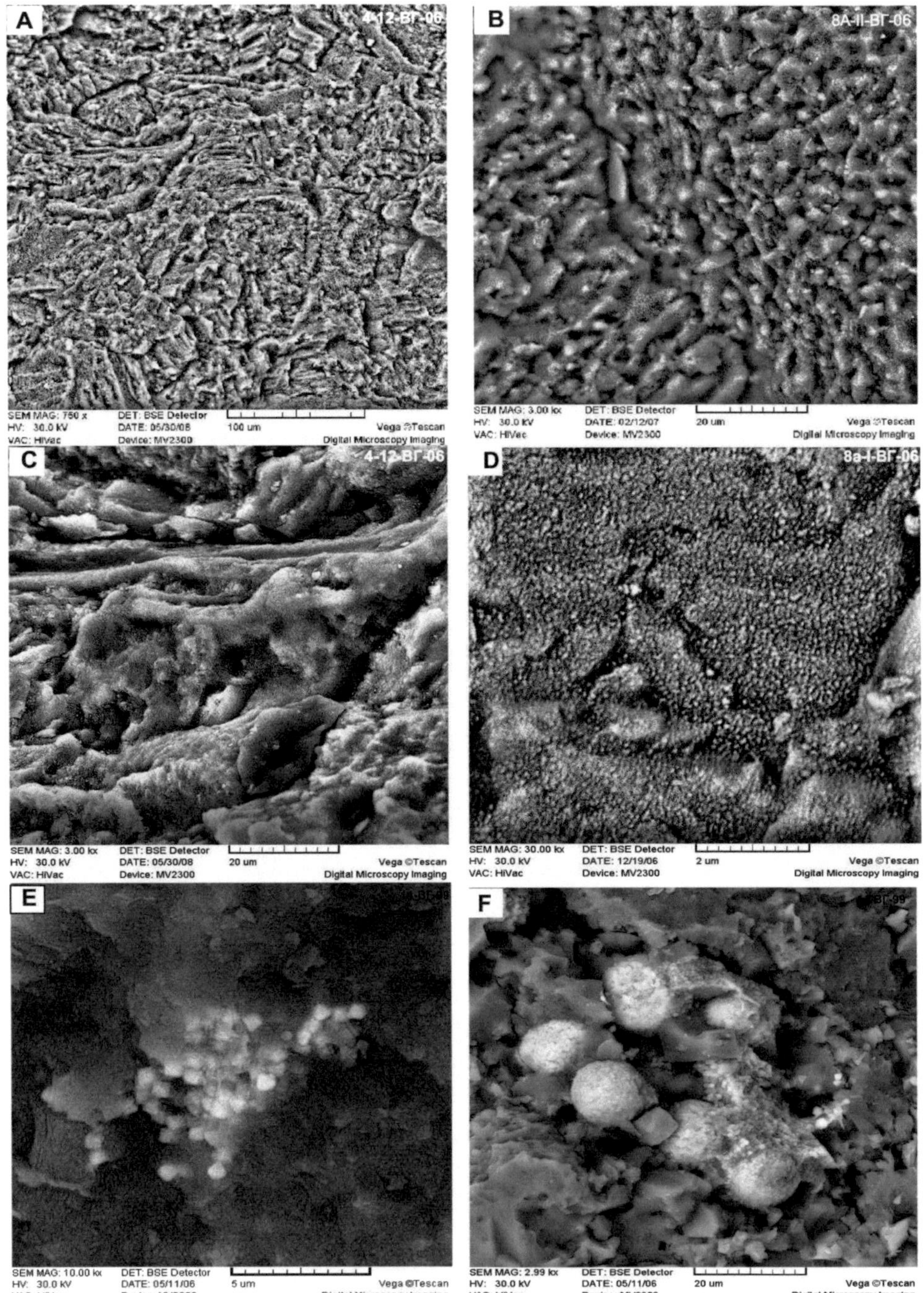

Figure 6.
(A–D) Bacteriomorph structures of the carbonatolites of the Magiveem formation; (E, F) Framboidal pyrite in the same rock; Lower permian, Magiveem formation, upper Paren river. Scanning electron microscope (SEM) photos at different magnifications.

are especially important for correct interpretation of these rocks. Pokrovsky [10] found that the bulk samples are characterized by very low values of $\delta^{13}C$, ranging from −9.4 to −26.4‰, and values of $\delta^{18}O$ ranging from 12.5 to 15.9‰ (**Figure 7**). Since brachiopod shells from the same rocks demonstrate isotopic ratios, typical for normal marine environment, it can be assumed that low $\delta^{13}C$ values in bulk samples should be explained by the fractionation of carbon by sulfate-reducing and methanotrophic prokaryotes. The intense sulfate reduction, as can be assumed, brought additional impact on the development of euxinia within the basin. According to B.G. Pokrovsky, low values of $\delta^{18}O$ can be explained by untypically intense fluid regime at the moment of rock formation, associated with hydrothermal activity. There is reason to believe that benthic organisms were rare due to rapidly changing redox conditions in the thin and mobile redox zone. As an analogy, we can turn to modern sedimentation in the Black Sea [11, 12]. Here, prokaryotes form an integrated symbiotic system in which methane oxidation is inextricably linked to sulfate reduction. Under anaerobic conditions in microbial mats, active sulfate reduction and anaerobic oxidation of methane occur due to the activity of archaea and sulfate-reducing bacteria. It was found that during the oxidation of the methane, part of the produced carbon is spent for the formation of carbon dioxide. The latter is fixed in carbonates that constitute nodules, bioherms, and continuous beds. At the same time, intensive sulfate reduction generates the huge amount of hydrogen sulfide. It can be suggested that the rarity of occurrence of fossil shelly benthos can be explained by unstable and quickly changing redox conditions within the thin and versatile redox zones.

The wide distribution of Middle-Upper Permian carbonates is associated with regional transgression, which began at the end of the Early Permian and ended by the end of the period. Carbonate crusts, bioherms, biostromes, as well as normally bedding limestones are ubiquitous all over the region (**Figure 8**). Bioherms and biostromes are normally 1–3 m high, rarely reaching 30 m, while their length

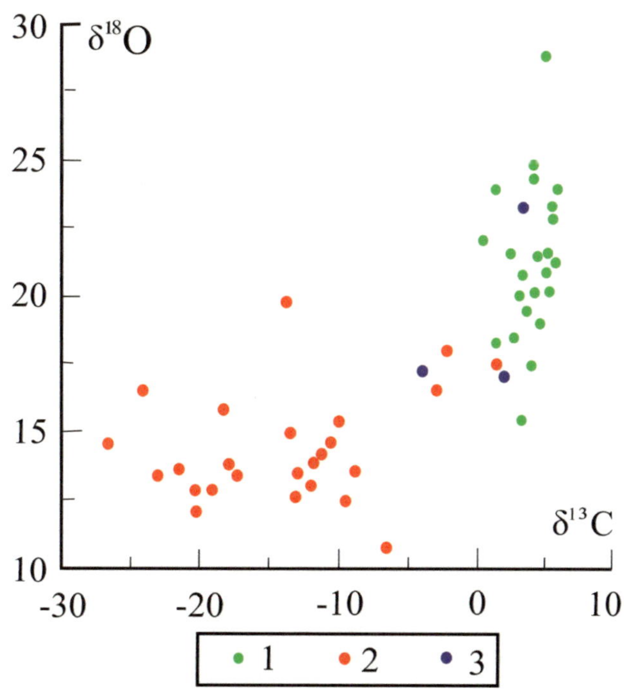

Figure 7.
Stable isotope ratios δ C (‰ PDB) and late Paleozoic carbonates of the Kolyma-Omolon basin. 1 middle-late Permian limestones, 2,3 late carboniferous-early Permian (2 bulk carbonate, 3 brachiopod shells).

Figure 8.
(A) Omolon massif. Bacterial limestones of the Omolon formation (1,2,3 indicate the members of the Omolon formation), middle Permian, northern Priokhotye, Khivach River basin (photo by I.L. Vedernikov). (B–E) Sugoi synclinorium. Bacterial limestones of the Taktay-Yuryakh formation. (B, C) Biostromes and bioherms of the Taktay-Yuryakh formation. D Biostromes overlying each other demonstrate strongly dissected upper surface, inverted bedding. (E) Bacterial crusts of prismatic calcite in limestones of the Taktay-Yuryakh formation. Middle Permian, middle reaches of the Kolyma River. (F) Omulevka uplift. Bacterial crusts of prismatic calcite in carbonaceous clay-silty rocks of the Rogachev formation. Upper perm, Kolyma river basin, Tuscan River. Photo by I.L. Vedernikov. (G) Gizhiga deflection. Bacterial crusts of prismatic calcite in carbon siltstone Fedorov formation. The second half of the lower perm; northern Priokhotye, upper course of the Paren' river. (H) for comparison, bacterial crusts among modern sediments of the Black Sea (A. Mazzini, M.K. Ivanov and others [9]).

extensively varies from the first meters to tens of meters and probably more. The contacts between individual bioherms are of the baselap type. Sometimes the top of bioherms demonstrates complex relief with column-like structures [column heights up to 1.5–2 m (**Figure 8D**)]. Carbonate crusts are especially common. They are present both in shallow facies (limestones) and in the deepwater facies—in the carbon-rich siliceous shales (**Figures 8D–G** and **9C**). A comparison of these crusts with similar bacterial formations in modern sediments of the Black Sea is indicated in **Figure 8H**. Limestones are always thin-layered (**Figure 9A–C**), fetid (with a pungent smell of hydrogen sulfide), and often saturated with finely dispersed pyrite and locally bituminous. Tubular fossils, composed of hematite and

Figure 9.
Structures of limestones of the Omolon series. (A–C) thin bedding of bacterial limestones of the middle Permian Omolon formation at Omolon massif: (A) upper member of the Omolon formation, zone 14 (Magadania bajkurica), Omolon River; (B) middle member of the Omolon formation, Khivach River; (C) bedding in the limestones of the Turin formation, Omulevka zone, Taskan River (photo by I.L. Vedernikov); (D, E) tubes in the bacterial limestones of the folk formation, the second half of the early Permian, zone 9 (Kolymaella-Bocharella), Omolon massif, Munugudzhak River (photo by T.V. Filimonova).

apparently formed during the oxidation of pyrite, can be met here (**Figure 9D, E**). In some samples of Permian limestones, quite numerous rounded transverse sections of presumable organisms with multilayered shell can be observed under SEM (**Figure 10B**). One can assume that these structures belong to organisms related to siboglinids—annelid worms, having the gut filled with chemosynthetic bacteria—serving as food for the host.

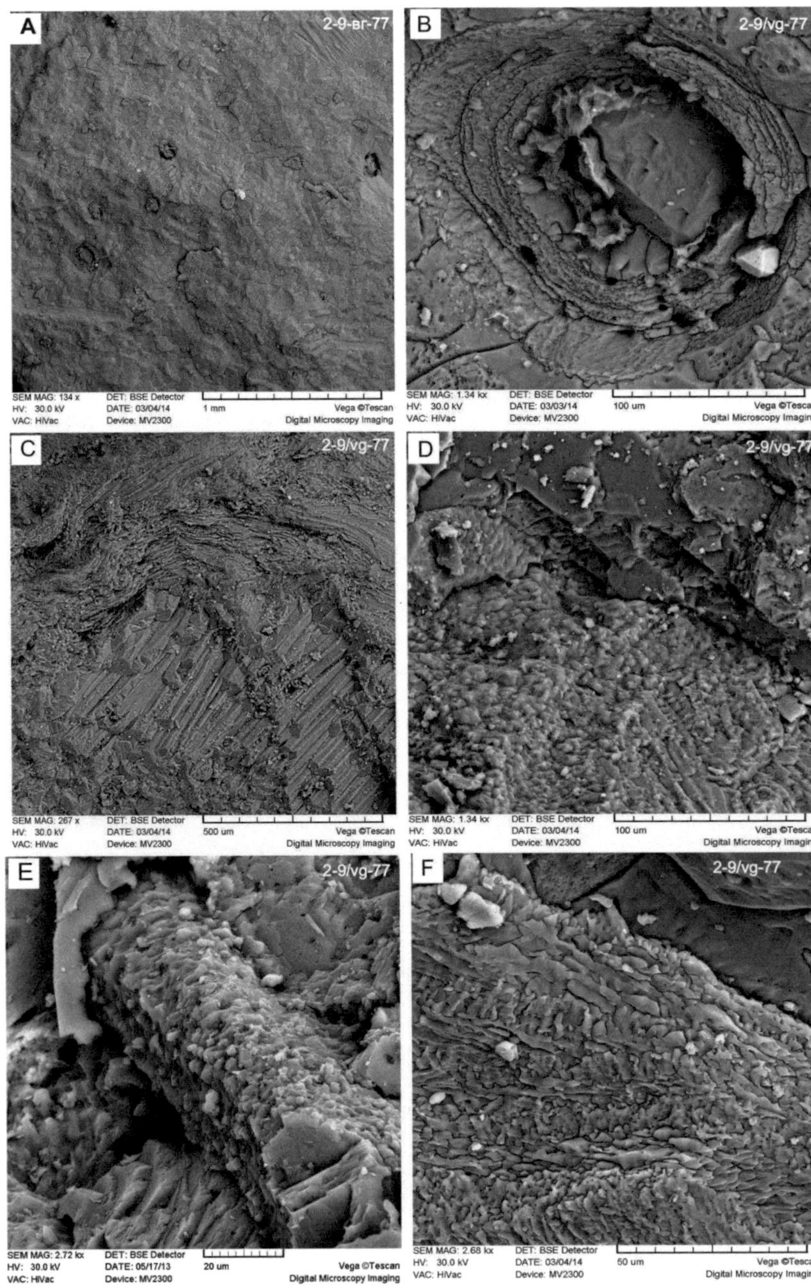

Figure 10.
Limestones of the Omolon formation (sample 2–9/vg-77) under SEM. (A) cross sections of microbial tubular structures; (B) the same, enlarged; (C) overlying prismatic crust and multilayered bacterial sheath; (D) another location of the same sample; (E) isolated prismatic colony, with no glycocalyx membrane and its cellular structure (with presumable cells >2 μm) is clearly visible; (F) differently oriented rod-shaped and spindle-shaped lithified cells forming elongated clusters; middle Permian, zone 12 (Terrakea borealis); Omolon massif, Omolon River.

The thickness of limestone bodies varies from centimeters to hundreds of meters. Microstructural features of these rocks (**Figure 11**), composed of the smallest calcite prisms, gave rise to an erroneous opinion about their detrital origin. Over the centuries, it was believed that calcite prisms are fragments of the destroyed prismatic layer of shells of inoceramid-like mollusks of the Colymiidae family, and therefore they are often called "Kolymiidae limestones" in the literature.

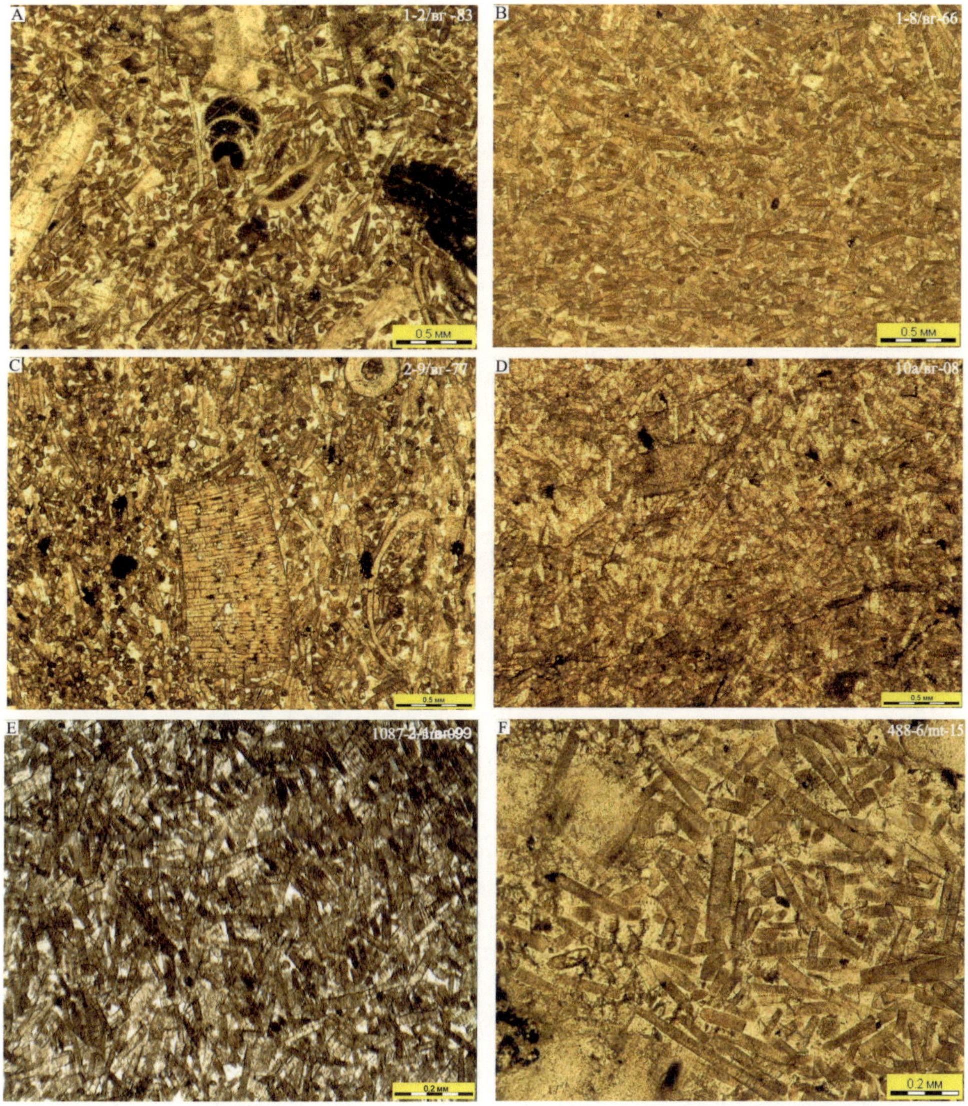

Figure 11.
Microstructures of limestone in transmitted light. (A) Ozerny formation, sample 1–2/вг-73; fragments of foraminifera and brachiopods are visible; upper lower Permian, Dzhigdali series, zone 9 (Megousia kuliki); Omulevka uplift, Kolyma river pool, upper course of the Zyryanka River. (B) Middle member of the Omolon formation, sample 1–8/вг-66, zone 13 (Terrakea korkodonensis), middle perm, Omolon massif, Khivach River basin. Sample 1–8/vg-66, zone 13 (Terrakea korkodonensis). (C) Middle member of the Omolon formation, sample 2–9/вг-76, zone 12 (Terrakea borealis), middle Permian, northern Priokhotye, Omolon massif, the left tributary of the Omolon River is the Russian River. (D) Taktay-Yuryakh formation, sample 10a/вг-08, middle Permian, the middle reaches of the Kolyma River. (E) Tupicovy formation, sample 1087–3/вш-09, the second half of the lower Permian, possibly the middle perm; Indigirka River basin (materials by V.S. Shulgina). (F) Mochov formation, sample 488–6/мт-15; fragments of shells of brachiopods and crinoid stems are present; upper lower Permian; Wrangel Island, the upper reaches of the Hishchnikov River (material by M.I. Tuchkova).

However, our detailed studies of these structures have undoubtedly shown the autigenicity of these rocks and their microbial nature. In addition to studying the morphology of these bodies, this is confirmed by a detailed study using SEM and thin sections, which leaves no doubt about this conclusion. Carbonate rocks have a well-preserved mineralized microbial structure. With a significant increase in SEM conditions, the breed looks like a set of fossilized colonies represented by prismatic and cup-shaped forms (**Figures 11, 12**). Prismatic colonies (up to 500 μm long and up to 30 μm wide) fit tightly against each other, and the matrix is completely absent. These colonies have a wrinkled and nostril surface—a membrane. It is formed, apparently, by the mineralized glycocalyx, which surrounds a cluster of mineralized cells 1–2 microns in size. Cells have a coccoidal, less often rod-shaped or ovoid shape (**Figures 12D, E, 13E**, and **10D, E**). When these structures are more strongly mineralized, the shell is covered with an even layer of porcelain calcite, and often prismatic colonies acquire a columnar appearance (**Figures 12D** and **10C**). In some places you can see fragmented cells, sometimes forming bundles or a complex interweaving of threads, branched, or rosette-like clusters (**Figures 13F** and **10F**). The rock is saturated with fine pyrite, which is often framboidal (**Figure 12E**). The surface of the colonies also has pores (gas vacuoles), with a diameter of up to 1 μm (**Figure 13C**).

Pokrovsky [10] established a significant difference in the isotopic composition of these rocks and limestones of the Middle Carboniferous-Early Permian era described above. The rocks of the Middle and Late Permian are characterized by a relatively high average content of $\delta^{13}C$ = 4.1 ± 1.4‰ (in the range from 1.5 to 5.9‰) and a wide range of $\delta^{18}O$ values: from 15.5 to 28.8 (in average 21.4 ± 2.8) (**Figure 7**). When interpreting these data, it should be borne in mind that the formation of rocks occurred during the widespread transgression. Therefore, it can be assumed that this violation caused the influx of deepwater sulfide water masses into the aerobic zone. Their mixing with oxygenated water masses, as can be assumed, caused changes in the composition of the bacterial biota. The methanotrophic anaerobic ecosystem of Middle Carboniferous-Early Permian was probably replaced by the chemotrophic sulfide ecosystem of Perm. It can be assumed that the trophic basis of this ecosystem was colorless sulfur bacteria and (possibly) archaea. They oxidized reduced sulfur compounds and produced carbon using seawater carbon dioxide. This explains the relatively high $\delta^{13}C$ for these limestones. It can be assumed that in addition to obligate aerobes, facultative prokaryotes were also present in the bacterial consortium. The latter explains the formation of rocks at different levels of the stratified basin. They are formed in the photic, oxygen zone, as well as in the deepwater aphotic zone with a low oxygen content.

2.2 Mixtites of the upper Paleozoic black shale series

Of no less interest than the carbonate rock, another characteristic member of the black shale series are the Late Permian mixtites, which form a belt along the Angarida margin. This belt extends from the Southern Verkhoyanie to Northern Priokhotye. Another belt of similar type extends along the southeastern margin of the Omolon massif and the Gizhiga trough. In Verkhoyanie and Priokhotye, mixtites are confined to thick (up to 1500 m) clay-shale sequences along the Angarida passive margin. Within the latter, the Late Paleozoic rocks are ore-bearing—they include large and very large gold stockwork-type mine fields. At the Omolon massif, the thickness of strata containing mixtites is 30–60 m, increasing up to 200–300 m in its marginal parts and areas surrounding the Gizhiga trough. A feature of stones is that they look like a drill: sand, gravel-pebble, and boulder materials are randomly dispersed in a dark gray and black clay-silty matrix, which is usually not layered but often has a characteristic of

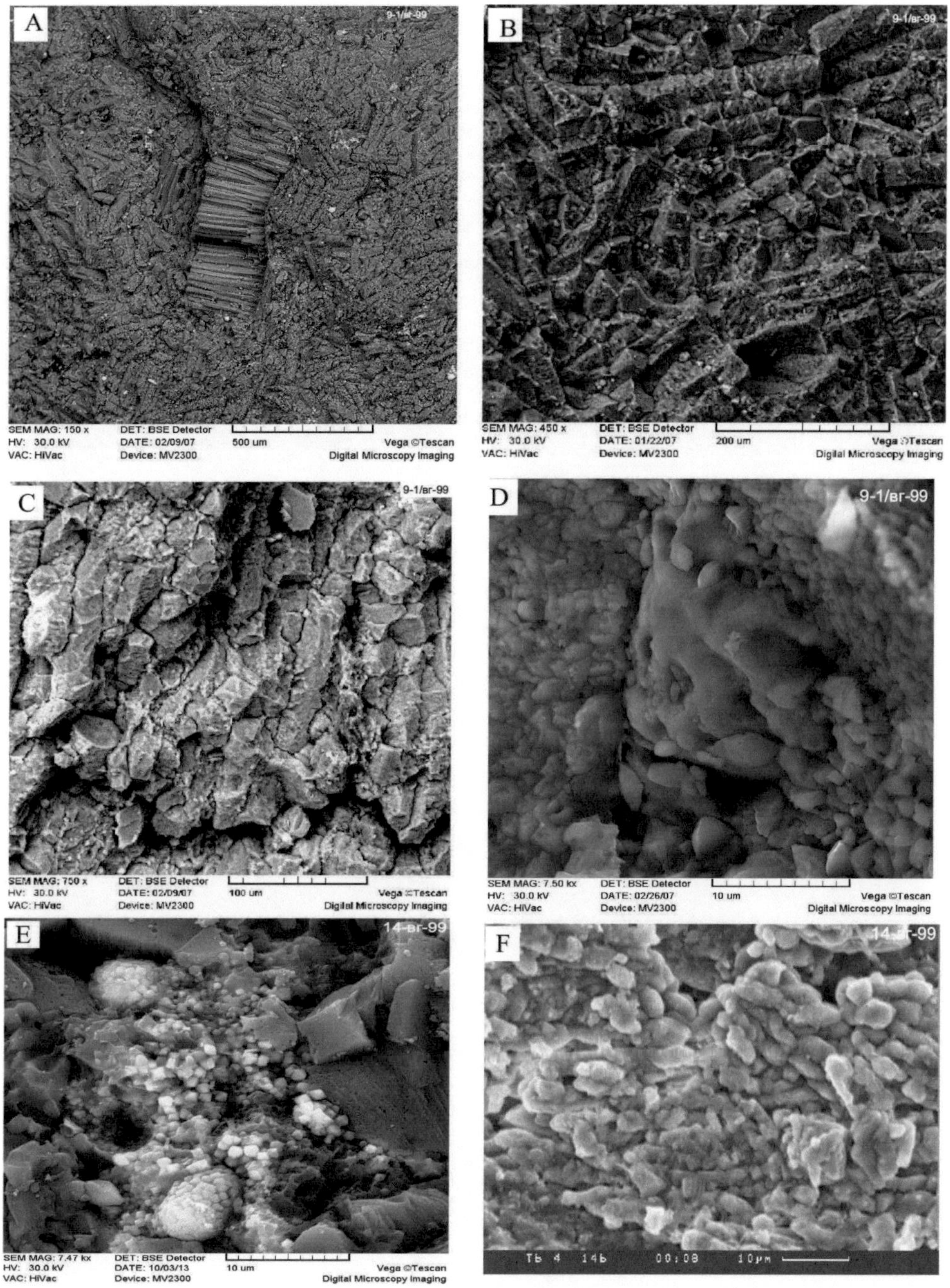

Figure 12.
Limestones under a SEM. (A-D) Rulon formation. (A–C) accumulation of prismatic microbial structures, at different magnifications (in a, a small carbonate crust is visible among clusters of prisms), (D) the accumulation of grains of bacterial carbonate and a whitened colony covered with porcelain calcite, sample 9–1/вг-99. Omolon massif, the upper half of the lower Permian, Dzhigdali series, zone 7 (Megousia aagardi). Left tributary of the Omolon River, the Munugujak River. (E, F) folk formation, sample 14–1/вг-99. (E) Framboidal pyrite, (F) rod-shaped microbial structures. Omolon massif, the upper half of the lower Permian, Dzhigdali series, zone 8 (Megousia kuliki). Same location.

concentrically layered parting (**Figure 14**). Another feature of the rocks is the predominantly volcanic composition of the clastic elements—andesit-dacite, less often liparite. Light, decayed, and altered fragments of volcanic rocks, scattered over the dark matrix, provide spotty appearance for the rocks. This served as a basis for their

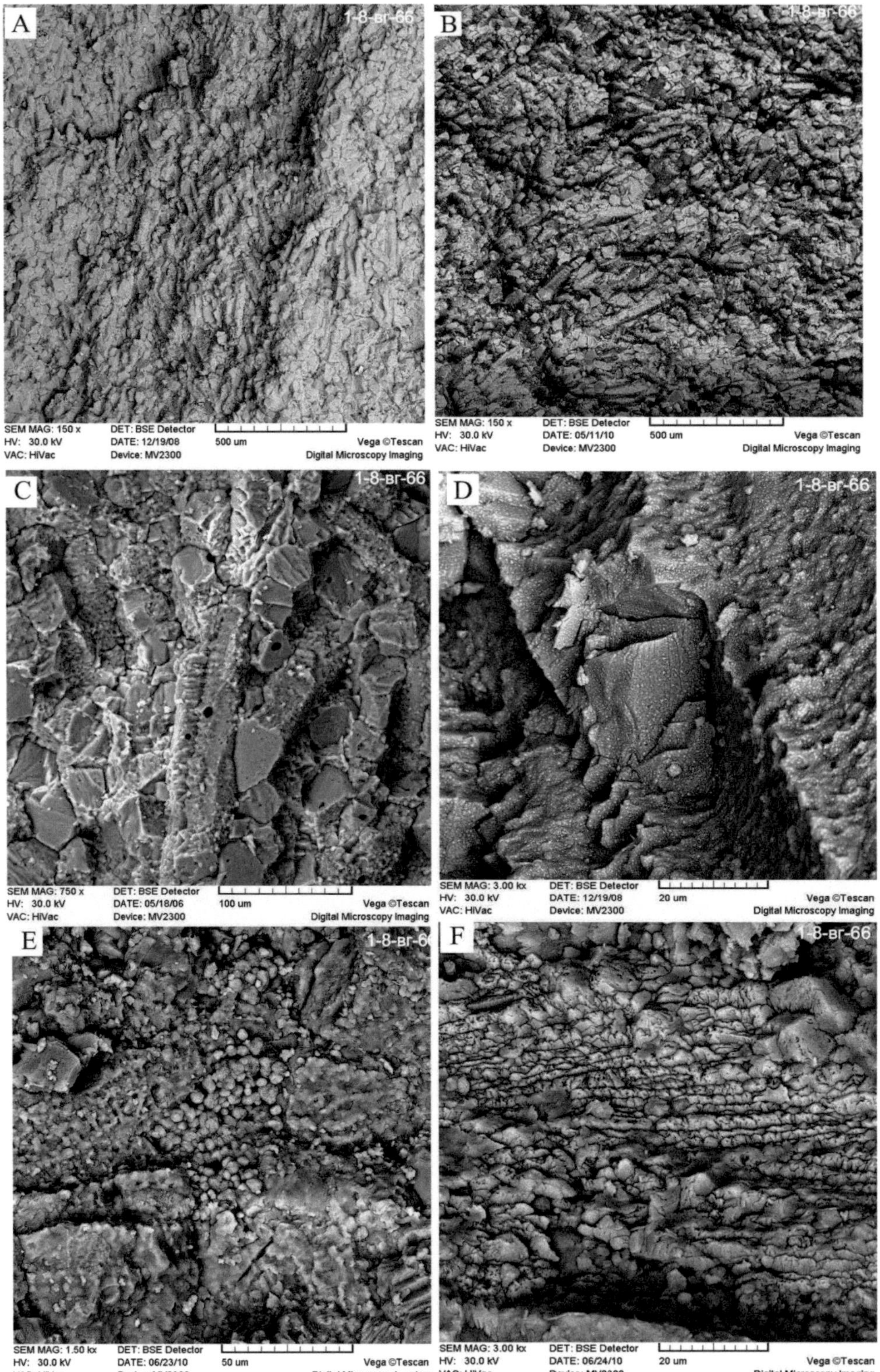

Figure 13.
(A–C) accumulation of prismatic microbial structures, depicted in different planes under different magnifications. (D) Surface of the prismatic structures, apparently covered by the mineralized glycocalyx. (E) Aggregates of microbial filaments formed by splitting cells and coated by a thickened mineralized glycocalyx. (F) Longitudinal section of a prismatic structure with distinct coccoid bacteriomorph structure; sample etched with HCl. Middle member of the Omolon formation. Middle Permian, zone 13 (Terrakea korkodonensis); Omolon massif, Khivach River.

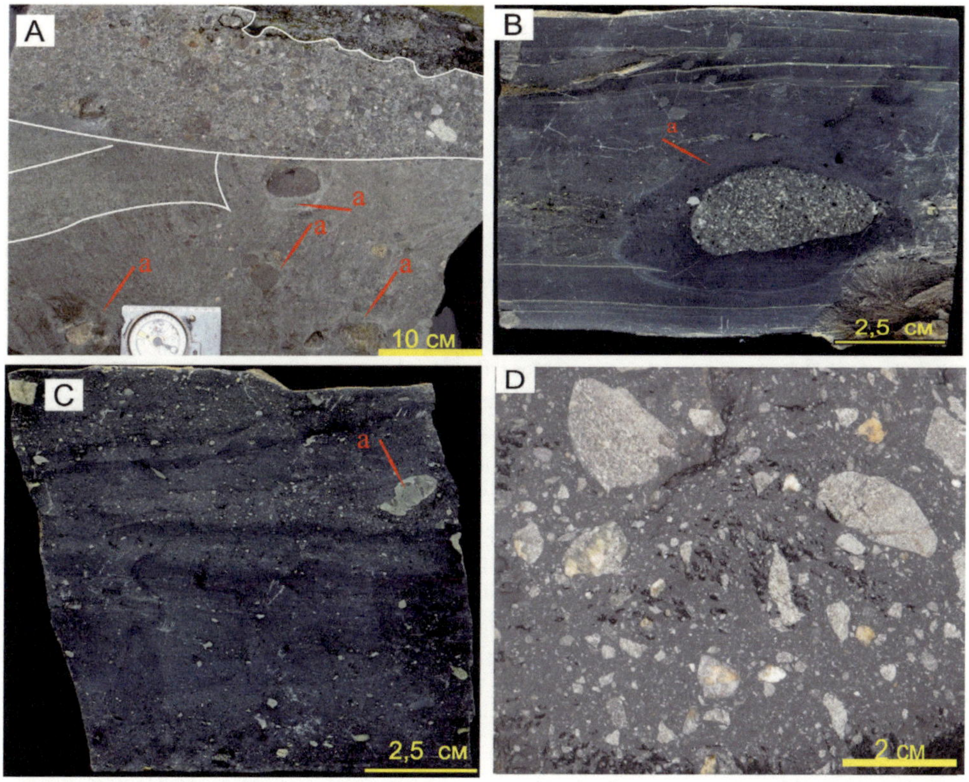

Figure 14.
Polished surfaces of Permian mixtites, showing their principal features. (A–C) Mixtites (diamictites) of the Permian of Northeast Asia: (A) fractal nature of diamictite (diamictite in diamictite; "a" letters indicate areas of mineral transformation around xenoliths); middle-upper Permian, Atkan formation; upper reaches of the Nelkoba River, basin of the upper reaches of the Kolyma River (materials by I.L. Vedernikov). (D) variable roundness of xenoliths; upper Permian, Atkan formation; Burgagylkan River, a tributary of the Chelomdzha River, the basin of the upper reaches of the Kolyma River (materials by A.S. Byakov).

naming in the geological slang as "hazel grouse rocks." The predominant size of the clasts is 0.5–5 cm, although fragments up to 20 cm can be observed. The clasts normally occupy 5–10% of the total rock volume, locally reaching 60–70%. Their roundness is variable, but often they are well-rounded. The silty-clayish matrix is unsorted, has chlorite-siliceous-clay composition, and apparently derives from the pyroclastic material (Mikhailov, 1971 — unpublished; Byakov, Vedernikov [13]). Intervals with clasts normally interbed with pure clayish-silty intervals, which locally demonstrate well-visible graded bedding and soft-sediment flow structures. Fluid textures are also noted. The mineralogical composition of the heavy fraction from the Omolon mixtites is represented by three types: (1) associated with rocks of acid and intermedium composition, (2) associated with rocks of basic and ultrabasic composition, and (3) deriving from metamorphic rocks (Mikhailov, 1971). The most peculiar thing is the presence of large amount of pyroclastic materials, while the corresponding effusive analogues remain unknown. This gave rise to different hypotheses on the origin of clasts. Since the late 1950s, there was a widespread opinion that the clasts were brought by the drifting seawater ice, while the source of clasts was from some hypothetical land in the Sea of Okhotsk [14]. Ustritsky [15] linked the transport of volcanogenic material with synchronous volcanism within his hypothetical "Shelikhovsky volcanogenic belt." Rejecting the ice transportation hypothesis, Byakov and Vedernikov, following Ustritsky, explained the origin of clasts by synchronous volcanism outside the territory, renaming the hypothetical "Shelikhovsky belt" into the equally hypothetical

"Okhotsk-Taigonos volcanic arc." These authors correctly marked the widespread presence of debris flows within the succession, which offer to be the main transportation agent. Recent special studies by Isbell et al. [16] have also rejected the hypothesis of the ice origin of these deposits. Obviously, the problem requires further study. However, several additional considerations should be made.

Noteworthy is the spatial distribution of the rocks along the continental margin, which was the zone of increased fluid permeability. For such areas fluid flows are one of the significant factors of sedimentogenesis. This suggests that the rocks under consideration may belong to the fluid-explosive type, the recognition and study of which receives more and more attention today. In recent years, the role of fluid systems has been recognized as the universal mechanism of the formation and transformation of the earth's crust, as well as of the localization of many currently known mineral resources. Fluid is a substantially aquaceous, water/gas, vapor or gas medium, enclosed or transported by the rocks within the lithosphere. The components of the fluid systems react with petrogenic, ore, and other elements [17]. Fluidolites are a resulting association of rocks, varying from breccias and conglomerates to aleuro-psammites and carbonates. Cases of erroneous interpretation of fluidolites as glacial formations are not rare at all [18]. The formation of such rocks was not associated with the magma intrusion. It happened due to drastic releases of gas-water solutions and more or less gradual intrusion of gas and water solutions into the crust. These solutions, having separated from their maternal extremely gas-saturated magma, brought various types of pyroclastic debris, from ash to psephites. The development of fluid systems was accompanied by a wide range of phenomena, such as mud volcanism.

The fluid-explosive-mud origin of the debris in mixtites of Northeast is confirmed by the structural position of these formations; sedimentation of black slate background; lack of a clear (and not hypothetical) surface source; sharp lateral disappearance, which can be explained by the presence of several foci of eruption; the presence of fluid textures; and the presence of stockwork type of mineralization. The interbedding of diamictites and pure silt-clay intervals, as we can assume, indicates the alternation of different phases: explosive (short-term explosive emissions of gases, debris, and dirt) and griffin (prolonged and quiet penetration of liquid mud). It can be assumed that the hazel grouse and the associated gold mineralization are syngenetic.

The characteristics of the Late Paleozoic sedimentation in Northeast Asia, considered above, suggest that a significant role in the Late Paleozoic riftogenesis was played by the endogenous activity, which generated ascending fluid flows of various compositions. The discharge of these flows resulted in the black shale sedimentation, the presence of peculiar mixtites within the succession, and the presence of carbonate bodies of bacterial genesis. The association of benthic fauna within bacterial carbonates indicates that the basin was characterized by autolithotrophic ecosystems, where the role of primary producers was played by methanotrophic and chemotrophic bacteria. Data indicating the significant role of similar processes in recent water reservoirs are quickly increasing ([11, 12, 19–22] and others). Lein described the problem as "life on methane and hydrogen sulfide" [23]. A summary on these processes is reflected in the Belenitskaya's concept of "fluid lithogenesis" [24] as a special type of rock formation processes (Yudovich [25]).

3. Stages of historical development of the benthic biota

The black shale "anomaly" of the Late Paleozoic sedimentation at the Northeast Asia is well correlated with the peculiarity of its benthic biota. The latter is very poor taxonomically and endemic for most of the Northern Hemisphere. Large animal groups such as fusulinids and trilobites are absent here. As a rule, corals are absent too, while conodonts are unknown. The explanations for this usually imply the paleogeographic position of the region—it was located in high palaeolatitudes and ocean waters here were cold [26–29]. However, earlier, the author has shown that the distribution of the so-called "boreal, cold-water" biota is not consistent with this idea. At the same time, a correlation between the occurrence of the benthic communities and the phases of riftogenesis and formation of deep black shale basins was revealed [2]. The analysis of the corresponding ecosystems revealed their lithotrophic nature, where the role of primary producers belongs not to plants but to autolithotrophic bacteria.

In the evolution of the Late Paleozoic biota of the northeast of Asia, three large stages can be distinguished, reflected in the successive change of three types of benthic associations—Verkhoyansk, Pechora-Kolyma, and Kolyma-Omolon types [9].

The Verkhoyansk type of benthic associations (*Verkhojania–Jakutoproductus* fauna) ranges from the mid-Bashkirian to the mid-Artinskian. Taxonomically these associations are very poor and sharply endemic. As a rule, fossil assemblages are confined to thin limestone bodies (bioherms, biostromes) with untypically low $\delta^{13}C$ values. Until the Early Artinskian, this type of associations remains extremely rare and mostly local.

Starting from the Late Artinskian and up to the end of the Early Permian, there was a biotic turnover caused by the beginning transgression. Monotypic associations of the Verkhoyansk type were replaced by polytypic association of Pechora-Kolyma type. In the newly established associations, the presence of the elements of Ural and, especially, of Pechora affinity is very noticeable. The diversity of foraminiferas from Lagenidae's family significantly increased after this turnover. Representatives of the genera *Waagenoconcha*, *Anidanthus*, *Striapustula*, and *Spiriferella* are the dominants of the brachiopod communities. Among mollusks, the most significant and noticeable event is the appearance of inoceramid bivalves.

The Kolyma-Omolon type of associations replaces the preceding Pechora-Kolyma type during the biotic crisis at the Early/Middle Permian transition. Its paleocenoses are confined to bacterial carbonates, forming thick bodies of limestones saturated with sulfides. This type is characterized by the highest taxonomic diversity. The foraminifera are very diverse but represented exclusively by lagenids. Among brachiopods, finely ribbed linoproductides are dominant (genera *Terrakea*, *Cancrinelloides*, and *Stepanoviella*), as well as the representatives of the family Lichareviidae. Among bivalves, the role and taxonomic diversity of inoceramid-like bivalves (the family Kolymiidae) grows. As a rule, these are large and very large (up to 70 cm) shells, suggesting endosymbiosis with bacteria, by analogy with recent *Bathymodiolus* and *Calyptogena*. The described type of associations existed throughout the rest of the Permian, completely disappearing near the Permian–Triassic boundary.

4. Analogues outside the Northeast Asia

The described features of the Late Paleozoic bio- and sedimentogenesis are not limited by the territory of Northeast Asia. Similar biotopes, as well as related biomes, are fragmentarily present along the Pamir-Himalaya belt, Inner Mongolia, and Primorsky Krai of Russia. But the most striking analogues can be found in the Southern Hemisphere, associated with sedimentary basins of Gondwana. In particular, benthic associations of the Middle Carboniferous-Early Permian of the Tepuel group in Patagonia [30] are almost identical with those of Verkhoyansk type. Fossils, similar to all dominants of the Verkhoyansk association, are known from Patagonia. Among them brachiopods of the genera *Verkhojania*, *Jakutoproductus*, *Lanipustula*, *Levipustula*, and *Costatumulus* are reported. Noteworthy is the black shale silt-clay composition of the sequence of the Tepuel group, containing several horizons of diamictites, interpreted here as a result of transportation by seawater ice.

The Perm system of New Zealand shows even more striking similarities. The thin-layer carbonate formation Wooded Peak on the South Island is built by lithified bacterial mats and consists of fetid hydrogen sulfide bacterial limestones rich in sulfides with characteristic carbonate crusts completely identical to similar rocks from Northeast Asia (**Figure 15**). Here, they are also mistakenly interpreted as clastic rocks—the result of fragmentation of bivalve mollusk shells ("Atomodesma limestone") [31, 32]. In the north of the southern island, in the Nelson region, limestones appear to be deep-sea and directly contact with ophiolite rocks (Dan Mountains and D'Urville Islands) [23]. In the flat facies of shallow water in the south of the same island, these rocks contain

Figure 15.
Limestones of the wooded peak formation (upper perm) at D'Urville Island, New Zealand, and recent bacterial formations in the Black Sea. (A-C) Limestones of the wooded peak formation. (A) Exposure of the wooded peak formation. (B) Bacterial carbonate crusts along bedding planes. (C) Thin sections in transmitted light. (D, E) comparison between recent and fossil bacterial structures: (D) modern bacterial mats on the Black Sea bottom (7) and (E) fossilized bacterial mats in the wooded peak formation.

shell benthic fossils, mostly similar with the Kolyma-Omolon association—finely ribbed linoproductides, inoceramid-like bivalve mollusks, etc.

Comparable associations of brachiopods, bivalves, and foraminifera are also characteristic of Perm Australia, especially eastern of fully occupied Australia (Bowen and Sydney basins, New England) [33]. The thick terrigenous sequence, similar to the Verkhoyansk terrigenous complex of Northeast Asia, also contains several mixtite horizons, interpreted here as glacial.

The considered materials allow us to see the similarity of some Late Paleozoic basins of Gondwana and Northeast Asia. The identical geology of these basins was the reason for the existence in these areas of peculiar ecosystems unknown in the previous eras.

5. Conclusion

The present brief overview allows to make some preliminary conclusions requiring further elaboration. The composition of the Late Paleozoic rocks in Northeast Asia, as well as the features of their fossil associations, suggests that the strongest agent which defined both sedimentation and development of life were the endogenous fluid processes. Those appeared, apparently, both as extrusions of cold fluids (seeps), confined to the system of regional faults (or fractures), and as mud-explosive activity caused by pressure gradients within the crust. This led to the saturation of marine waters by reduced compounds—methane and hydrogen sulfide. The result was the formation of the Late Paleozoic sulfide-rich sedimentary basins and their typical sulfide biota. This defines the peculiarity of the Late Paleozoic biosphere, where lithotrophic ecosystems played an important role along with "normal" phototrophic associations.

Acknowledgments

The author thanks Dr. Bruce Waterhouse, the best expert on the Permian of New Zealand, for the opportunity to visit some sections in that country. I am also deeply grateful to Dr. Arturo Caesar Taboada for organizing a tour to the outcrops of the Carboniferous and Permian in Patagonia.

Author details

Victor Gdal'evich Ganelin
Geological Institute of Russian Academy of Sciences, Moscow, Russia

*Address all correspondence to: vigdal@yandex.ru

References

[1] Ganelin VG, Kotlyar GV. Zoning and general characteristics of the Perm system in the USSR. In: The main features of the stratigraphy of the Perm system of the USSR. Leningrad: Nedra; 1984. pp. 15-21. (In Russian)

[2] Ganelin VG. Boreal benthonic biota of the late Paleozoic World Ocean. Stratigraphy and Geological Correlation. 1997;**5**(N3):231-242

[3] Lychagin PP, Lyckman VB. Late Paleozoic belt of basic volcanism. In: Erupt rocks and questions of their geological mapping in the Northeast of the USSR. Magadan: SVKNII; 1988. pp. 132-145. (In Russian)

[4] Ganelin VG. Taimyr-Kolyma subregion. In: The main features of the stratigraphy of the Perm system of the USSR. Leningrad: VSEGEI; 1984. pp. 111-142. (In Russian)

[5] Ganelin VG, Chernjak GE. Mariner basins of the Northeast Asia. In: The Carboniferous of the World. Vol. 3. The former USSR, Mongolia, Middle Eastern Platform, Afghanistan, & Iran. General Editor Carlos Martinez Diaz. Instituto Tecnologio GeoMinero de Espano. Madrid; 1996. pp. 207-234, tables 51-57

[6] Kheraskov NP. Geology and Geomorphology of the Western Verkhoyansk Proceedings of the All-Union Scientific Research Institute of Mineral Raw Materials. Materials of the Yakut Expedition. 1938;**116**(3):116. (In Russian)

[7] Ganelin VG, Solovyova NA. Some features of the late Paleozoic sedimentogenesis and ecogenesis of Northeast Asia. In: Upper Paleozoic of Russia. In: Materials of the 2 All-Russian Meeting. Kazan. KSU; 2009. pp.46-48. (In Russian)

[8] Ganelin VG, Biakov AS. The Permian biostratigraphy of the Kolyma–Omolon region, Northeast Asia. Journal of Asian Earth Sciences. 2006;**26**:225-234

[9] Ganelin VG. Brachiopods in the late Paleozoic benthic biota of Northeast Asia. In: 6th International Brachiopod Congress. Melbourne: Abstracts. 2010. pp. 43-44

[10] Ganelin VG, Byakov AS, Vedernikov IL, Pokrovsky BG, Solovyova NA. Authigenic carbonates of the Late Paleozoic basins of the North-East of Asia. In: Materials of the All-Russian lithological meeting "Reefs and Carbonate Psephytolites"; 2010. pp. 47-49, Syktyvkar. (In Russian)

[11] Lein AY, Pimenov NV, Ivanov MV. The isotopic composition of methane and products of its anaerobic microbial oxidation in the Black sea. In: Past and present water column Anoxia. Proc. NATO ARW 4-8 October, 2003. Sevastopol; 2003. pp.53-54

[12] Lein AY, Ivanov MV. The largest methane reservoir on earth. Nature. 2005;**2**:19-25. (In Russian)

[13] Byakov AS, Vedernikov IL, Akinin VV. Perm diamictites of Northeast Asia and their probable origin. In: The bulletin of the research Center of the far Eastern Branch of the Russian Academy of Sciences. Vol. 1. Magadan: SVKNI; 2010. pp. 14-24. (In Russian)

[14] Kalugin HI. Stratigraphy of Permian deposits of the western part of the Okhotsk- Kolyma watershed. In: Materials on geology and minerals north-east of the USSR. Release 13. Magadan: SVTGU; 1958. pp. 35-48. (In Russian)

[15] Ustritsky VI. The history of the development of the north-east of the USSR in the late Paleozoic. In: Upper Paleozoic of the north-east of the USSR. Leningrad: NIIGA; 1975. pp. 54-75. (In Russian)

[16] Isbell JL, Biakov AS, Vedernikov IL, Davydov VI, Gulbranson EL, Fedorchuk ND. Permian diamictites in northeastern Asia: Their significance concerning the bipolarity of the late Paleozoic ice age. Earth-Science Reviews. 2016;**154**:279-300

[17] Letnikov FA. Fluid facies of the continental lithosphere and ore formation problems. In: Bulletin of the OGGGN RAS, Irkutsk. Vol. 4(10). 1999. p. 99. (In Russian)

[18] Kasack AP, Kopylova NN, Tolmacheva EV, Jacobson KE. Fluid – Explosive Formations in Sedimentary Complexes. St. Petersburg: VSEGEI; 2008. p. 38. (In Russian)

[19] Lein AY. Authigenic carbonate formation in the ocean. Lithology and Mineral Resources Journal. 2004:3-35. (In Russian)

[20] Mazzini A, Ivanov MK, Parnell J, Stadnitskaia A, Cronin BT, Poludetkina E. Methane-related authigenic carbonates from the Black Sea: Geochemical characterization and relation to seeping fluids. Marine Geology. 2004;**212**(1-4): 153-181

[21] Reitner J, Peckmann J, et al. Concretionary methane-seep carbonates and associated microbial communities in Black Sea sediments. Palaeogeography, Palaeoclimatology, Palaeoecology. 2005;**227**:18-30

[22] Van Weering TCE. Methane-related authigenic carbonates from the Black Sea: Geochemical characterization and relation to seeping fluids. Marine Geology. 2004;**212**(1-4):153-181

[23] Lein AY. Life on hydrogen sulfide and methane. Nature. 2003;**10**:1-13

[24] Belenitskaya GA. The role of deep fluids in sedimentary rock and ore formation. In: Volcanogenic Sedimentary Ore Formation. St. Petersburg: VSEGEI; 1997. pp. 155-159. (In Russian)

[25] Yudovich YE. Fluid mineral formation - an alternative to lithogenesis? Overview. Ural Geological Journal. 2009;**4**:31-80. (In Russian)

[26] Ustritsky VI. The climate of Perm. News of the USSR Academy of Sciences. Series Geology. 1972;**4**:267-274

[27] Ustritsky VI. On the bipolarity of the Late Paleozoic fauna. Paleontological Journal. 1974;**2**:33-37

[28] Ustritskii VI, Stepanov DL. The Permian Paleobiogeography and Climate of Eurasia. In: Paleontologiya, Morskaya geologiya (Paleontology, Marine Geology). Moscow: Nauka; 1976. pp. 61-71

[29] Chumakov NM, Zharkov MA. Climate during the Permian–Triassic biosphere reorganizations. Article 2. Climate of the late Permian and early Triassic: General inferences. Stratigraphy and Geological Correlation. 2003;**11**(4):361-375

[30] Taboada AC, Pagani MA. The coupled occurrence of Cimmeriella-Jakutoproductus (Brachiopoda: Productidina) in Patagonia: Implications for Early Permian high to middle paleolatitudinal correlations and paleoclimatic reconstruction. Geologica Acta. 2010;**8**(4):513-534

[31] Waterhouse JB. Permian Brachiopods of New Zealand. New Zealand Geological Survey Paleontological Bulletin. 1964;**5**:287

[32] Waterhouse JB. Permian Stratigraphy and Faunas of New Zealand. New Zealand Geological Survey Paleontological Bulletin. 1964;**72**:101

[33] Briggs DJC. Permian Productidina and Strophalosiidina from the Sydney – Bowen Basin and New England Orogen: Systematic and Biostratigraphic Significance. In: Association of Australasian Paliontologists. Mem; 1998. p. 258

Chapter 10

Structural Differentiation and Sedimentary System of the Permian Sichuan Cratonic Basin

Haofu Zheng and Bo Liu

Abstract

The Sichuan Basin located in the western region of the Yangtze block was a stable craton basin in the Permian. The structural differentiation caused by the Dongwu movement and the Emei rifting activity controlled the sedimentary system and the Permian carbonate gas reservoirs in the Sichuan Basin. In this study, we have investigated the stratigraphic characteristics of each Permian formation, studied the depositional systems of each period of the Permian stage, and discussed the overall tectono-sedimentary evolution of the sedimentary basin. During the Permian, the Sichuan Basin experienced an intense tectonic activity, controlling the variations of the sedimentary environments occurring in the basin. The depositional systems of the basin were controlled by the tectonic setting of the intra-cratonic depression and marginal rifts during the period. Therefore, this is an important period in the tectono-sedimentary evolution of the study area, which can be divided into the following stages: (1) From the Late Carboniferous to the Early Permian, the Sichuan Basin was dominated by tectonic uplift and denudation. In the Middle Permian, a regional transgression occurred in the whole upper Yangtze region, and the sedimentary environments of the Sichuan Basin and its adjacent areas gradually changed to the carbonate platform. (2) In the early phase of the Late Permian (the Wujiaping period), being influenced by the Emei rift, the Sichuan Basin and its adjacent areas formed a complex pattern of structural highs and adjacent depressions, controlled by a differential subsidence. (3) In the late phase of the Late Permian (the Changxing period), with the cessation of the volcanic activity and the enhancement of the regional extension, the pattern of structural highs and depressions is more obvious, and the relatively calm structural environment makes the carbonate sedimentary environment tending to dominate.

Keywords: Yangtze block, Permian, structural differentiation, sedimentary system, tectono-sedimentary evolution

1. Introduction

The Yangtze block is located in South China, with the Qinling-Dabie-Sulu orogenic belt in the north and the Songpan-Ganzi fold belt in the west [1].

The Sichuan cratonic basin located in the western region of the Yangtze block has experienced multiple tectonic movements during its evolution. During the

Permian, the structural differentiation caused by the Dongwu movement and the Emei rifting activity controlled the sedimentary system [2].

Moreover, the depositional systems controlled the distribution of the Permian carbonate gas reservoirs in the area, where several new gas fields have been discovered in recent years. The platform margin reef-shoal facies of the Upper Permian Changxing Formation is one of the favorable sedimentary facies belts for the formation of favorable reservoirs in the Puguang and Yuanba gas fields [3, 4]. Besides, the high-energy granular beach sedimentary facies belt of the Middle Permian is also considered as a reasonable basis for the development of natural gas reservoirs with the discovery of several high-yield gas fields in the northwestern and central part of the Sichuan Basin [5, 6]. On the other hand, shales with high organic matter content widely developed in the Wujiaping period of Upper Permian in the Sichuan Basin are the suitable hydrocarbon-generating beds [7]. Therefore, it is of considerable significance for natural gas exploration to analyze and study the structural differentiation and sedimentary system of the Permian Sichuan cratonic basin.

Although many different opinions have been put forward on the genetic mechanisms of the structural differentiation pattern of the Sichuan Basin [2, 5, 8, 9], these opinions lack a complete analysis on the basin sedimentary filling processes. The previous study lack a detailed discussion on the relationships between the structural differentiation and the depositional processes in the basin, which is precisely the basis and key of structural sedimentary evolution in the Sichuan Basin.

In this study, we focus on the structural differentiation, on the sedimentary systems, and on the tectono-sedimentary evolution of the Sichuan cratonic basin during the Permian.

2. Geological setting

2.1 Tectonic evolution of the Yangtze block and the formation of the Sichuan Basin

The Sichuan Basin is a superimposed basin developed on the upper Yangtze craton [8] (**Figure 1**). However, the craton block has long been in the transitional position between Gondwana and Laurasia [10, 11], showing a vigorous tectonic activity. During the later stage there are many regional unconformities in the craton. The cratonic margin was involved in orogenic deformation and was strongly re-shaped. The processes of basin formation and evolution were quite complicated.

At the end of Mesoproterozoic (1000 Ma), the island arc and accreted continental crust in the margin of the Yangtze Paleocontinent were spliced onto the Yangtze block. The Yangtze, Cathaysia, and North China blocks were combined to form a part of Rodinia ancient land [11, 12]. Since 850 Ma, it has experienced (1) the early cracking, the formation of rifted sag, passive continental margin basin, and composite basin of intra-cratonic depression in 850–460.9 Ma (Nh-O_2), and (2) the late convergence, the establishment of intra-continental foreland basin and large-scale tectonic uplift in 460.9–416 Ma (O_3-S) [12]. In the Tethys Ocean evolutionary stage from Sinian to Silurian, the inner plate tension between the Yangtze block and the Cathaysia block resulted in the formation of the Xianggui continental rift basin and in the internal and marginal rifting of the middle and upper Yangtze craton. Then the compression orogeny in the Caledonian stage resulted in the formation of the South China continent. In the Late Paleozoic, it turned into the Paleo-Tethys Ocean evolution stage.

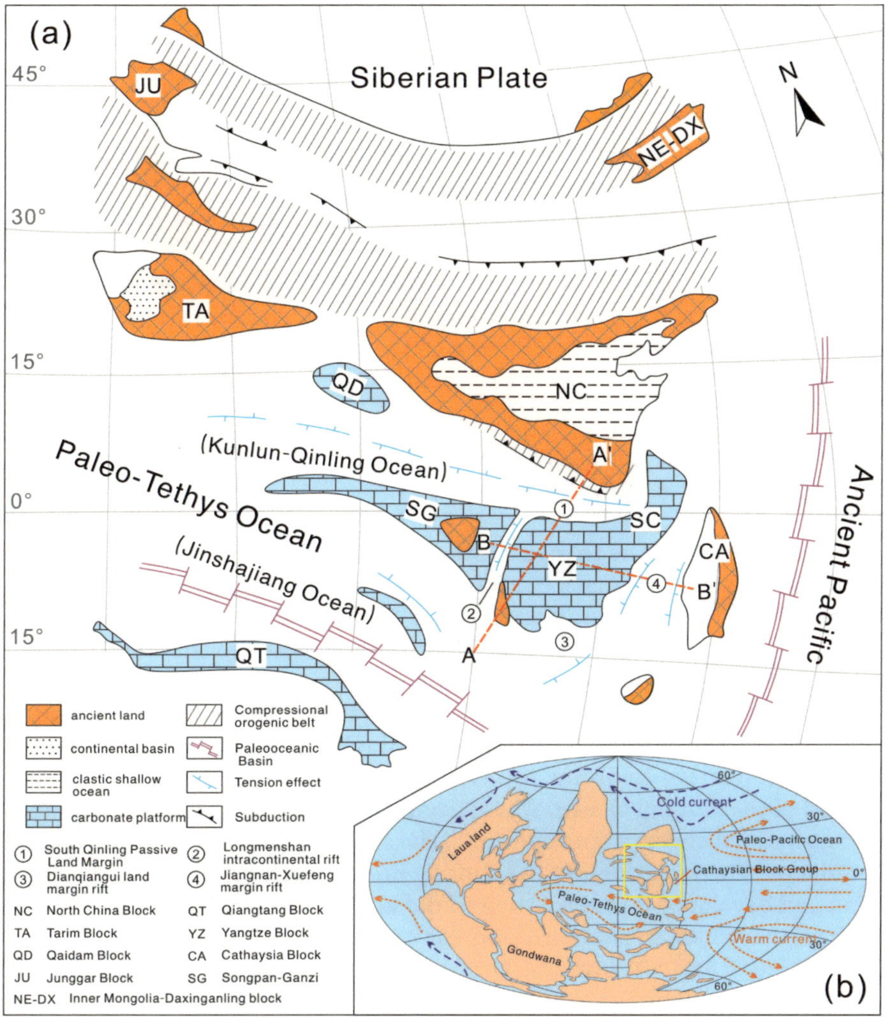

Figure 1.
(a) Location of the Yangtze block and its relationships with other tectonic units in Permian; (b) the restoration of Permian global paleogeography (modified from Huang et al. [45]).

From the Late Paleozoic to the Middle Triassic, the formation and evolution of the Sichuan Basin was genetically related to the geological process of South China block splitting and drifting from the northern margin of the Gondwana continent. The southwestern and northern margin of the Yangtze block became passive continental margins facing on different branches of the Paleo-Tethys Ocean. The middle and upper Yangtze block experienced a relatively short convergence and compression process from Devonian to Permian, forming the inner depression, the marginal depression, the passive continental margin, and the foreland basin of the middle and upper Yangtze craton [12–15].

After the development of the foreland basin in the Late Triassic, the middle and upper Yangtze were involved in an extensional and convergent cycle, closely related to the development of the Neo-Tethys Ocean, which included (1) the short-term extension from the Early Jurassic to the Early-Middle Jurassic (199.6–167.7 Ma) and (2) the long-term compression and transformation from the Late-Middle Jurassic to the Quaternary (167.7–0 Ma) [12, 16].

Above all, the basin prototypes of the Sichuan Basin were controlled by the activities of the surrounding plates in different phases. The basement of the basin was formed in the Pre-Nanhua Period; the rift basin was established in the Nanhua Period; the cratonic margin rift and the cratonic inner depression were created in the Sinian-Ordovician; the cratonic inner depression and the peripheral foreland basin were developed in the Silurian; the cratonic margin rift and the cratonic inner depression were developed from the Devonian to the depositional period of the third member of the Xujiahe Formation in the Late Triassic (D-T_3x^3); the foreland basin was established in the depositional period of the upper Xujiahe Formation (T_3x^{4-6}); the large-scale cratonic depression was developed in the Early to the Middle Jurassic; the compressional basin was developed in the Late Jurassic-Early Cretaceous; the depression basin was established from the Late Jurassic to the Early Cretaceous; and from the Late Cretaceous to the Quaternary, a transitional compressional foreland basin was formed. As a whole, the Sichuan Basin shows the characteristics of an alternating development of extensional basin (Z-O, D-P-T_3x^3, J_{1-2}) and compressional basin (S, T_3x^{4-6}, J_3-K_1) [12].

2.2 The tectonic stages of the Permian

In the Early Permian the striped continents composed of the Cimmerian block, Qiangtang block, and Sibumasu block were separated from the northern margin of Gondwana. Then the back part was extended to form the new Tethys Ocean. In the Middle Permian the transgression reached its maximum in South China, leading to the development of a wide southward-dipping carbonate platform. At that time, the Jiangnan-Xuefeng area was a submerged structural high and was separated by the southern and northern sedimentary regions. On the north side of the Yangtze block, the sedimentary environments were the shallow slope and deepwater basin southward of the Qinling Ocean, where deepwater dark limestones with nodular cherts and bedded cherts were deposited. The southern Qinling Ocean crust was subducted northward under the Qinling micro-block at the end of the Early Permian and developed corresponding island arc volcanic rocks, while the passive continental margin on its southern side was still growing (**Figure 1**).

In the Sichuan Basin, the Middle Permian strata include the Liangshan Formation (P_2l), the Qixia Formation (P_2q), and the Maokou Formation (P_2m), with a thickness of 400–500 m. In the early transgression stage of the Middle Permian sandstones, mudstones, marls, and marshes were deposited. In the middle period, the deposits evolved to shallow platform limestones and shaly limestones interlayered with sandy limestones. Massive limestones, dolomites, and black shales developed in the late Middle Permian, during which Leshan-Luzhou biological shoals grew.

At the end of the Middle Permian characterized by the Emei rift movement, the region was uplifted, and the Maokou strata were denuded in different degrees. The event of the Emei rift may be related to the subduction of the Jinshajiang-Mojiang ocean basin from the south to the north, and a large-scale extension occurred in the back of the arc (**Figure 1**). Longmenshan, Kaijiang-Liangping, and Chengkou-Exi rifts developed in the Yangtze craton block [12, 17].

3. Structural differentiation of the Sichuan cratonic basin during the Permian

3.1 Tectonic characteristics of plate margin rift

During the Permian, the Yangtze block was generally located in the low-latitude area near the equator [18, 19], in the transitional position between the

Gondwana and the Laurasia supercontinents. From the Early Permian to the Late Permian, the Yangtze platform was surrounded by the Paleo-Tethys Ocean and the Paleo-Pacific Ocean [20]. It rotated mainly anticlockwisely and formed the South China block, together with the Cathaysia block in the southeast (**Figure 2**). In the Late Permian the North China block was located northward of the South China block.

The paleomagnetic data showed that both the North China block and the South China block had the trend of northward movement and their latitudinal changes showed a certain degree of synchronicity, which may be related to the first collision of the two blocks in the east [21]. The South Qinling Ocean was located between the two blocks, opening toward the west in the form of scissors, with an angle of 70–80° [21]. Because of the continuous northward subduction, the North Qinling orogenic belt was formed at the southern margin of the North China block, while a passive continental margin emplaced at the north margin of the Yangtze block [22].

The southwest margin of the South China block was Simao-Indosinian block, which was located in the low-latitude area near the equator [23]. Between the two blocks was Jinshajiang Ocean, a branch of the Paleo-Tethys Ocean. The ocean basin rapidly expanded from the Early Permian to the Late Permian [24]. The subduction of the Jinshajiang Ocean to the Simao-Indosinian block in the southwest reflected a strong tectonic compression, while the South China block was a passive continental margin in the overall extensional setting [25]. The South China block was connected with the Songpan-Ganzi Ocean in the west, while the Jiangnan-Xuefeng intra-continental rifting zone was connected with Youjiang rifting zone, mainly in deepwater shelf environment, with isolated carbonate platform sporadically developed [26]. During the Permian, the tectonic setting of South China block was mainly of regional extension, showing the characteristics of a strong tension at the margins of the block.

3.2 Characteristics of block internal structure differentiation

In the Permian period, the tectonic setting of the South China block was mainly of regional extension, showing the characteristics of strong extension at the margin of the block and of a weak tension within the block and developing a set of marginal rift basins, intra-continental depression basins, and rift basins. In the Kangtien ancient land area, located on the southwestern margin of the block,

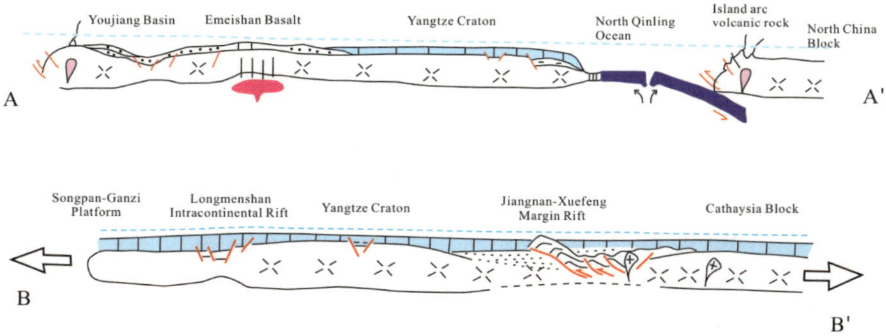

Figure 2.
Structural differentiation profile of the middle and upper Yangtze plate in the Permian (A-A', NW-SE direction; B-B', NE–SW direction, location in Figure 1(a); modified from He et al. [12]).

a basalt eruption was considered to be a stratigraphic marker of the extensional process. The exposed area of basalt could reach 250,000 square kilometers, with a wide range of influence [9, 27]. During the Permian, the upper Yangtze block had the characteristics of a structural high to the south and of a depression to the north. The Kangtien ancient land was the main source area supplying terrigenous clastic in the upper Yangtze craton [28, 29]. To the north, it was dominated by the shallow sea carbonate environment and graded into the passive continental margin basinal environment.

The Sichuan Basin is a typical cratonic basin located in the western margin of the Yangtze block, which has recorded many geological events during the Permian. Several geological events occurred in the Permian, including the Emeishan Large Igneous Province [30–32], the Paleo-Tethys Ocean expansion and evolution [19, 33–35], and the end of the Permian biological extinction, have been widely of concern by scholars [36–39]. During recent years, according to the field outcrop data and to the stratigraphic records of drilling data, it was found that from the southwest to the northeast of the basin, the Sichuan Basin has deposited a wide range of marine carbonate rocks during the Permian. However, in the central and northern sectors of the basin, there were many deepwater sedimentary areas toward the carbonate platform margins. This unique sedimentary filling pattern reflects that the Sichuan Basin had the unique structural differentiation characteristics during the Permian [40].

4. Sedimentary system of the Permian

Permian strata widely developed in the eastern sectors of the Sichuan Basin during the early depositional stages, when carbonate rocks are dominant. During late stages, the difference between the eastern and the western sectors increased. In the Panxi area, the continental basic volcanic rocks were dominant, while the eastward transition was represented by continental clastic rocks. On the contrary, in the eastern Sichuan Basin, thick marine carbonate rocks were dominant [41]. In the Sichuan Basin, the Permian strata can be divided into the Middle Permian Liangshan Formation, Qixia Formation, Maokou Formation, and Upper Permian Wujiaping Formation and Changxing Formation (**Figure 3**).

4.1 Stratigraphic characteristics of the Liangshan formation

In the late Early Permian, the tectonic movement uplifted the northwestern region of the upper Yangtze to land, controlling variable degrees of denudation of the Paleozoic strata [42]. Until the Middle Permian, transgression occurred from south to north, and the Liangshan Formation was the product of the land-sea transformation at the beginning of this transgression [41, 43, 44]. The Liangshan Formation, widely developed in the Sichuan Basin and its adjacent areas, which is a coal-bearing deposit dominated by clastic rocks, and the contact between Liangshan Formation and underlying strata (e.g., Carboniferous or older layers) are disconformable [41, 43–45].

In the Sichuan Basin and Panxi areas, the lithology and the thickness of the Liangshan Formation considerably vary [41, 44]: (1) The sandstone content of the Liangshan Formation deposited in the western region is relatively large, with a thickness generally ranging from 10 to 42 m and finally reaching 88 m (e.g., Gan Luo). (2) The sedimentary thickness of the Liangshan Formation rapidly decreases

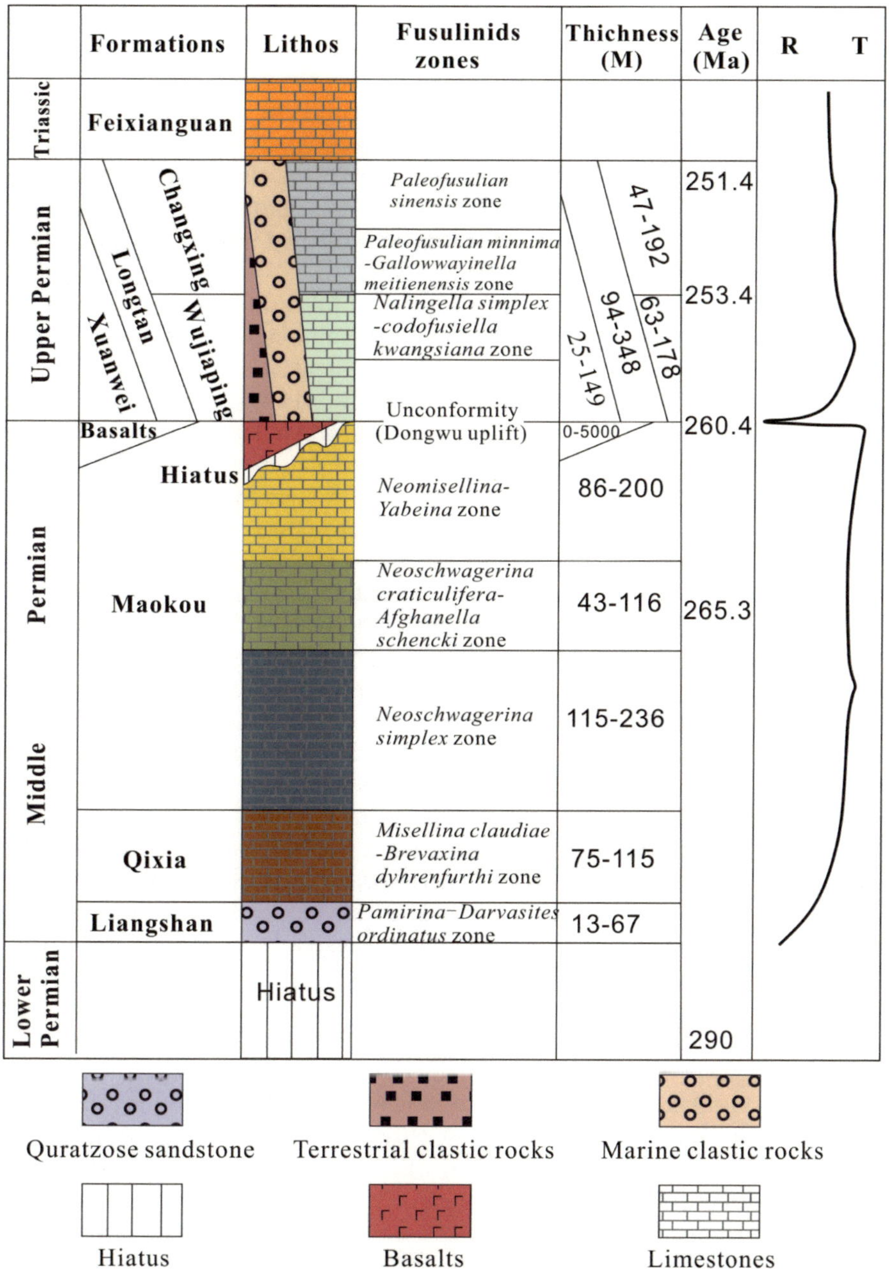

Figure 3.
Generalized Permian stratigraphy and division of fusulinid zones in SW China (after He et al. [54]).

eastward to 5–15 m in the area of Emeishan and Leshan, which is often dominated by carbonaceous shales. (3) The Liangshan Formation in the south of Sichuan is 4–17 m thick and is mainly composed of carbonaceous shale and clay rock, containing bauxite and hematite. (4) The Liangshan Formation in eastern Sichuan is dominated by coal-bearing claystone and sandstone and occasionally contains oolitic and bean-shaped hematite, with a thickness of 4–8 m, reaching 21 m. (5) The Liangshan Formation in the north of Sichuan and Longmenshan is 3–30 m thick, thinning eastward, and is mainly composed of aluminum clay rocks, bauxite, and weak coal seams. The Liangshan Formation contains plant fossils (e.g., *Lepidodendron,*

Problemocumnum wongii, Taeniopteris multinervis), brachiopods (e.g., *Orthotichia indica*), and bryozoans [44].

4.2 Stratigraphic characteristics of the Qixia formation

During the deposition period of the Qixia Formation, the crustal subsidence was stable and seawater intruded on a large scale. As a result, the early sedimentary environment dominated by clastic rocks was transformed into carbonate platform sedimentary environment [41, 45].

Qixia Formation is widely distributed in the middle and eastern sectors of the Sichuan Basin. It is mainly composed of dark gray-black limestone, with a massive and micrite structure, locally mixed with bioclastic limestone, siliceous limestone, siliceous bands, and siliceous concretion [41, 44]. The limestones of the Qixia Formation generally contain high asphaltene and siliceous components and show dolomitization, and abundant eyeball-shaped structures (e.g., Huayingshan area) occur locally [46]. The Qixia Formation is interlayered above the Liangshan Formation, and its stratigraphic thickness ranges from tens of meters to more than 300 meters, gradually thickening from west to east [41].

According to the observation results of the outcrop in the wild, the Qixia Formation can be divided into two types [44]. One is called "White Qixia," which is distributed in the northern section of Micangshan and Longmenshan. It is mainly composed of light gray-black limestone with dolomitic limestone and dolomite and with shale at the bottom. The other is called "Black Qixia," which is distributed in other areas of the Sichuan Basin, with shale and siliceous layer at the bottom, dark gray thick layer of biological limestone, micritic shell limestone in the lower part, and light gray biological limestone in the upper part.

The Qixia Formation contains many types of fossils, mainly including fusulinids (e.g., *Nankinella orbicularia*, *N. nankingensis*, *N. discoides*, *N. regularis*, *Pisolina excessa*, *Schwagerina tshernyschewi*), corals (e.g., *Hayasakaia yunnanensis*, *Wentzellophyllum denticulatum, Polythecalis chinensis*), brachiopods, and conodonts [41, 44].

4.3 Stratigraphic characteristics of the Maokou formation

In the Sichuan Basin, the lithology of the Maokou Formation is relatively uniform with shallow marine, light gray, thick micritic fossiliferous limestone, including siliceous concretions and thin siliceous layers, ranging in thickness from 50 m to 600 m. Due to the influence of the Dongwu tectonic movement, the Maokou Formation was involved by various degrees of erosion, and the integrity of the strata gradually improved from west to east. In the southern part of the Sichuan Basin, basalts erupted in the middle sedimentary period of the Maokou Formation. The Maokou Formation can be divided into two members in the northeastern Sichuan Basin and into three members in the central Sichuan Basin.

The lower member of the Maokou Formation is composed of dark gray muddy micritic limestones, bioclastic limestones with black calcareous shales, and a thin siliceous layer at the top. Microbial rocks and storm rocks can be seen in southern Sichuan, eastern Sichuan, and Longmenshan areas. The Maokou Formation contains brachiopods (e.g., *Cryptospirifer omeishanensis*, *C. striatus*) and fusulinids (e.g., *Schwagerina quasibrevipola*, *S. declinata*, *Chusenella sinensis*, *Neoschwagerina*, *Pseudodolina*).

The middle member of Maokou Formation consists of light gray and dark gray thick layer massive micritic bioclastic limestone and micritic limestones with siliceous concretions. In the northwest and south of Sichuan, tempestite is relatively developed. There are abundant organisms, including fusulinids (e.g., *Neoschwagerina craticulifera*, *N. colaniae*, *N. sphaerica*, *Verbeekina heimi*, *Pseudodoliolina ozawai*, *Chusenella conicocylindrica*), corals (e.g., *Wentzelella elegans*), brachiopods, and conodonts.

The upper member of the Maokou Formation consists of gray-white micritic limestone with siliceous concretion, gray-black micritic limestone, and bioclastic limestone, including fusulinids (e.g., *Yabeina*, *Neomisellina*) and ammonoids (e.g., *Altudoceras*, *Paraceltites*, *Shouchangoceras*).

The Maokou Formation in Dabashan, Wushan, and southeastern Sichuan only remains lower and middle members. Parts of the lower, middle, and upper members of the Maokou Formation are preserved in Micangshan, Longmenshan, and Huayingshan. The Maokou Formation is well preserved in southern Sichuan (such as Gongxian).

4.4 Stratigraphic characteristics of the Wujiaping formation

The Wujiaping Formation is mainly distributed in the northeastern Sichuan Basin and can be subdivided into two members according to lithology differences. The lower member (formerly known as Wangpo shale) is a coal-bearing stratum at the intersection of land and sea. Its lithology is an aluminous clay rock, carbonaceous shales with coal seam or coal lines, oolitic hematite, and monohydrallite.

The upper member of the Wujiaping Formation (limestone section) has little change in lithology, which is micritic limestone, limestone with calcareous, siliceous, carbonaceous shale, and coal lines, with a siliceous layer at the top. From west to east, the dolomite content of deposits increased. In Mianzhu and Youyang, there are thin micritic limestone, limestone with shale, and multilayer coals. To the west of the line of Mianzhu-Daxian-Nanchuan-Gulin, Wujiaping Formation gradually changed into Longtan Formation.

In the Sichuan Basin, Wujiaping Formation consists of fusulinids (e.g., *Codonofusiella*), brachiopods (e.g., *Dictyoclostus*), and corals (e.g., *Waagenophyllum*).

4.5 Stratigraphic characteristics of the Changxing formation

As a lithostratigraphic unit at the top of the Permian, the Changxing Formation usually refers to the carbonate formation of platform facies sedimentary under the Lower Triassic in the Sichuan Basin, which is roughly equivalent to the sedimentary period of the Dalong Formation, and the difference between them is the sedimentary environment [47].

The Changxing Formation is mainly composed of shallow water carbonate, while the Dalong Formation primarily consists of deepwater siliceous rock and shale. Therefore, the sedimentary area of the Dalong Formation is also called "siliciclastic rock basin" [48].

The Dalong Formation is defined as the layer dominated by black and gray-black thin-layer siliceous rocks and siliceous shales in Sichuan and Chongqing area, with relatively stable sedimentary thickness, generally 15–42 m. The Dalong Formation is interlayered with the underlying Wujiaping Formation and the overlying Feixianguan Formation, and there is no apparent stratigraphic division between them [41]. In the Sichuan Basin, the distribution of the Dalong Formation is strictly controlled by the paleogeographic pattern, which is distributed in the north deep-water trough or basin facies area, roughly along the line of Guangyuan- Wangcang-Chengkou-Wushan. The lithology of the Dalong Formation is dominated by siliceous rock, siliceous shales, and siliceous limestones with tuff, mudstone, shales, and siltstones, and the siliceous composition is gradually reduced from the bottom to the top. There are abundant fossils, mainly including fusulinids (e.g., *Palaeofusulina*, *Codonofusiella*), ammonoids (e.g., *Pseudotirolites*, *Pseudogastrioceras*), brachiopods (e.g., *Spinomarginifera*), conodonts (e.g., *Clarkina changxingensis*, *C. meishanensis*), and radiolarians (e.g., *Neoalbaillella*, *Albaillella*).

The lithology of the Changxing Formation is mainly composed of medium-thick bioclastic limestones, micritic limestones, reef limestone, and dolomites, containing siliceous bands and concretions. The sedimentary thickness of the Changxing Formation varies from tens of meters to more than 100 m, and in some places, it can be as thick as 200–300 m. The fossils of the Changxing Formation are extremely rich, including algae (e.g., *Tubiphytes*, *Archaeolithoporella, Permocalculus*), fusulinids (e.g., *Palaeofusulina*, *Codonofusiella*), foraminifers (e.g., *Nodosaria*, *Colaniella*, *Pseudoglandulina*, *Pachyphloia*, *Geinitzina*), brachiopods (e.g., *Oldhamina*, *Enteletina*, *Orthothetina*, *Leptodus*), gastropods, bivalves (e.g., *Aviculopecten*), coral (e.g., *Lophophyllidium*, *Plerophyllum*, *Waagenophyllum*, *Huayunophyllum*), bryozoon (e.g., *Fistulipora*, *Fenestella*, *Polypora*), sponge (e.g., *Sphinctozoa*, *Inoza*, *Sclerospongiae*), trilobite (e.g., *Pseudophilipsia*), and conodonts (e.g., *Clarkina changxingensis, C. meishanensis, C. yini*).

Coral reefs and sponge reefs in the Changxing Formation is very well developed. Due to the tectonic control of faults within the platform, the shallow water area around the trough is deposited to form a platform margin reef. The most developed reef is located eastward of Sichuan, such as Wujiti, Huanglongchang, Damaoping, Gaofeng, and Laolongdong. The high-quality reservoir formed by the Changxing reef also provides favorable conditions for the formation of a reef gas reservoir [4].

5. Tectono-sedimentary evolution

After the small-scale transgression in Late Carboniferous, the Sichuan Basin experienced tectonic uplift in the Early Permian, controlling a wide stratigraphic gap, as shown by the lacking of Lower Permian strata in the basin filling. At the beginning of the Middle Permian, a new transgression occurred in the Sichuan Basin, which extended to the west of Hunan. As a result, the pre-existing ancient land around the basin was submerged by seawater. Only the Kangtien ancient land in the southwest of the basin and the Xuefeng ancient land in the east of the basin remained, which controlled the source supply in the basin and around the basin [41].

5.1 Liangshan period

During the Middle Permian, the Liangshan Formation, the Sichuan Basin, and its surrounding areas were generally characterized by clastic shore deposits, as controlled by the relatively shallow water body and by the high siliciclastic supply [45]. In the area around the paleo-uplift, the sediments were mainly medium-grained or fine-grained quartz sandstone, which represented the near-source coastal sedimentary environment. In the central Sichuan area, due to its high structural position, the lithology was mainly composed of sandy carbonate sediments, which showed a sedimentary environment of sand flat. In other areas of the basin, due to the relatively deepwater and lower water energy, the sedimentary environment was mainly muddy flat (**Figure 4**).

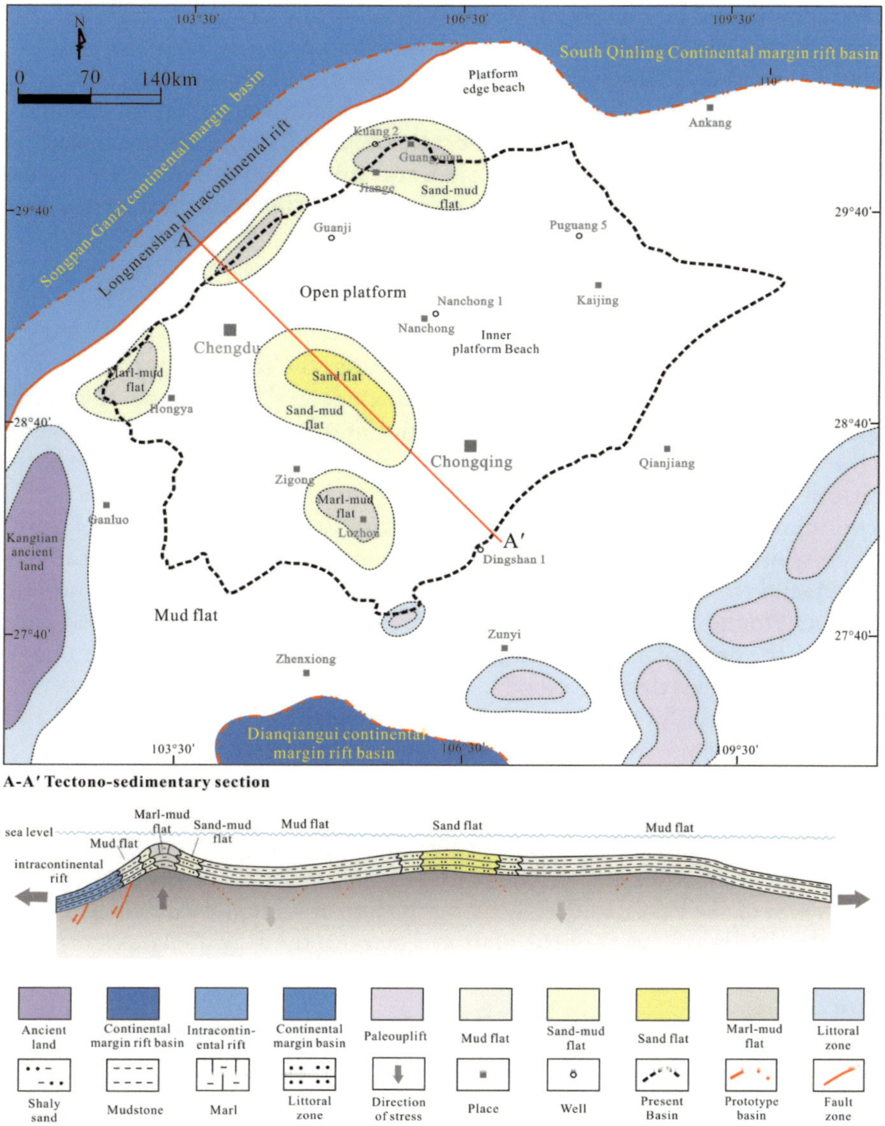

Figure 4.
Tectono-sedimentary system of the Liangshan period in the Sichuan Basin and its adjacent areas (modified from Huang et al. [45]).

5.2 Qixia period

During the deposition of the Qixia Formation, the sea level gradually decreased. At this time, the overall topography of the basin was high in the west and low in the east. With the relative shallowness of the water depth, as well as the subsidence of the depression in the basin and the extensional rifting of the basin margin, the sedimentary environment changed accordingly [45].

In the periphery of the Kangtien ancient land, it was still shoring facies. In the southwest of the basin, the lithology was mainly characterized by dolomites, dolomitic limestones, sandy limestones, and marls, with a low biological content, showing a typically restricted platform facies deposition. To the east of the restricted platform was an open platform facies area with good circulation of seawater, dominated by micrites and locally developed bioclastic limestones. The platform margin was established in the northwestern Sichuan Basin, roughly along the line of Mianzhu-Guangyuan. The sediments of the platform margin facies were mainly composed of thick bioclastic limestones, which were rich in species and high in abundance, and dolomitization locally occurred. In the northern part of the Sichuan Basin, influenced by Longmenshan ancient fault in the west, the terrain was sharply reduced, and the sediments were mainly mudstone, limestones, and shales, representing slope deposits (**Figure 5**).

5.3 Maokou period

During the deposition of the Maokou Formation, the southwestern margin of the upper Yangtze block was the Kangtien ancient land, and the northeastern margin was the passive continental margin environment, which was connected with the South Qinling continental margin basin. The northern part of the basin was mainly extensional. In contrast, the southern part of the Kangtien ancient land was continuously uplifting, and the basin was distributed in the pattern of uplift and depression [49]. From the southwest to the northeast, the seawater gradually deepened, and the sedimentary environment slowly changed from the continental environment to the marine one. The terrigenous shore facies, the restricted platform facies, the open platform facies, the platform margin facies, and the slope deepwater shelf facies then developed.

The restricted platform facies is distributed in Leshan-Emeishan-Zigong and other places in the southwest of the basin, and bioclastic beach facies is locally developed. The open platform facies area is located in the vast area to the east of the restricted platform facies area, where bioclastic banks are widely deposited in the higher ground, and micrite representing a low-energy environment was widely deposited in other places. The platform margin shoals were developed in the northwest of the basin, generally along the line of Qionglai-Anxian-Jiangyou-Guangyuan. In the area west of the platform margin, the seawater was steeply deepened, and the sedimentary environment also changed into slope deep shelf (**Figure 6**).

5.4 Wujiaping period

In the Late Permian the Sichuan Basin was located eastward of the Paleo-Tethys Ocean, with a high relief in the southwest and a low relief in the northeast. During the deposition of the Wujiaping Formation, the seawater transgressed from the northeast of the basin, and the provenance was mainly from the Kangtien ancient land [41, 50]. From the southwest to the northeast, due to the deepening of the ocean, the Sichuan Basin and its surrounding areas successively developed terrigenous shore facies, restricted platform facies, open platform facies, platform margin facies, and slope facies.

Shoreline facies extended from the southwest to the southeast of the basin, being composed mainly of coal-bearing terrigenous clastic deposits. In the area of Chengdu, Suining, Guangan, Chongqing, Nanchuan, and Zunyi, the poor circulation of seawater controlled the deposition of restricted platform facies. The area located eastward of the restricted platform was an open platform facies, where micrite and bioclastic limestones deposited. In the north and east Sichuan, a series of bioclastic shoals were developed along the line of Guangyuan-Dazhou-Wanyuan-Shizhu, which together form the platform margin. On the west side of the range of Guangyuan-Dazhou-Wanyuan-Shizhu,

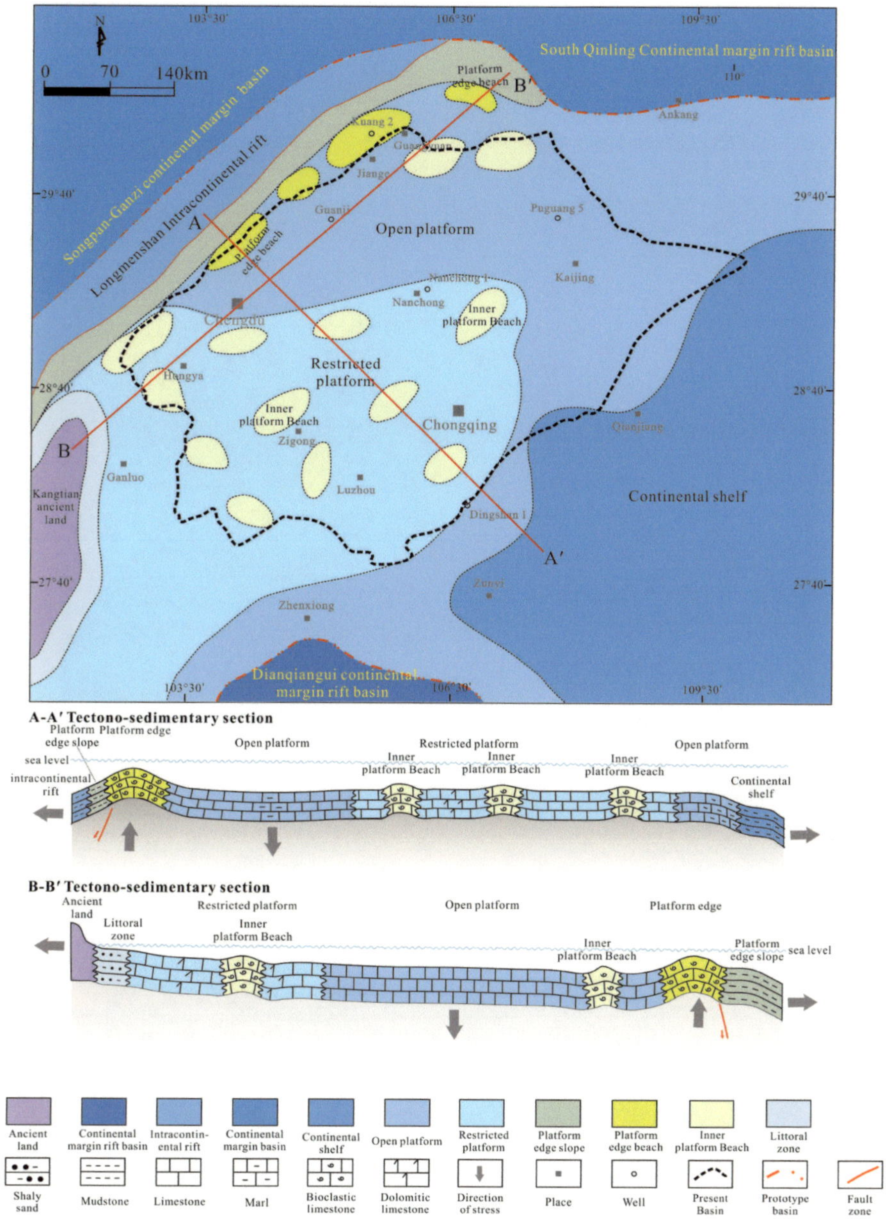

Figure 5.
Tectono-sedimentary system of the Qixia period in the Sichuan Basin and its adjacent areas.

there were slope and deepwater shelf sedimentary environments, and the sediments are mainly mudstone and siliceous rock (**Figure 7**).

5.5 Changxing period

During the deposition of the Upper Permian Changxing Formation, the paleogeographic pattern of the Sichuan Basin and its surrounding areas was controlled by the Longmenshan-Kangdian ancient land in the west [51]. Its structural lithofacies paleogeographic pattern features can be summarized as strong tectonic activity (e.g., "Emei ground fissure movement" [27]), obvious lithofacies

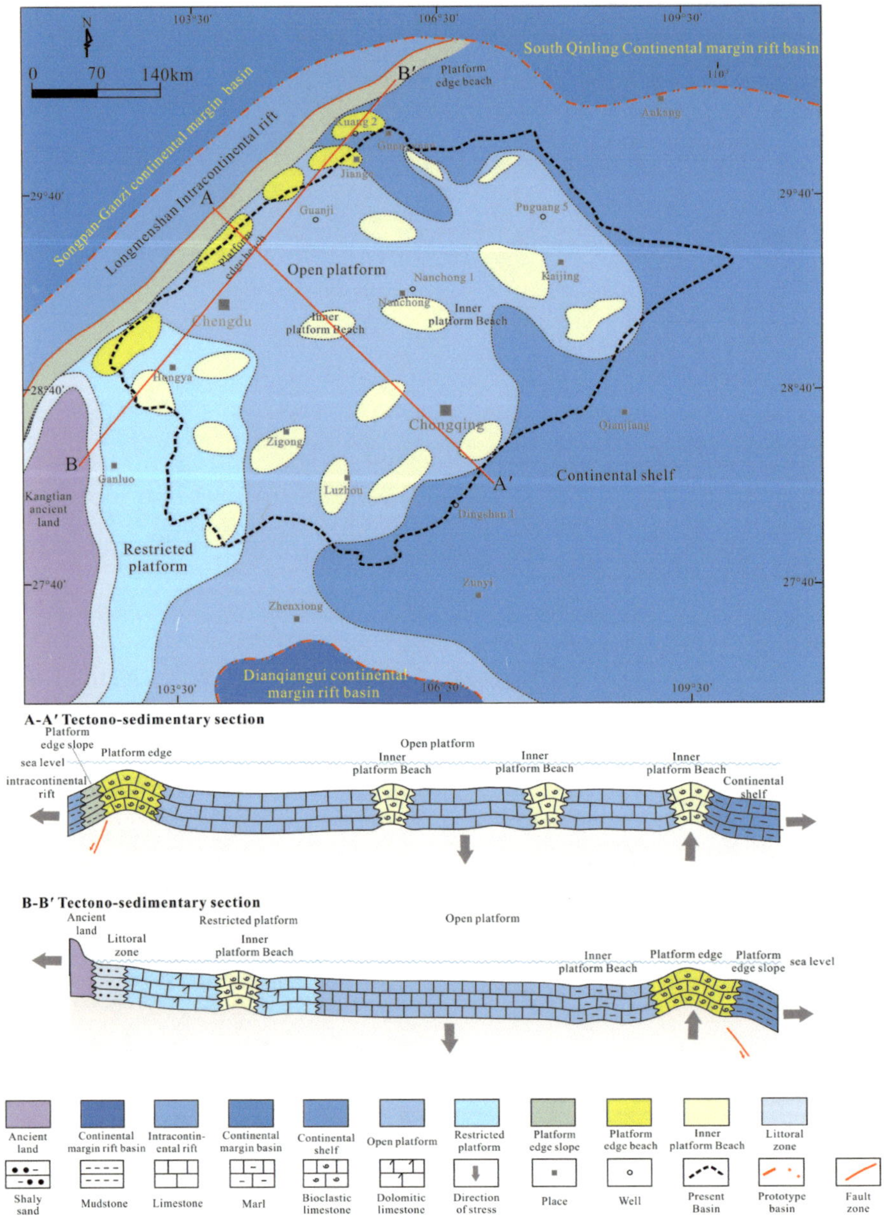

Figure 6.
Tectono-sedimentary system of the Maokou period in the Sichuan Basin and its adjacent areas.

differentiation (e.g., siliceous clastic lithofacies and shallow marine carbonate lithofacies), and relatively rich and complex paleogeomorphic types (e.g., trough and uplift) [50].

During the Late Permian, the eruption of Emei basalts reached its peak due to the strong ground fissure movement, which was called the "Emei ground fissure movement" [27]. This volcano-tectonic activity had a profound impact on the paleogeographic pattern of the Late Permian and Early Triassic in the upper Yangtze region. Wang et al. believed that a series of the northwest and southeast deepwater troughs developed in the Late Permian in the upper Yangtze region, such as "Guangyuan-Wangcang trough," "Kaijiang-Liangping trough" in the west, and "Chengkou-Exi trough" in the east. The formation of these

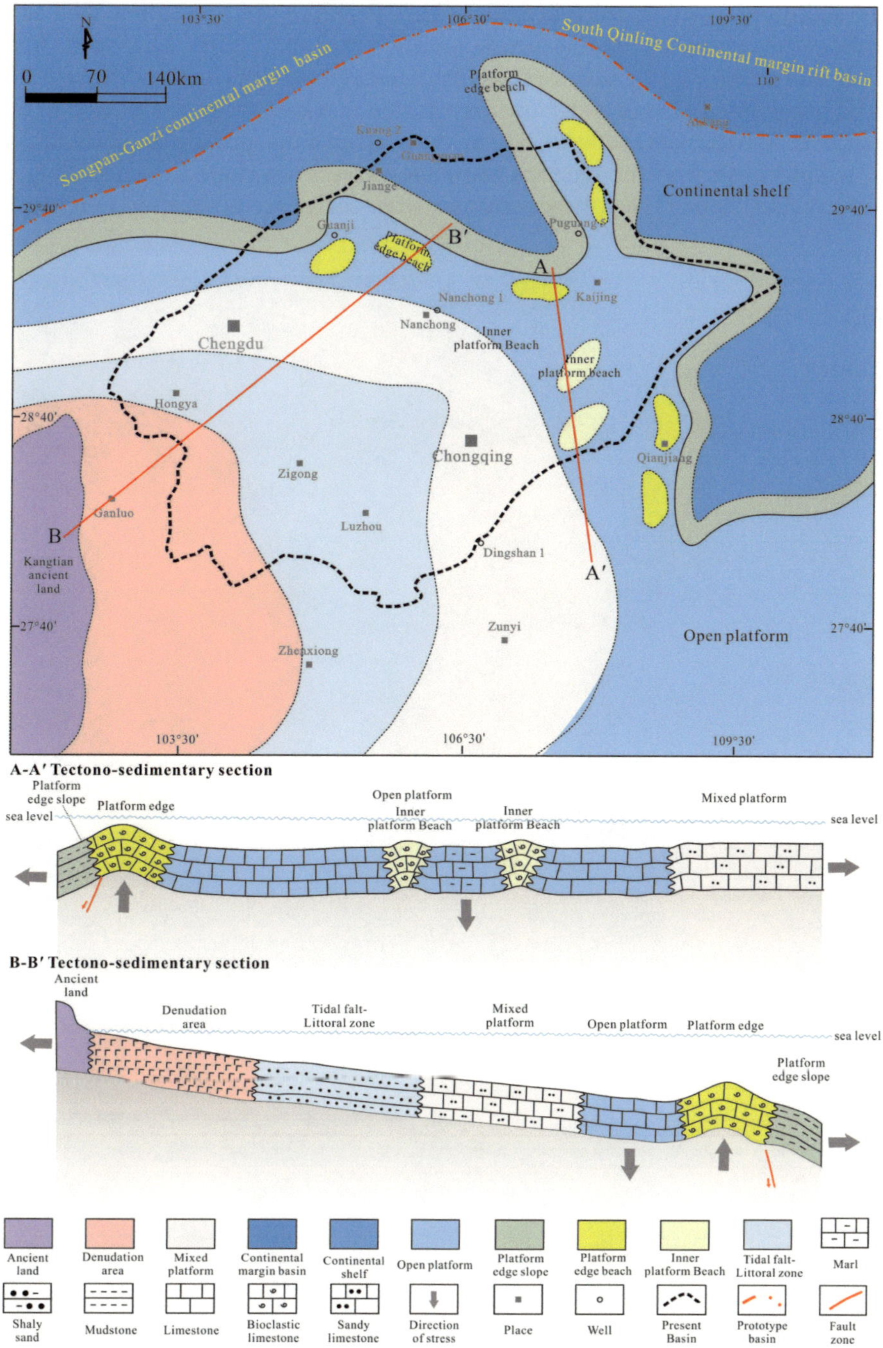

Figure 7.
Tectono-sedimentary system of the Wujiaping period in the Sichuan Basin and its adjacent areas.

troughs was controlled by the extensional movement of the southern Qinling Ocean in the north and the "Emei rift movement" in the west [27, 52]. In the Late Permian, the sedimentary facies belts in the upper Yangtze area were obviously different. The dark thin siliceous rocks, siliceous mudstones, and siliceous limestones were mainly deposited in the deepwater trough facies area, and the bioclastic limestones and reef limestones were mainly deposited in the shallow water platform facies area [53].

The sedimentary characteristics of the Late Permian can be summarized as follows: the sedimentary facies belt was generally distributed in the east–west direction and was controlled by the north–south direction structure and differentiated; reef and beach deposits are generally distributed in a belt along the trough to form the platform margin. On the west side of the platform margin, open platform, restricted platform, and terrigenous shore facies were developed, respectively, and on the east side, slope facies and deepwater shelf facies are developed (**Figure 8**).

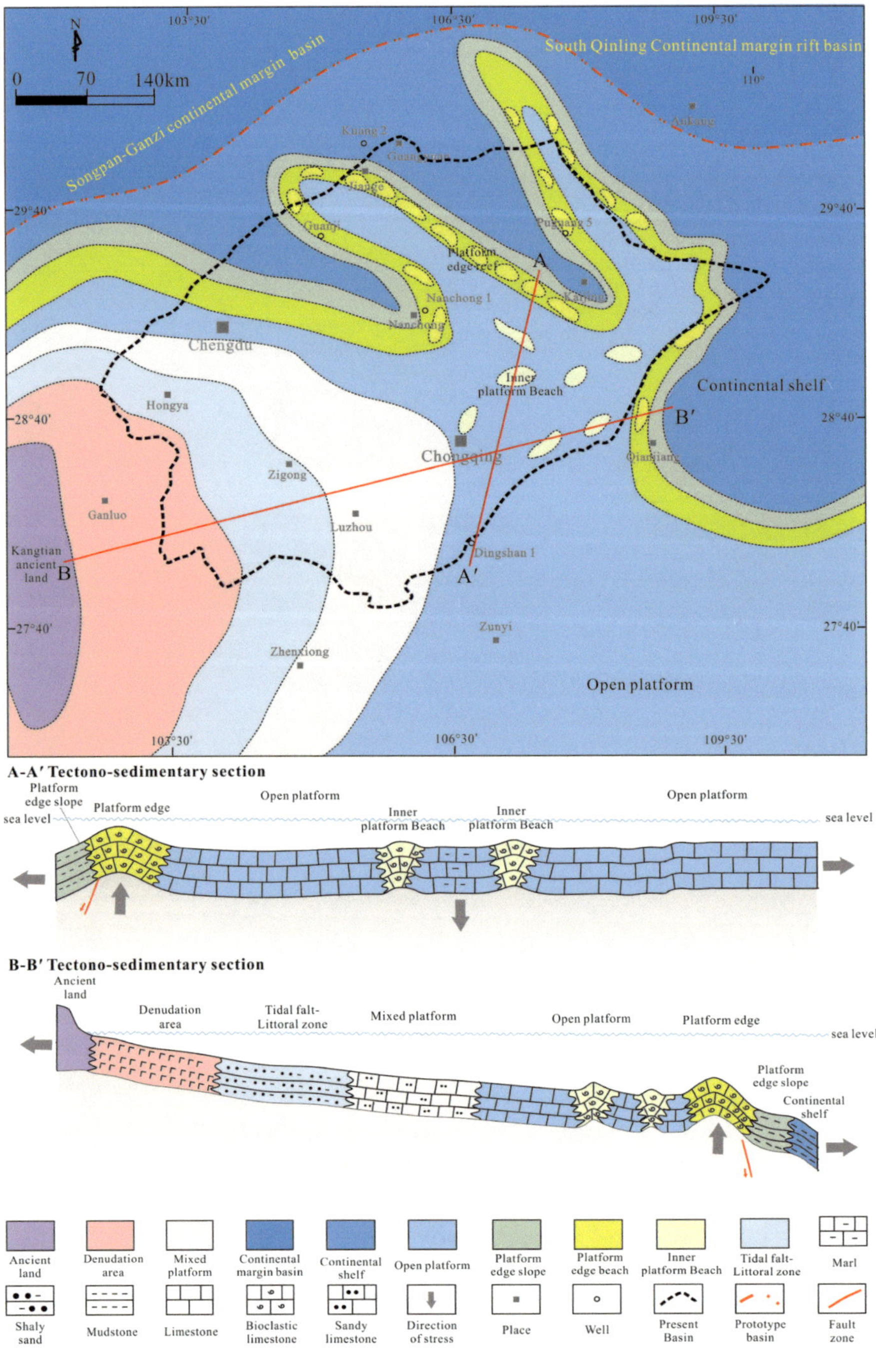

Figure 8.
Tectono-sedimentary system of the Changxing period in the Sichuan Basin and its adjacent areas.

6. Conclusions

The northern and western margins of the Yangtze block were passive continental marginal environments, while the southern and southeastern margins were continental marginal rift basins with deepwater settings.

During the deposition of the Liangshan and Qixia formations, the northern and western margins of Sichuan cratonic basin were discrete passive continental margin environments, while the south and southeast margins were continental margin rift basins with deepwater sedimentary characteristics. The sedimentary environment in this period gradually changed from the early shore tidal flat to the stable carbonate platform, which was generally characterized by east–west differentiation and north–south gradual change.

The tectonic setting of the basin was relatively stable in the Maokou period. The Kangtien ancient land in the south of the basin showed a trend of continuous uplift. The basin presented a pattern of alternating structural highs and depressions. The development of sedimentary facies belts in each sedimentary stage had obvious inheritance and migration. From southwest to northeast, the sedimentary facies included shore facies, limited platform facies, open platform facies, continental shelf facies, platform margin facies, and slope facies. These are the characteristics of a carbonate platform system.

Being influenced by the regional extension controlled by the Emei rifting, the tectonic subsidence of the basin in the Late Permian gradually increased from southwest to northeast. In the early stage, the basin was controlled by the thermal effect of deep materials, and then the northern part of the basin was mainly affected by the superimposition of regional extensional phases and finally formed the sedimentary pattern of platform shelf structural differentiation. At the same time, the sea level changes and the development of reefs and beaches at the platform edge had an important impact on the filling sequence. In the study area transgression gradually occurred from NE to SW, forming a sequence of shelf, slope, platform margin, carbonate platform, mixed platform, tidal flat, and volcanic facies.

Author details

Haofu Zheng[1] and Bo Liu[2,3]*

1 College of River and Ocean Engineering, Chongqing Jiaotong University, Chongqing, China

2 School of Earth and Space Sciences, Peking University, Beijing, China

3 School of Earth Sciences, Yunnan University, Kunming, China

*Address all correspondence to: bobliu@pku.edu.cn

References

[1] Guo Z, Deng K, Han Y. Formation and Evolution of the Sichuan Basin. Beijing: Geological Publishing House; 1996

[2] Mei Q, He D, Wen Z, Li Y, Li J. Geologic structure and tectonic evolution of Leshan-Longnvsi paleo-uplift in Sichuan Basin, China. Acta Petrolei Sinica. 2014;**27**(8):1427-1438. DOI: 10.7623/syxb201401002

[3] Ma Y. Puguang Gas Field. Marine Oil and Gas Exploration in China. Beijing: Springer; 2020. pp. 239-264

[4] Yuanba MY, Field G. Marine Oil and Gas Exploration in China. Beijing: Springer; 2020. pp. 265-283

[5] Hu M, Hu Z, Wei G, Yang W, Liu M. Sequence lithofacies paleogeography and reservoir potential of the Maokou formation in Sichuan Basin. Petroleum Exploration and Development. 2012;**39**(1):51-61

[6] Zheng H, Ma Y, Chi G, Qing H, Liu B, Zhang X, et al. Stratigraphic and structural control on hydrothermal Dolomitization in the middle Permian carbonates, southwestern Sichuan Basin (China). Minerals. 2019;**9**(1):32

[7] Saitoh M, Isozaki Y, Yao J, Ji Z, Ueno Y, Yoshida N. The appearance of an oxygen-depleted condition on the Capitanian disphotic slope/basin in South China: Middle–upper Permian stratigraphy at Chaotian in northern Sichuan. Global and Planetary Change. 2013;**105**:180-192

[8] Wang Zecheng ZW, Lin Z, Shixiang W. The Structural Sequence of the Sichuan Basin and the Natural Gas Exploration. Beijing: Geological Publishing House; 2002

[9] Luo Z. The determination of Emei taphrogenesis and its significance. Geological Journal of Sichuan. 1989;**9**(1):1-16

[10] Ren J. The continental tectonics of China. Acta Geosicientia Sinica. 1996;**13**(3):197-204

[11] Li X, Li Z, Sinclair JA, Li W, Carter G. Revisiting the "Yanbian Terrane": Implications for Neoproterozoic tectonic evolution of the western Yangtze block, South China. Precambrian Research. 2006;**151**(1-2):14-30

[12] He D, Li D, Zhang G, Zhao L, Fan C, Lu R, et al. Formation and evolution of multicycle superposed Sichuan Basin, China. Dizhi Kexue/ Chinese Journal of Geology. 2011;**46**(3):589-606

[13] Zhang G, Meng Q, Lai S. Tectonics and structure of Qinling orogenic belt. Science in China (Scienctia Sinica) Series B. 1995;**11**(38):1379-1394

[14] Zhang G, Meng Q, Yu Z, Sun Y, Zhou D, Guo A. Orogenesis and dynamics of the Qinling orogen. Science in China Series D: Earth Sciences. 1996;**39**(3):225-234

[15] Zhang G, Zhang B, Yuan X. Qinling Orogenic Belt and Continental Dynamics. Beijing: Science Press; 2001

[16] Hu Guangcan XY. Carboniferous Reservoirs in the High and Steep Structure, Eastern Sichuan Basin. Beijing: Petroleum Industry Press; 1997. pp. 63-130

[17] Yigang W, Yingchu W, Haitao H, Maolong X, Shujun S. Dalong formation found in Kaijiang-Liangping oceanic trough in the Sichuan Basin. Natural Gas Industry. 2006;**26**(9):32

[18] Xianyun M. Researches on the paleomagnetism of the strata from

carboniferous to triassic in southeastern Hubei. Hubei Geology. 1995;**9**:41-53

[19] Yin Hongfu WS, Yuansheng D, Peng Y. South China defined as part of Tethyan Archipelagic Ocean system. Earth Science. 1999;**24**:1-12

[20] Wu H, Lv J, Zhu R, Bai L, Guo B. Phanerozoic geomagnetic pole-shift curve and block motion characteristics of Yangtze block. Science in China (Series D). 1998;**28**(S1):71-80

[21] Huang Baochun ZY, Rixiang Z. Discussions on Phanerozoic evolution and formation of continental China, based on paleomagnetic studies. Earth Science Frontiers. 2008;**15**(3):348-359

[22] Zhang G, Guo A, Wang Y, Li S, Dong Y, Liu S, et al. Tectonics of South China continent and its implications. Science China Earth Sciences. 2013;**56**(11):1804-1828

[23] Wan TF. The Tectonics of China. Beijing: Higher Education Press; 2011

[24] Chen F, Wang Q, Yang S, Zhang Q, Liu X, Chen J, et al. Space-time distribution of manganese ore deposits along the southern margin of the South China block, in the context of palaco-Tethyan evolution. International Geology Review. 2018;**60**(1):72-86

[25] Zhang Q, Qian Q, Wang Y, Xu P, Han S, Jia X. Late Paleozoic basic magmatism from SW Yangtze massif and evolution of the Paleo-Tethyan Ocean. Acta Petrologica Sinica. 1999;**15**(4):576-583

[26] Cai J, Zhang K. A new model for the Indochina and South China collision during the Late Permian to the middle Triassic. Tectonophysics. 2009;**467**(1-4):35-43

[27] Luo Z, Jin Y, Zhu K. On emei taphrogenesis of the upper yangtze platform. Geological Review. 1988;**34**(1):15-28

[28] Liang X, Zhou Y, Jiang Y, Wen S, Fu J, Wang C. Difference of sedimentary response to Dongwu movement: Study on LA-ICPMS U-Pb ages of detrital zircons from upper Permian Wujiaping or Longtan formation from the Yangtze and Cathaysia blocks. Acta Petrologica Sinica. 2013;**29**(10):3592-3606

[29] Shao L, Gao C, Zhang C, Wang H, Guo L, Gao C. Sequence-palaeogeography and coal accumulation of Late Permian in southwestern China. Acta Sedimentologica Sinica. 2013;**31**(5):856-866

[30] Zhang Z, Wang F. Sr, Nd and Pb isotopic characteristics of Emeishan basalt province and discussion on their source region. Earth Science - Journal of China University of Geosciences. 2003;**28**(4):431-439

[31] Shi R, Hao Y, Huang Q. Comment on Re-Os isotopes constrain the formation of the Emeishan large Igneous Province. Acta Petrologica Sinica. 2008;**24**(11):2515-2523

[32] Li H, Zhang Z, Santosh M, Linsu L, Han L, Liu W, et al. Late Permian basalts in the northwestern margin of the Emeishan Large Igneous Province: Implications for the origin of the Songpan-Ganzi terrane. Lithos. 2016;**256**:75-87

[33] Liu H. Paleo-Tethyan basin evolution and basin-bound feature of Mn-bearing formation in the west margin of Yangtze plate. Journal of Mineralogy and Petrology. 2001;**21**(3):105-113

[34] Gao C, Ye D, Huang Z, Liu G, Ji R, Qin D. Two Paleo-oceans in the Late Paleozoic and their control to basins in China. Petroleum Geology and Expeximent. 2005;**2**

[35] Shaocong L, Qin J, Wenjuan Z, Li X. Geochemistry of the Permian basalt and its relationship with east Palaeo-Tethys evolution in Xiangyun area,

Yunnan Province. Journal of Northwest University (Natural Science Edition). 2009;**39**(3):14

[36] Huang S. A study on carbon and strontium isotopes of Late Paleozoic carbonate rocks in the upper Yangtze platform. Acta Geologica Sinica. 1997;**1**:45-53

[37] Lai X, Wang W, Wignall P, Bond D, Jiang H, Ali J, et al. Palaeoenvironmental change during the end-Guadalupian (Permian) mass extinction in Sichuan, China. Palaeogeography Palaeoclimatology Palaeoecology. 2008;**269**(1-2):78-93

[38] Chen L, Lu YC, Guo TL, Deng LS. Growth characteristics of Changhsingian (Late Permian) carbonate platform margin reef complexes in Yuanba gas Field, northeastern Sichuan Basin, China. Geological Journal. 2012;**47**(5):524-536

[39] Xie S, Algeo TJ, Zhou W, Ruan X, Luo G, Huang J, et al. Contrasting microbial community changes during mass extinctions at the Middle/ Late Permian and Permian/Triassic boundaries. Earth and Planetary Science Letters. 2017;**460**:180-191

[40] Ma Y, Mou C, Tan Q, Yu Q. A discussion on Kaijiang-Liangping ocean trough. Oil and Gas Geology. 2006;**27**(3):326-331

[41] Gu XD, Liu XH. Multiple Classification and Correlation of the Stratigraphy of China (51): Stratigraphy (Lithostratic) of Sichuan Province. Wuhan: China University of Geosciences Press; 1997

[42] Wang LT, Lu YB, Zhao SJ, Luo JH. Permian Lithofacies, Paleogeography and Mineralization In South China. Beijing: Geological Publishing House; 1994

[43] Zhang QM, Jiang XS, Qin JH, Cui X. Lithofacies palaeogeography of the early middle Permian Liangshan formation in northern Guizhou-southern Chongqing area and its bauxite ore-forming effect. Geological Bulletin of China 2012;**31**(4):558-568

[44] The Sichuan Geology and Mineral Bureau. Regional Geology of Sichuan Province. Beijing: Geological Publishing House; 1991

[45] Huang H, He D, Li Y, Wang B. The prototype and its evolution of the Sichuan sedimentary basin and adjacent areas during Liangshan and Qixia stages in Permian. Acta Petrologica Sinica. 2017;**33**:1317-1337

[46] Luo JX, He YB. Origin and characteristics of Permian eyeball-shaped limestones in middle-upper Yangtze region. Geological Review. 2010;**056**(005):629-637

[47] Shen SZ, Zhang H, Zhang YC, Yuan DX. Permian integrative stratigraphy and timescale of China. Science China Earth Sciences. 2019;**62**(01):158-192

[48] Feng ZZ, Yang YQ, Jin ZK, He YB. Lithofacies paleogeography of the Permian of South China. Acta Sedimentologica Sinica. 1996;**14**(2):1-11

[49] Li R, Hu MY, Wei Y, Liu M. Sedimentary facies model and favorable reservoir distribution of the middle Permian in Sichuan Basin. Oil and Gas Geology. 2019;**40**(2):369-379

[50] Ma YS, Chen HD, Wang GL. Sequence Stratigraphy and Paleogeography in South China. Beijing: Science Press; 2009

[51] Li YC. Stable Isotope Cyclostratigraphy of the Permian/ Triassic Limestones from South China: An Indexes and Stratigraphic Correlations and Palaeoenvironmental Implications. Hefei: University of Science and Technology of China Press; 2003

[52] Du Y, Yin HF, Wang ZP. The late Caledonian-early Hercynian basin's framework and tectonic evolution of Qinling orogenic belt (in Chinese with English abstract). Earth Science. 1997;**22**:401-405

[53] Wang YG, Wen YC, Hong HT, Xia ML, Fan Y, Wen L, et al. Carbonate slope facies sedimentary characteristics of the Late Permian to early Triassic in northern Sichuan Basin. Journal of Palaeogeography. 2009;**11**(2):143-156

[54] He B, Xu Y-G, Guan J-P, Zhong Y-T. Paleokarst on the top of the Maokou formation: Further evidence for domal crustal uplift prior to the Emeishan flood volcanism. Lithos. 2010;**119**(1): 1-9. DOI: 10.1016/j.lithos.2010.07.019

Chapter 11

Late Neo-Proterozoic Tectono-Sedimentary Evolution of the Tarim Block, NW China

Kaibo Shi, Bo Liu, Weimin Jiang, Jinxing Yu, Yue Kong, Tong Li and Changhai Li

Abstract

The study of the late Neo-Proterozoic tectono-sedimentary evolution of the Tarim Basin is a key to unravel the tectonic setting, the intracontinental rift formation mechanism, and the sedimentary filling processes of this basin. Since in the Tarim Basin, the late Neo-Proterozoic to early Cambrian sedimentary successions were preserved, this basin represents an excellent site in order to study the Precambrian geology. Based on the outcrop data collected in the peripheral areas of the Tarim Basin, coupled with the intra-basinal drill sites and seismic data previously published, the late Neo-proterozoic tectono-sedimentary evolution of the Tarim Basin has been investigated. These data show that there were two individual blocks before the Cryogenian Period, namely, the north Tarim Block and the south Tarim Block. In the early Neo-Proterozoic (ca. 800 Ma), the amalgamation of two blocks resulted in the formation of the unified basement. During the late Neo-Proterozoic, the Tarim Block was in an extensional setting as a result of the Rodinia supercontinent breakup and then evolved into an intracontinental rift basin. The tectono-sedimentary evolution of the basin may be divided into three stages: the rifting stage (780–700 Ma), the rifting to depression transitional stage (660–600 Ma), and the post-rift depression stage (580–540 Ma). In the rifting stage, intracontinental rifts (i.e., the Awati Rift, the North Manjar Rift, and the South Manjar Rift) were formed, in which coarse-grained clastic sediments were deposited, generally accompanied by a massive volcanic activity due to an intensive stretching. In the rifting-depression transitional stage and in the post-rift depression stage, the paleogeography was characterized by uplifts to the south and depressions to the north. Three types of depositional association (i.e., clastic depositional association, clastic-carbonate mixed depositional association, and carbonate depositional association) were formed. The distribution of the lower Cambrian source rock was genetically related to the tectono-sedimentary evolution during the late Neo-Proterozoic. The lower Cambrian source rock was a stable deposit in the northern Tarim Basin, where the late Ediacaran carbonate was deposited, thinning out toward the central uplift. It was distributed throughout the entire Mangar region in the east and may be missing in the Magaiti and the southwestern Tarim Basin.

Keywords: Tarim Block, late Neo-Proterozoic, tectono-sedimentary evolution, intracontinental rift, lower Cambrian source rock

1. Introduction

The Neo-Proterozoic to early Cambrian was a significant period of geological history. Several global geological events occurred in this time interval, for example, the assembly and breakup of the Rodinia supercontinent, the Snowball Earth, and the global sea-level rise and anoxic events [1–13]. The Tarim Basin is a large superimposed basin that underwent multiple phases of tectonic deformation from the Precambrian to the Cenozoic [14]. The late Neo-Proterozoic sedimentary succession was preserved in the peripheral areas of the Tarim Basin, which recorded convergence breakup cycles of the Rodinia supercontinent, multi-glacial events, multiphase volcanism, and evolution of continental rift [15–25]. The study of the late Neo-Proterozoic tectono-sedimentary evolution of the Tarim Basin is a key to unravel the tectonic setting, the continental rift formation mechanisms, and the sedimentary filling processes.

Additionally, being the largest petroliferous basin in northwest (NW) China, the lower Cambrian source rock was a focus of scientific debate [17, 18, 26–28]. In recent years, the results of oil and source rock correlation and the exploration discovery of primary oil and gas in the Cambrian subsalt dolomite reservoirs have shown that the lower Cambrian Yurtusi Formation is the most important source rocks in Tarim Basin [26, 29]. However, there are few stratigraphic data about the lower Cambrian source rock. Due to its small thickness and the deep burial, it is difficult to identify and trace the seismic horizons corresponding with the lower Cambrian source rocks on the seismic sections. Therefore, there are a lot of controversies over the distribution of the lower Cambrian source rocks, especially in the western sectors of the Tarim Basin [17, 18, 26–28]. These issues directly affect the evaluation and the selection of the target areas of deep oil and gas exploration. An accurate prediction of the distribution of the lower Cambrian source rock is essential for the deep oil and gas exploration in the Tarim Basin. The late Neo-Proterozoic tectono-sedimentary evolution of the Tarim Basin was reconstructed based on a comprehensive analysis of a large number of field outcrop data, drilling data, and high-resolution seismic profiles. Moreover, we have discussed how the early basin tectonic background influenced the sedimentary characteristics during the early Cambrian period, with a view to provide new ideas for the distribution of the lower Cambrian source rock.

2. Geological setting

The Tarim Block has an ancient crystallization basement and was separated from the Rodinia supercontinent during the late Neo-Proterozoic. The Tarim Basin, covering an area of approximately 56×10^4 km^2, was the stable area of the Tarim Block (**Figure 1**). As one of the three major continental blocks in China, the Tarim Block experienced several stages of tectonic evolution since its formation, having both similarities and dissimilarities to the North and South China Blocks [30]. The continental crust evolution of trondhjemite, tonalite, and granodiorite (TTG) during the late Neoarchean [31–34] and two orogenic events at the end of the Paleo-Proterozoic and the late Meso-Proterozoic to early Neo-Proterozoic, respectively, occurred. During the early period of the Neo-Proterozoic (ca. 900 Ma), the Tarim Block, that was a part of the Rodinia supercontinent, collided with the Australian Plate [4, 35–39]. Since ca. 800 Ma, the Tarim Block was separated from the Australian Plate, as a result of the breakup of the Rodinia supercontinent, resulting in the late Neo-Proterozoic cover deposits [1, 2]. During the extensional phase, an intense continental rifting, magmatic events, and sedimentary processes subsequently occurred, both in the interior and periphery of the Tarim Block, ranging in age from Cryogenian to Ediacaran [1–13, 15–22, 30, 40, 41].

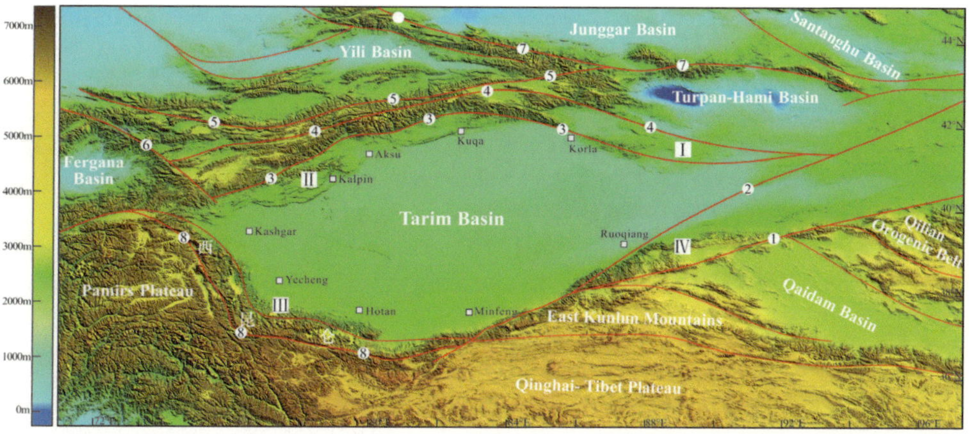

Figure 1.
Tectonic characteristics of the Tarim Basin and its peripheral areas. I-Kuruktag Quruqtagh uplift belt; II-Kalpin uplift belt; III-Tiekelike uplift belt; and IV-Altun uplift belt. 1-Altun strike-slip fault; 2-Xingxingxia strike-slip fault; 3-North Tarim fault; 4-south marginal fault of the central Tianshan; 5-Nikolaev-Nalat fault belt; 6-Talass-Fergana strike-slip fault; 7-north Tianshan suture belt; and 8-Kangxiwar fault belt.

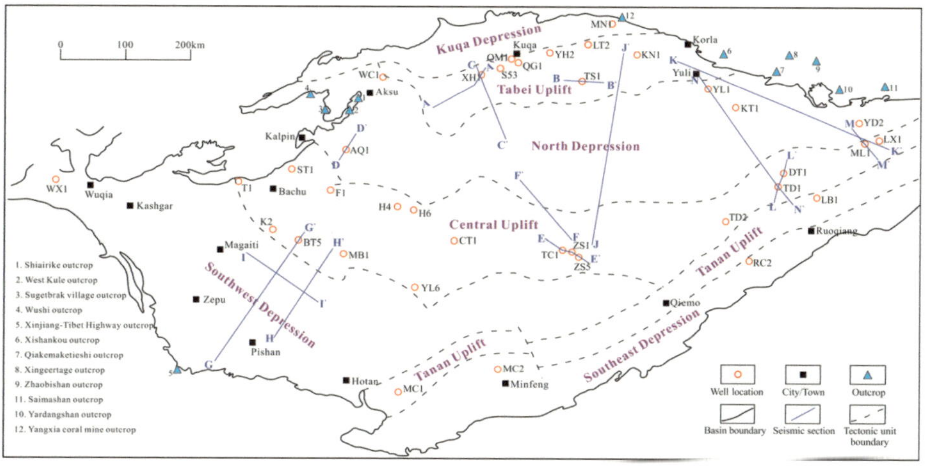

Figure 2.
Tectonic units division and key well locations in the Tarim Basin.

Currently, four uplift belts, that is, Quruqtagh in the northeast, Kalpin in the northwest, Tiekelike in the southwest, and Altun in the southeast (SE), are distributed on the margin of the Tarim Basin, while the hinterland of the basin is covered by desert areas (**Figure 1**). Based on the top of the basement and the regional characteristic of large-scale faults, the Tarim Basin was divided into seven first-order tectonic units, with three structural uplifts and four depressions, that is, the Kuqa Depression, the Tabei Uplift, the North Depression, the Central Uplift (Bachu Uplift, Tazhong Uplift, and Gucheng Uplift), the Southwest Depression, Tanan Uplift, and the Southeast Depression (**Figure 2**) [24].

3. The late Neo-Proterozoic sedimentary distribution

The hinterland of the Tarim Basin was covered by desert. The Neo-Proterozoic outcrops were distributed along the basin margin, mainly in the Aksu-Kalpin area in the northwestern margin, the Quruqtagh area in the northeastern margin, and the Tiekelike area in the southwestern margin, while they are lacking in the Altun area in the southeastern margin of the basin. Within the basin, several wells (including well XH1, well WC1, well QG1, well LT1, well YL1, well DT1, well TD1, and well TD2,) were drilled into the late Neo-Proterozoic strata and some other wells (including well ST1, well T1, well F1, well H4, well BT5, well MB1, well CT1, well TC1, well ZS1, well MC1, and well YD2,) were drilled into the Precambrian basement or into volcanic rocks (**Figure 2**).

3.1 Periphery of the basin

3.1.1 The northeastern margin

The Neo-Proterozoic outcrops in the northeastern margin of Tarim Basin were mainly located in the Quruqtagh area. The Quruqtagh area was separated into northern and southern regions by the Xingdi fault and preserved intact Neo-Proterozoic sedimentary successions.

The Cryogenian sequence was subdivided into the Baiyisi Formation, the Zhaobishan Formation, the Altungol Formation, and the Tereeken Formation from the bottom to the top (**Figures 3** and **4**). In the northern Quruqtagh region, the Baiyisi Formation comprised of diamictites in its lower part and was overlain by volcanic rocks. The Zhaobishan Formation was composed of sandstones, siltstones, and shales in the northern region and was lacking in the southern region. In the northern region, the Altungol Formation developed diamictites in the lower part and consisted of siltstones, sandstones, and volcanic rocks in its upper part. In the southern region, the Altungol Formation was dominated by diamictites and covered by cap dolomite with negative $\delta^{13}C$ values [42]. In the northern region, the Tereeken Formation consisted of diamictites separated by several layers of siltstones, mudstones, and volcanic rocks and was covered by a 10-m thick cap dolomites, characterized by negative $\delta^{13}C$ values [42]. In the southern region, a shallowing upward sequence crops out, which is composed of shales, siltstones, sandstones interlayered with carbonates (**Figure 4**).

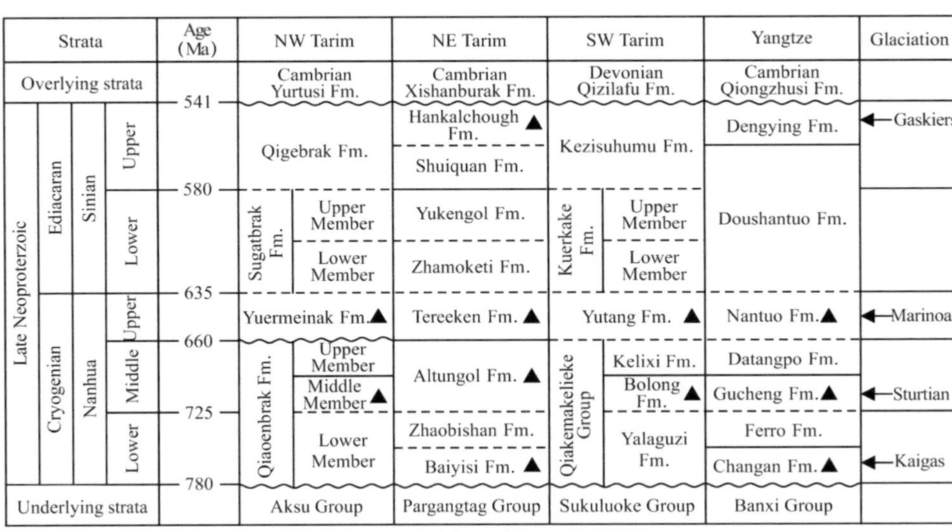

Figure 3.
Stratigraphic classification and correlation of Cryogenian-Ediacaran in the Tarim Basin.

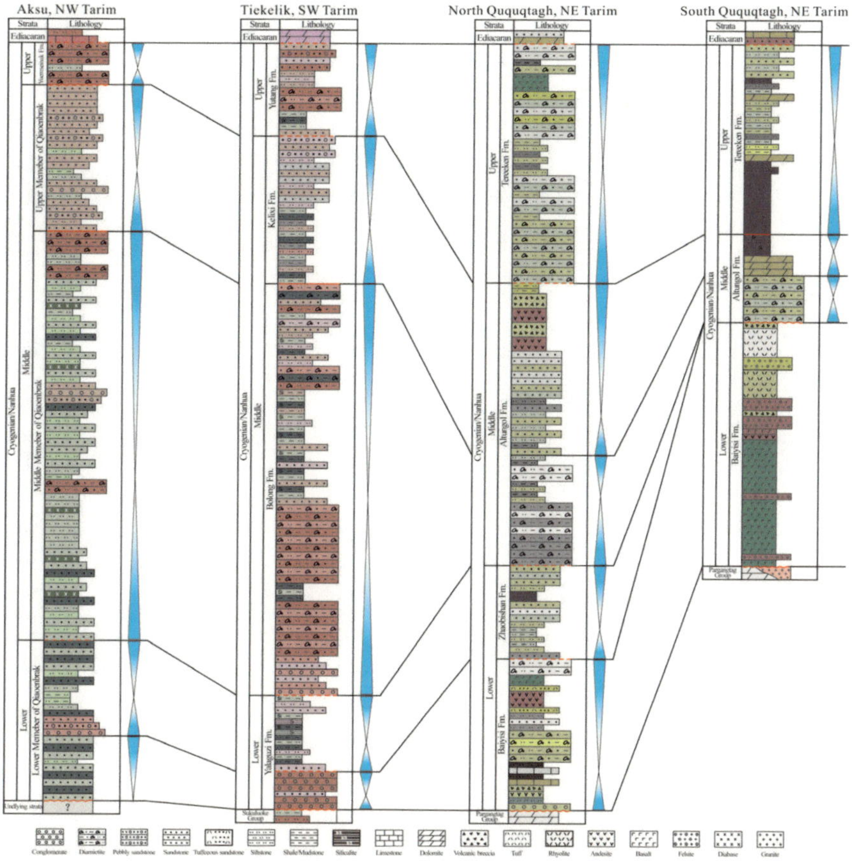

Figure 4.
Stratigraphic correlation of Cryogenian in periphery of the Tarim Basin.

The Ediacaran sequence was disconformably underlain by the Cryogenian strata and was unconformably overlain by the Cambrian Xishanbulake Formation (**Figure 5e**). It was subdivided into the Zhamoketi Formation, the Yukengol Formation, the Shuiquan Formation, and the Hankalchough Formation from the bottom to the top (**Figures 3** and **6**).

In the northern region, the Zhamoketi Formation were comprised mainly of fine-grained clastic deposits and volcanic rocks developed on the figures. The Yukengol Formation was composed of shales, siltstones interlayered with carbonates (**Figures 5e** and **6**). In the southern region, the Zhamoketi Formation and Yukengol Formation have similar sedimentary characteristics, which are mainly sandstones and siltstones (**Figures 5e** and **6**). There are weak differences between the southern and northern regions in the upper Ediacaran and the carbonate deposition gradually increased. The Shuiquan Formation was principally composed of carbonate rocks with a positive $\delta^{13}C$ ratio (**Figures 5g** and **6**) [42]. The youngest Hankalchough Formation in the top part of Ediacaran was comprised of diamictites and cap dolomites, characterized by negative $\delta^{13}C$ values in both northern and southern regions (**Figure 6**) [42].

3.1.2 The northwestern margin

The Neo-Proterozoic outcrops in the northwestern margin of the basin were mainly distributed in the Aksu-Wushi area. The Neo-Proterozoic units are

Figure 5.
Characteristics of the Cryogenian-Ediacaran sequence boundary in periphery of the Tarim Basin.

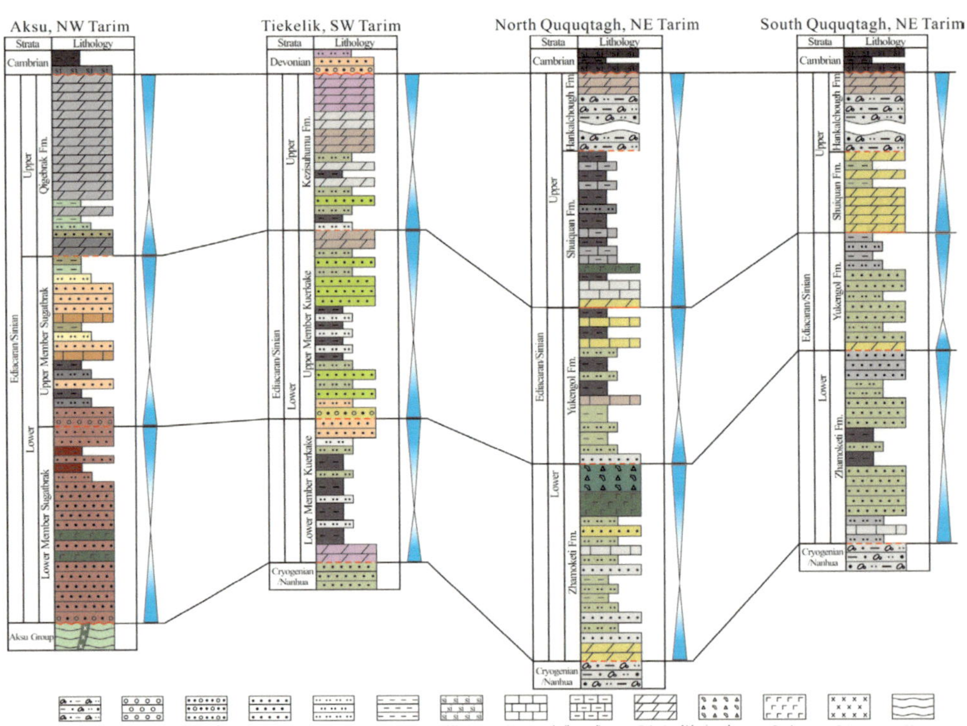

Figure 6.
Stratigraphic correlation of Ediacaran in periphery of the Tarim Basin.

composed of the metamorphic Aksu Group and the unmetamorphosed Cryogenian and Ediacaran sequences. The Aksu Group was mainly comprised of pelitic, psammtie, and mafic schists, which underwent green-schist to blue-schist facies metamorphism (**Figure 5a**) [42–44]. In addition, a series of NW-trending mafic dykes intruded the Aksu Group, with given zircon U-Pb ages of 757 ± 8.9 Ma [5] and 759 ± 7 Ma [45]. The Aksu Group was unconformably overlain by the Cryogenian-Ediacaran sedimentary strata, including the Qiaoenbrak and Yuermeinak Formations of the Cryogenian and the Sugetbrak and Qigebrak Formations of the Ediacaran (**Figures 3**, **4**, and **6**).

The Qiaoenbrak Formation was subdivided into the lower, middle, and upper members (**Figure 4**). The lower member was mainly composed of feldspar sandstones, feldspar-quartz sandstones, and siltstones. The middle member was characterized by thick rhythmic gray-green sandstones and siltstones (**Figure 4**). Several mafic dykes intruded into the middle member, with given zircon U-Pb age of 633 ± 7 Ma. The upper member was mainly calcareous sandstones and coarse-grained feldspars sandstones. There is a clear angular unconformity between the Qiaoenbrak Formation and the overlying Yuermeinak Formation. The Yuermeinak Formation was locally exposed and consisted of thick diamictites and sandstones, which were interpreted as glacial deposits. The Yuermeinak diamictite is generally correlated with the Tereeken diamictite in the northeastern Tarim Basin.

The Sugetbrak Formation was unconformably underlain by the Yuermeinak Formation or the Aksu Group (**Figure 5a**). It consisted of two members. The lower member was composed of red conglomerates and sandstones, which were deposited in an oxidized environment (**Figure 5a** and **b**). The mafic dykes intruded into the lower member, with given zircon U-Pb age of 615 Ma [3]. The upper member was characterized by mixed deposits of fine-grained clastic and carbonate rocks. The Qigebrak Formation was composed of bedded dolomite and stromatolites developed. There is a weathering crust with a thickness of ca. 30–50 m in the uppermost of the Qigebrak Formation, which was unconformably overlain by the Cambrian Yurtusi Formation (**Figure 5d**).

3.1.3 The southwestern margin

The Neo-Proterozoic strata in the southwestern margin of the basin were mainly distributed in the Tiekelike area and the outcrops along the Xinjiang-Tibet Highway were complete. The late Neo-Proterozoic succession has been divided into the Qiakemakelieke Group (including Yalaguzi, Bolong, and Kelixi Formations) and the Yutang Formation of the Cryogenian, Kuerkake, and Kezisuhumu Formations (Ediacaran; **Figures 3**, **4**, and **6**).

The Yalaguzi Formation was composed of conglomerates and unconformably underlain by the Tonian Sukuluoke Group (**Figure 4**). The upper part was composed of gray-green laminated silicalite and siliceous mudstones. The Bolong Formation was mainly composed of two sets of thick diamictites, which were separated by layers of laminated siliceous mudstones, siltstones, and sandstones. The Bolong diamicitie can be generally correlated with the Altungol and Qiaoenbrak diamictites (**Figure 4**). The Kelixi Formation was a shallowing upward sequence. It was composed of mudstones, siltstones, sandstones, and conglomerate-bearing sandstones from the bottom to the top (**Figure 4**) [46]. There was another set of diamicites in the lower part of the Yutang Formation, which was contemporaneous with the Tereeken and Yuermeinak diamictites. The upper part was composed of siltstones and sandstones (**Figure 4**).

The Kuerkake Formation includes two members (**Figure 6**). The lower member was composed of black and dark-gray mudstones intercalated with siltstones, while the upper member consisted of sandstones and siltstones interlayered with dark-gray mudstones. The Kezisuhumu Formation was composed of mudstones, siltstones interlayered with dolomites in its lower part and thick dolomites in the upper part, which was unconformably covered by Devonian or Carboniferous strata.

3.2 The areas within the basin

3.2.1 The northern Tarim Basin (the Tabei area)

In the northern part of Tarim Basin, only well WC1, well XH1, well QG1, and well LT1 have drilled in the Ediacaran strata, while no well drilled in the Cryogenian strata. The Sugetbrak Formation was characterized by fine-grained clastic sediments and limestones, with a thickness of ca. 70–90 m. These features were similar to those in the upper member of Sugetbrak in the Aksu area. The Qigebrak Formation has a thickness of ca. 160–180 m and is composed of dolomites. The well XH1 has drilled phyllite, quartz schist, and granite gneiss beneath the Sugetbrak mudstones. The granite gneiss yielded a zircon U-Pb age of 832 ± 4 Ma [30]. The well WC1 has drilled the chlorite schist and quartz schist with detrital zircon ages clusters at ca. 800 Ma [30, 47]. In the Yangxia section, the sericite quartz schist developed and the detrital zircon ages clusters were at ca. 800 Ma [48]. This metamorphic basement has also been drilled in wells YH2, well LT2, and well MN1. It might be correlated with the Aksu Group in Aksu area accordingly to the detrital zircon ages and to the degree of metamorphism. In addition, Precambrian basement granite (ca. 1.8–1.9 Ga [30, 47, 49]) was revealed in many boreholes. These lithologic and chronological characterisitics suggest that the northern Tarim Basin developed a metamorphic basement which might be corresponded to the Aksu Group, and the Paleo-Proterozoic crystalline basement locally occurred.

As shown on seismic profiles, the Ediacaran strata were distributed stably in the southern area of the Tabei uplift (**Figure 7a–c**). Toward the north, the Ediacaran strata were pinched out due to uplifting and denudation during Paleozoic, and hence the Precambrian metamorphic basement was directly covered by Mesozoic strata (**Figure 7c**).

3.2.2 The Bachu-Tazhong area

In the Bachu uplift, only well T1 has drilled the late Neo-Proterozoic strata with a thickness of ca. 200 m. The lithology is composed of mudstones, sandstones, and volcanic rocks, and the underlying andesite (zircon U-Pb age of 755 ± 3 Ma [30]) is intercalated with mudstones. The youngest detrital zircon age of tuffaceous sandstone just below the Cambrian carbonate rocks of well T1 is 707 ± 8 Ma, which was interpreted as the maximum sedimentary age [50]. In addition, some wells (e.g., well ST1, well F1, and well H4) directly drilled in the mafic volcanic rocks just below the Cambrian carbonate rocks (ca. 26–224 m), which were supposed to correspond to the eruption in the period of the late Neo-Proterozoic based on the zircon U-Pb dating [47].

In the northern Bachu uplift, there was no borehole drilled in the Neo-Proterozoic strata. The seismic interpretation has shown that the Cryogenian depositional distribution was controlled by faults and was characterized by intracontinental rift deposition (**Figure 8a**). The Ediacaran strata have stable distribution over the rift sedimentary system (**Figure 8a**).

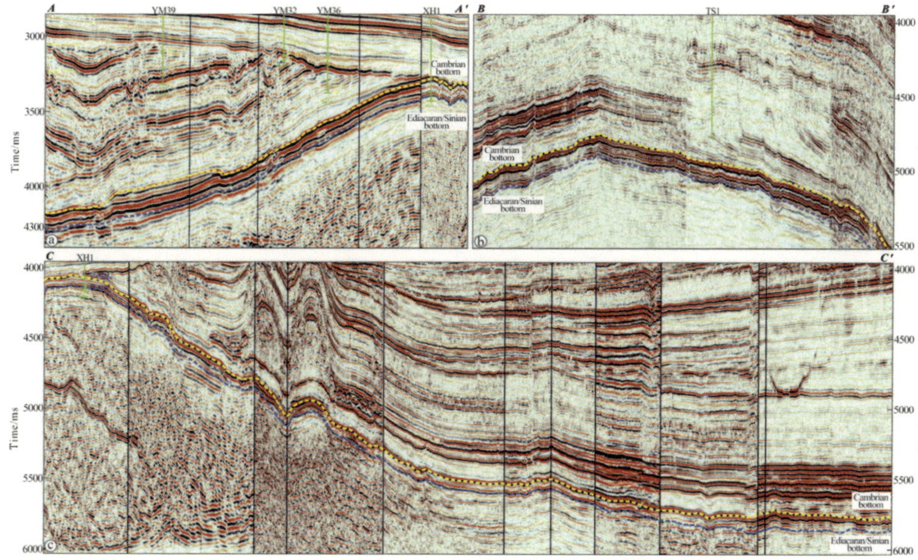

Figure 7.
*Seismic reflection characteristics of Cryogenian-Ediacaran in the northern Tarim Basin (profiles location as shown in **Figure 2**).*

In the southern Bachu uplift, well BT5 has revealed a set of breccias below the Cambrian strata. The breccias mainly consisted of basalt/diabase, indicating the deposits were near the source. The well MB1 drilled the granite gneiss (zircon U-Pb age of 1920 ± 14 Ma [50]) just below the Cambrian carbonate rocks. It was suggested that the Paleo-Proterozoic crystalline basement occurred in the Maigaiti area. The well YL6 drilled a set of marble below the middle Cambrian [50], which might be corresponded to the Bochatetage Formation in the Tiekelike area. According to detrital zircon dating, the Bochatetage Formation was deposited in the middle Neo-Proterozoic [11].

In the Tazhong uplift, there was no borehole drilled in the Neo-Proterozoic sedimentary successions. The well ZS1 drilled the olivine-bearing granite (zircon U-Pb ages of 1895 ± 1 Ma [47] and 1915 ± 5 Ma [30]) and well TC1 revealed diorite and granodiorite (zircon U-Pb age of 757 ± 6 Ma [51]) below the Cambrian carbonate rocks. In the northern region of Tazhong uplift, there were obvious seismic reflection characteristics of the Cryogenian intracontinental rift deposition with large thickness and the Ediacaran depression sedimentary successions (**Figure 8c**). However, in the southern area, the late Neo-Proterozoic sedimentary successions were lacking, and the Paleo-Proterozoic crystalline basement was directly overlain by the Cambrian carbonate rocks (**Figure 8b**). The above features implied that the northern part of the Tazhong uplift was a deposition area, while the southern part was the structural high without deposition during the late Neo-Proterozoic.

3.2.3 The eastern Tarim Basin (the Tadong area)

In the Tadong area, several boreholes (e.g., well TD2, well TD1, well DT1, and well YL1) have drilled in the Ediacaran strata. In the Tadong low uplift zone, well TD1, well TD2, and well DT1 drilled in Shuiquan carbonate rocks with a thickness of ca. 28–800 m. The Paleo-Proterozoic granite crystalline basement (zircon U-Pb ages of 1908 ± 9 Ma [51], 1908 ± 9 Ma, and 1908 ± 9 Ma [47]) was revealed in well

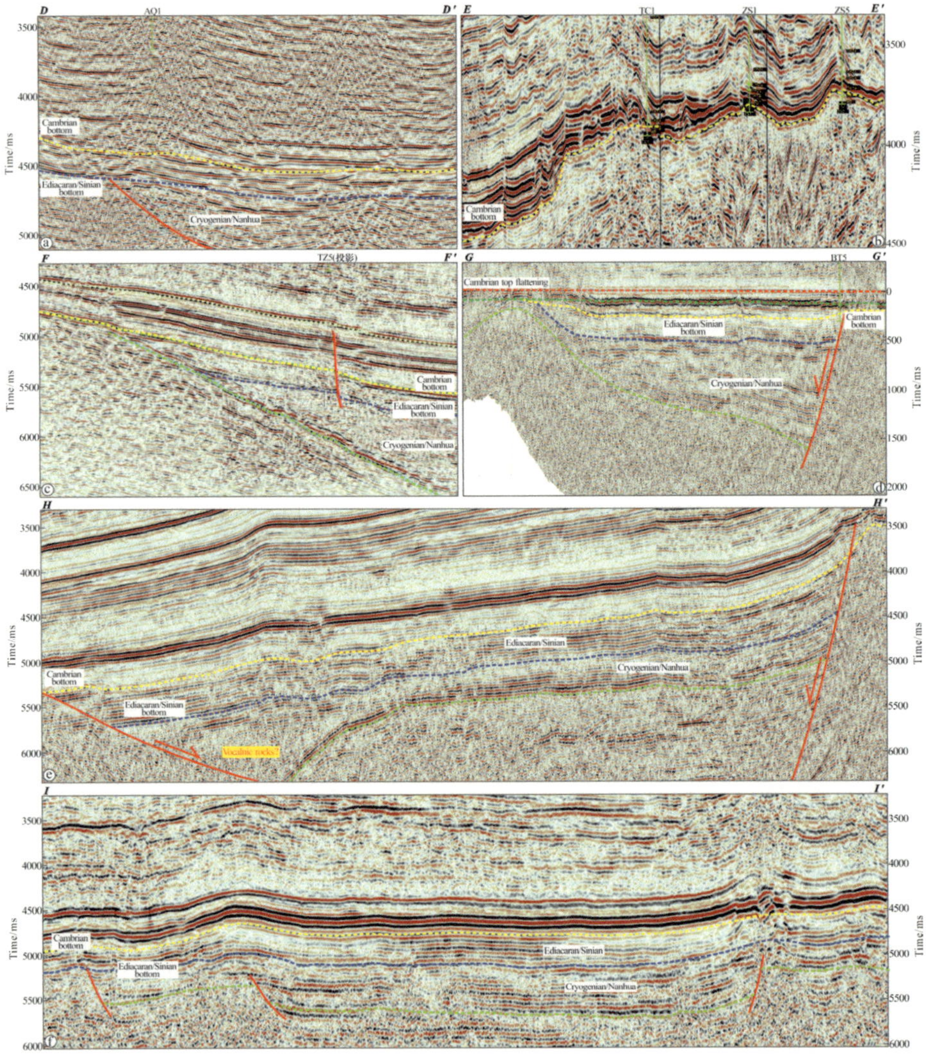

Figure 8.
*Seismic reflection characteristics of Cryogenian-Ediacaran in the western Tarim Basin (profiles location as shown in **Figure 2**).*

TD2. The well YD2 drilled the granite (zircon U-Pb ages of 750 ± 7 Ma [47]) and was overlain by the Cambrian black shale. The Ediacaran sedimentary successions including the Zhamoketi, Yukengol, and Shuiquan Formations were drilled by well YL1 in the Manjar sag, which composed of mudstones and siltstones in the lower part as well as micrite and argillaceous limestones intercalated with mudstones in the upper part.

As shown on seismic profiles, the Cryogenian was characteristic of the intracontinental rift deposition with a great thickness variation, which was controlled by faults (**Figure 9a** and **b**). The Ediacaran was the post-rifting depression deposition and stably distributed in the Manjar area (**Figure 9a**, **b**, and **e**). It was implied that the Tadong low uplift was a structural high, which might be the volcanic eruption center or the rift flank, hence it directly deposited the Shuiquan dolomite. The thickness of the Shuiquan dolomite pinched out toward the south (**Figure 9c** and **d**). The Shuiquan dolomite was missed in well YD2, and this might be attributed to denudation due to uplifting at the end of the Ediacaran.

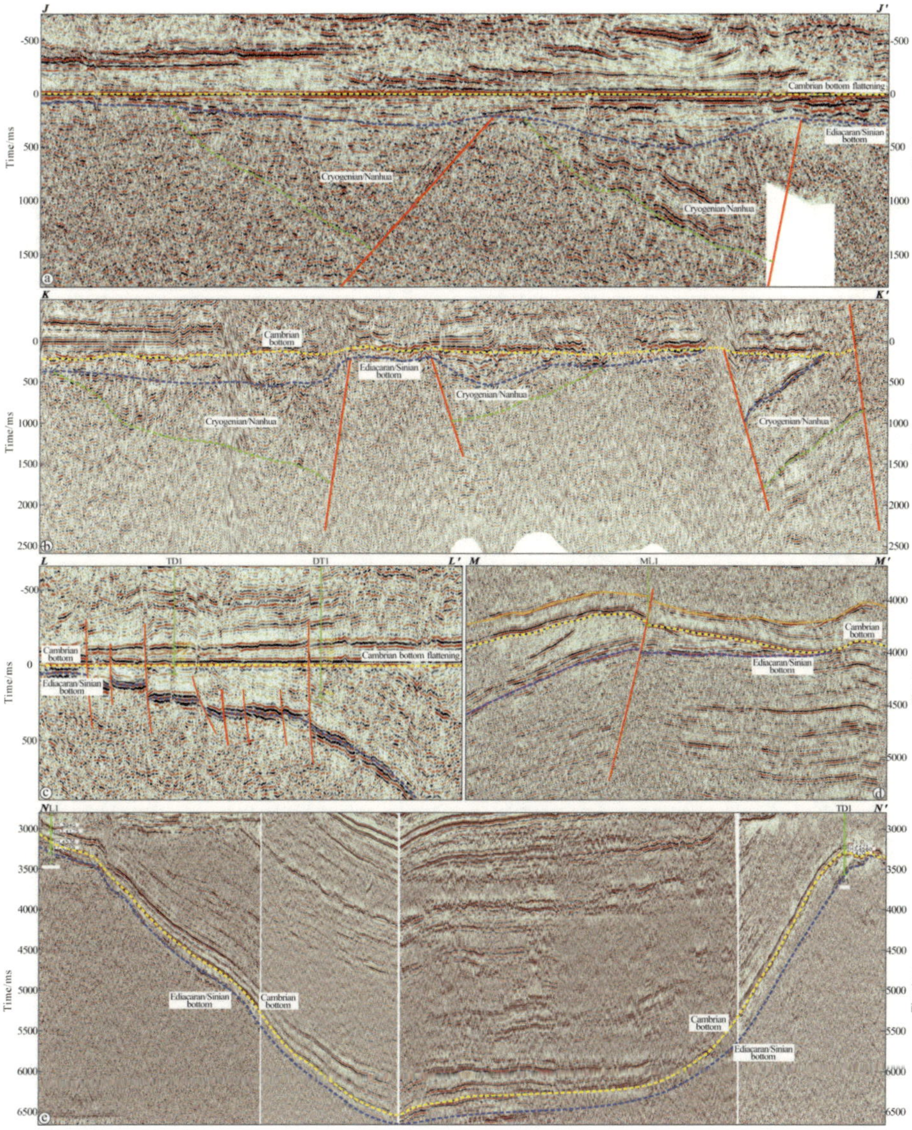

Figure 9.
*Seismic reflection characteristics of Cryogenian-Ediacaran in the eastern Tarim Basin (profiles location as shown in **Figure 2**).*

3.2.4 The southwestern to southeastern Tarim Basin

Although no Cryogenian-Ediacaran strata have drilled in the southwestern Tarim Basin due to large buried depth, the distinct seismic reflection signatures could be identified on the seismic section (**Figure 6d–f**). The Cryogenian developed intracontinental rift sedimentary successions with a large thickness and the Ediacaran was characteristic of depression sedimentary successions (**Figure 6d–f**). The 3D resistivity inversion results displayed that a depth range of 6–15 km exhibited low resistivity in the Magaiti area [52], and the areas of low-resistivity decrease with depth. Thus, it was inferred that the Cryogenian-Ediacaran developed in the southwest (SW) Tarim Basin. The northeast (NE)-SW trending aeromagnetic anomalies belts might indicate the basement difference and were not the identification mark of the Cryogenian-Ediacaran sedimentary successions.

In the southwestern Tarim Basin, no Cryogenian-Ediacaran strata have drilled. The wells MC1 and MC2 drilled in the low green-schist facies metamorphic rocks, with detrital zircon age clusters at ca. 750–850 Ma and 1.8–1.9 Ga [30, 47]. According to the detrital zircon ages, metamorphic grade, and petrological characteristics, it might be equivalent to the Ailiankate Group in the Tiekelike area. The Ailiankate Group was considered as Paleo-Proterozoic, but recent studies revealed it was Neo-Proterozoic [11, 53, 54]. Moreover, no Cryogenian-Ediacaran sedimentary successions were observed in the Altun outcrops, and the Cambrian/Ordovician strata were underlain by the Tonian. Thus, it was concluded that the southeastern Tarim Basin was the uplift area without deposition during the period of Cryogenian-Ediacaran.

4. Discussion

4.1 Tectono-sedimentary evolution

4.1.1 The pre-rifting stage (ca. 880–800 Ma)

Recently, a large number of chronological, geochemical, and geophysical studies on the Precambrian basement were carried out [1–13, 30, 34, 47, 51, 53–62]. These results suggested that there were two individual blocks before the Cryogenian Period, namely, the north Tarim Block and the south Tarim Block (**Figure 10**). The Neo-Proterozoic granitie (ca. 930–910 Ma [60, 63, 64]) was widely distributed in the Altun area, southeastern margin of Tarim Basin. Geochemical analysis indicates that the granite was formed in a collision orogenic tectonic background and interpreted to syn-collision granite [60, 64, 65], which was a result of the Rodinia supercontinent convergence. Thus, it is inferred that the south Tarim Block converged to the northern margin of the Australia Plate and collision orogenesis at ca. 930–910 Ma. This tectonic movement resulted in the formation of extensive

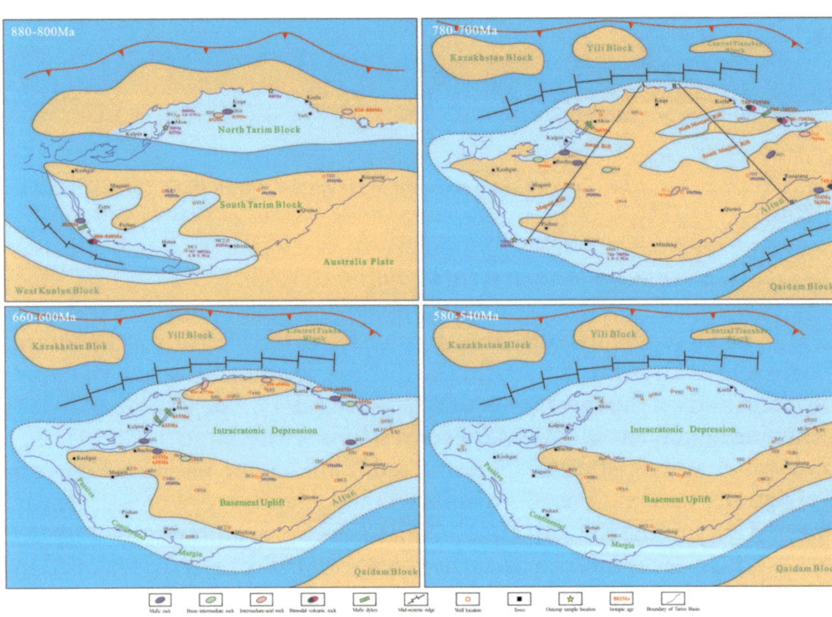

Figure 10.
Tectonic-sedimentary characteristics of the Tarim block during Cryogenian to Ediacaran.

syn-collision granite in the southeastern margin of Tarim Basin. The Sailajiazitage bimodal volcanic rocks were identified in the southwestern margin of Tarim Basin, which were composed of basalts and rhyolites [12]. The geochemical and chronological studies have suggested that these bimodal volcanic rocks formed in the intracontinental rift setting and erupted during the early Neo-Proterozoic (ca. 880–840 Ma) [12, 54, 66].

These different characteristics suggest that the western and eastern parts of south Tarim Block had different tectonic settings during the early Neo-Proterozoic. After the south Tarim Block converged to the northern margin of Australia Plate, the western part of the south Tarim Block was in an extensional tectonic setting and the West Kunlun Ocean gradually opened. The NE-SW trending rifts began to develop at ca. 880 Ma (**Figure 10**) and then deposited the early Neo-Proterozoic successions (Sailajiazitage Group, Bochatetage Formation, Sumalan Formation, and Sukuluoke Formation). This extensional tectonic setting might be last to the end at ca. 800 Ma, which was demonstrated by the ca. 800 Ma basalt and mafic dyke swarms in the Candilik-Xuxugou area [12, 54, 66]. During this period (ca. 880–800 Ma), the Kazakhstan, Yili, and Central Tianshan Blocks were not yet separated from the north Tarim Block (**Figure 10**). Due to subduction of the Tianshan Ocean, a back-arc basin formed and the Aksu Group volcanic-clastic rocks deposited [9, 20]. At ca. 800 Ma, the north and south Tarim Blocks amalgamated together to form the uniform Tarim Block. At the same time, the low green-schist facies metamorphism of the Ailiankate Group in the southwestern margin of Tarim Block and ultra-pressure blue-schist facies metamorphism of the Aksu Group in the northern margin of Tarim Block occurred.

4.1.2 The rifting stage (ca. 780–700 Ma)

In the late Neo-Proterozoic, the Rodinia supercontinent had an initial breakup due to the activity of the super-mantle plume. As a result, the Tarim Block was in an extensional setting and evolved the intracontinental rift basin. Tectono-sedimentary evolution of the basin may be divided into three stages: the rifting stage (780–700 Ma), rifting to depression transitional stage (660–600 Ma), and post-rifting depression stage (580–540 Ma) (**Figures 10–12**).

In the periphery of Tarim Block, with the opening of Altyn Ocean, the Qaidam Block was separated from the southeastern edge of Tarim Block and associated intracontinental bimodal volcanic rocks (ca. 760–750 Ma) [60] were developed in the Altun region, while the Altun Block was not separated from the Tarim Block (**Figure 10**) [67]. In the southwestern margin of Tarim Block, the NW-SE trending rifts had been died out by reason of amalgamation of the north and south Tarim Blocks during the early Neo-Proterozoic, and they evolved into the passive continental margin basin during the late Neo-Proterozoic. This tectonic setting continued, at least, till Cambrian, forming the late Neo-Proterozoic to Cambrian unmetamorphosed sedimentary succession. Due to the continuous subduction, the South Tianshan Ocean was opened, and the Kazakhstan, Yili, Central Tianshan, and other blocks were separated from the northern edge of Tarim Block, forming a strong extensional setting in the northern margin of the present Tarim Basin (**Figure 10**). Large-scale rift-related bimodal volcanic rocks (ca. 760–710 Ma) [1–3, 47] distributed in the northeastern margin of the present Tarim Basin, and abundant mafic dykes (ca. 760 Ma) [5, 60] intruded into the Aksu Group in the northwestern margin of the present Tarim Basin.

In the Bachu area, several boreholes have directly drilled the volcanic rocks just below the Cambrian carbonate rocks. The geochemical characteristics suggested that these volcanic rocks erupted in an intracontinental rift setting. In addition,

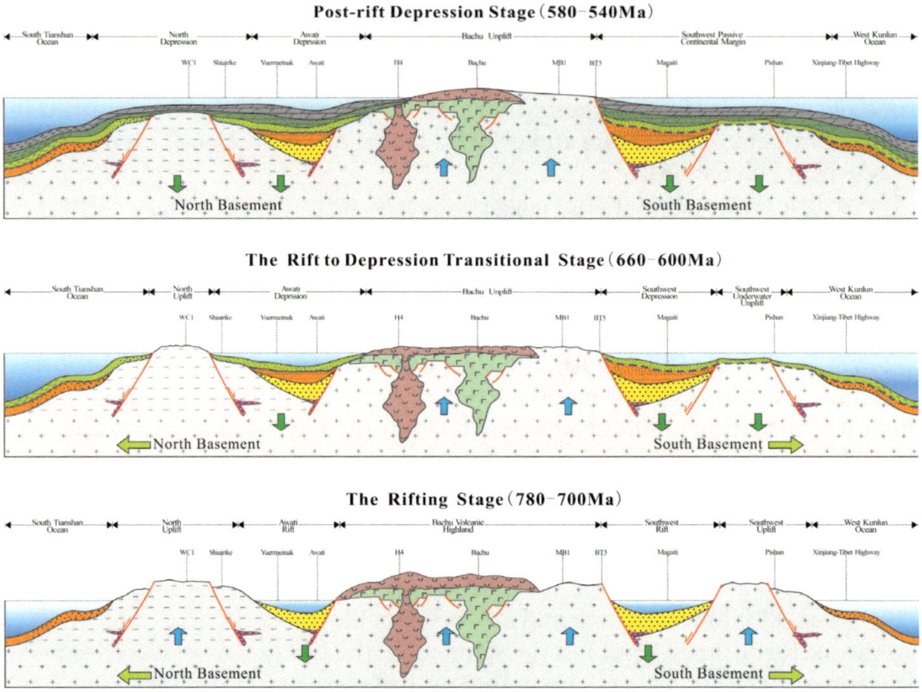

Figure 11.
Tectonic-sedimentary evolution profile of the Cryogenian-Ediacaran in the western Tarim Basin.

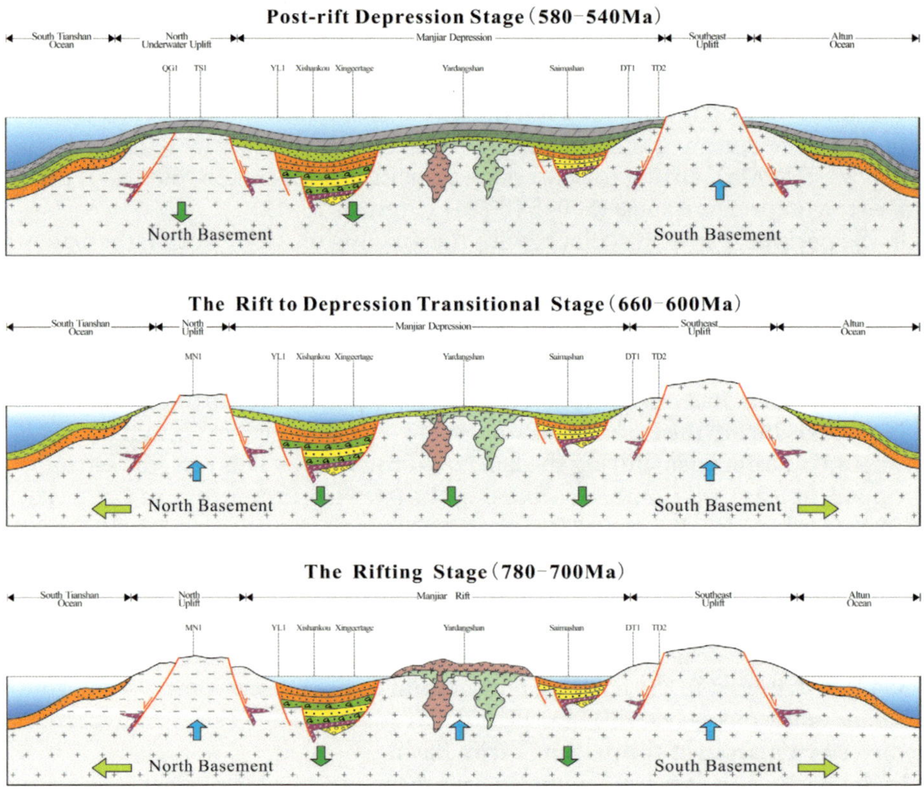

Figure 12.
Tectonic-sedimentary evolution profile of the Cryogenian-Ediacaran in the eastern Tarim Basin.

seismic profiles show obvious characteristics of intracontinental rift in the Awati-Kalpin, Manjar-Kuruktag, and Magaiti areas (**Figures 7–9**).

In summary, the Tarim Block was in a strong extensional setting during the period of ca. 780–700 Ma, and three intracontinental rifts (i.e., the Awati Rift, the North Manjar Rift, and the South Manjar Rift) formed in the northern part of Tarim Block (**Figures 10–12**). Rift shoulders were the uplift area without deposition. The Bachu area was the volcanic eruption center and resulted in the formation highland, which was a provenance denudation zone without deposition (**Figure 11**). The southern part was generally supposed to basement uplift and intracontinental rift developed in the Magaiti area. The southwestern margin was a passive continental margin (**Figures 11** and **12**).

4.1.3 The rifting to depression transitional stage (ca. 660–600 Ma)

At the end of the middle Cryogenian, a tectonic uplifting movement occurred, which resulted in the angular unconformity contract between the Qiaoenbrak Formation and the overlying strata in the Aksu area, and in the parallel unconformity contract between the Altungol Formation and the overlying Tereeken Formation in the Quruqtagh area.

After this phase of tectonic uplifting, the north part evolved into the transformation stage from rifting to depression and still was in an extensional tectonic setting. Episodic magmatic thermal events occurred in the northeast margin, such as rhyolite (ca. 655 Ma) [8] at the top of the Altungol Formation, mafic volcanic and pyroclastic rocks (ca. 635 Ma) [40] at the upper part of the Tereeken Formation, mafic dykes intruded into Zhamoketi Formation (ca. 615 Ma) [2, 8], and large area distribution granite (ca. 660–630 Ma) [13]. In the northwestern margin, ca. 633 Ma and ca. 615 Ma [3] mafic dykes intruded into the Qiaoenbrak Formation and Sugetbrak Formation, respectively.

In the northern part of the Tarim Block, because of continuous stretching and transgression, the isolated rifts gradually interconnected (Awati Rift, North Manjar Rift, and South Manjar Rift), the depositional area continued to expand and the residual paleo-uplift was existed in the Tabei area (**Figures 10–12**). The Bachu-Tazhong-Tadongnan area was still a highland, which was a provenance denudation zone without deposition (**Figure 10**). The seawater was from the South Tianshan Ocean with NE and NW directions of transgression, and the shore sedimentary was in development. In Manjar, grayish green mudstone intercalated with sandstone and siltstone was deposited in the shelf environment.

In the southwestern part of Tarim Block, the intensity of tectonic activity weakened gradually and was mainly post-rifting the thermal subsidence. Pishan palaeohigh was submerged, the continental rift (developed in Magaiti area) connected with the southwest margin and formed a passive continental margin (**Figures 10** and **11**). Influenced by the West Kunlun Ocean, the transgression from the southwest, the shore sedimentary developed in Magaiti and toward southwest shelf to deepwater deposition.

4.1.4 The post-rift depression stage (ca. 580–540 Ma)

During this period, the Tarim Block was in a stable tectonic setting without significant magmatic events and evolved into a post-rift depression stage. The northern part was the intra-cratonic depression basin and was characterized by thermal subsidence. The residual paleo-uplift was gradually flooded by seawater, but the Bachu-Tazhong-Tadongnan uplift still existed (**Figures 10–12**). Due to the decrease of terrigenous clastic and seawater evaporation and concentration, the carbonate content increased. Two deepwater sedimentary environments were developed in the Manjar and the northwest part of the Aksu-Kuqa area, where carbonate rocks and dark mudstone were deposited (**Figure 12**). A tidal-flat environment was developed in the southeastern part of the Manjar to the Kalpin-Awat area and deposited bedded dolomite with abundant stromatolite (**Figure 11**). At the same time, carbonate depositional systems prevailed in the southwestern part of Tarim Block.

At the end of Ediacaran, Tarim Block experienced a tectonic uplifting and the Ediacaran strata suffered from long-term weathering. Hence, a thick weathered crust (ca. 30–50 m) was formed at the top of Qigebrak dolomite, and meanwhile the lower Ediacaran strata were absent or partially absent in the southwestern of Tarim Basin.

4.2 Implication for the distribution of Cambrian source rocks

4.2.1 The lower Cambrian source rock

In the western region of the Tarim Basin, the lower Cambrian source rock developed in the Yurtusi Formation, which composed of silicalites, siliceous shales, and black shales with argillaceous dolomite. In the Aksu area, the Yurtusi source rock was in stable distribution with the thickness of ca. 10–15 m [26]. The total organic carbon (TOC) content was 2–16% [26], up to a maximum 22.39% [28], which is the highest quality marine hydrocarbon source rock in China [26]. In the Tabei uplift, only two boreholes (well XH1 and well LT1) have revealed the Yurtusi Formation whose sedimentary characteristics were similar to the Aksu outcrops. In well XH1, the thickness of the Yurtusi source rock was ca. 33 m, with a TOC of 1.0–9.43% (5.5% on average) [28]. In the Bachu-Tazhong uplift, many boreholes drilled through the Cambrian strata and entered into Precambrian, but no well revealed the Yurtusi source rock. In Magaiti, no well drilled and revealed the Yurtusi source rock. It is probably of large burial depth or was absent in the early Cambrian deposition.

In the eastern region of the Tarim Basin, the lower Cambrian source rock developed in Xishanbrak and Xidashan Formations. In north Quruqtagh, the Xishanbrak Formation developed in siliceous mud rocks with the thickness of ca. 12.5 m and TOC of 1.53%. The Xidashan Formation consisted of black mud rocks whose thickness was ca. 15 m and the TOC was of 1.39–2.17% (1.78% on average). In south Quruqtagh, the Xidashan Formation developed ca. 20 m black mud rocks with a TOC of 0.17–0.92% (0.46% on average). In Manjar area, several boreholes have drilled in the lower Cambrian, but it is difficult to divide between the Xishanbrak Formation and the Xidashan Formation. In well TD1, the thickness of black mud rocks was ca. 55 m with a TOC of 0.70–5.52% (2.3% on average) [27]. In well TD2, the thickness of black mud rocks with limestone was ca. 85 m and the TOC was more than 1%. In well YL1, the lower Cambrian composed of shales and argillaceous limestone, and the thickness was ca. 65 m with a TOC of more than 0.5% (1.56% on average) [27].

4.2.2 The distribution of the lower Cambrian source rock

In recent years, the results of oil and source rock correlation and the exploration discovery of primary oil and gas in the Cambrian subsalt dolomite reservoirs have shown that the lower Cambrian is the most important source rocks in the Tarim Basin [26, 29]. However, there are few stratigraphic data about the lower Cambrian source rock. Due to its small thickness and the deep burial, it is difficult to identify and trace the seismic horizons corresponding with the lower Cambrian source rocks on the seismic sections. Therefore, there are a lot of controversies over the distribution of the lower Cambrian source rocks, especially in the western sectors of the Tarim Basin [17, 18, 26–28].

Based on the oil and gas exploration discovery of Ediacaran-Cambrian in the Sichuan Basin, paleogeography in the late Neo-Proterozoic controlled the distribution of the lower Cambrian source rock [68–70]. Similar to the Sichuan Basin, the sedimentary characteristics of the lower Cambrian were related to the tectono-sedimentary in the late Neo-Proterozoic and uplifting movement at the end of Ediacaran. Based on outcrops and drilling data, in the western of the Tarim Baisn, the Yurtusi source rock existed in an area where the late Ediacaran carbonate deposited. The Yurtusi source rock did not develop in the area where the basement uplift existed in the late Neo-Proterozoic.

Therefore, we suggest that paleogeography in the late Neo-Proterozoic controlled the transgression in the early Cambrian and distribution of the lower Cambrian source rock. During the early Cambrian, the Tarim Basin had the paleogeographical characteristics of uplift in the south and depression in the north. The lower Cambrian source rock was composed of stable deposits in the northern Tarim Basin, where the late Ediacaran carbonate was deposited and thinning out toward the central uplift. It was distributed throughout the entire Manjar region in the east, and its overall thickness was thicker than that in the northern Tarim Basin. The lower Cambrian source rocks may be missing in the Magaiti and the southwestern Tarim Basin.

5. Conclusions

1. During the late Neo-Proterozoic, the Tarim Block was in an extensional setting as a result of the Rodinia supercontinent breakup and then evolved into an intracontinental rift basin. The tectono-sedimentary evolution of the basin may be divided into three stages: the rifting stage (780–700 Ma), the rifting to depression transitional stage (660–600 Ma), and the post-rift depression stage (580–540 Ma).

2. During the different stages of tectonic evolution, there were different paleogeographic characteristics and sedimentary association. In the rifting stage, intracontinental rifts (i.e., the Awati Rift, the North Manjar Rift, and the South Manjar Rift) were formed, in which coarse-grained clastic sediments were deposited, generally accompanied by a massive volcanic activity due to an intensive stretching. In the rifting-depression transitional stage and in the post-rift depression stage, the paleogeography was characterized by uplifts to the south and depressions to the north. Three types of depositional association (i.e., clastic depositional association, clastic-carbonate mixed depositional association, and carbonate depositional association) were formed.

3. The distribution of the lower Cambrian source rock was genetically related to the tectono-sedimentary evolution during the late Neo-Proterozoic. The lower Cambrian source rock was a stable deposit in the northern Tarim Basin, where the late Ediacaran carbonate was deposited, thinning out toward the central uplift. It was distributed throughout the entire Manjar region in the east and may be missing in the Magaiti and the southwestern Tarim Basin.

Acknowledgements

This study was supported by the National Natural Science Foundation of China (No. U19B6003) and Sinopec Science and Technology Major Project (P19022-7). The authors appreciate the Academic Editor Gemma Aiello for the detailed and constructive comments on the early version of manuscript. The authors also extend great thanks to Xiaoqiao Gao, Hongguang Liu, and Hangyu Liu for their field assistance.

Author details

Kaibo Shi[1], Bo Liu[1,2]*, Weimin Jiang[1], Jinxing Yu[1], Yue Kong[1], Tong Li[1] and Changhai Li[1]

1 School of Earth and Space Sciences, Peking University, Beijing, China

2 School of Earth Sciences, Yunnan University, Kunming, China

*Address all correspondence to: bobliu@pku.edu.cn

References

[1] Xu B, Jian P, Zheng H, Zou H, Zhang L, Liu D. U–Pb zircon geochronology and geochemistry of Neoproterozoic volcanic rocks in the Tarim block of Northwest China: Implications for the breakup of Rodinia supercontinent and Neoproterozoic glaciations. Precambrian Research. 2005;**136**(2):107-123. DOI: 10.1016/j.precamres.2004.09.007

[2] Xu B, Xiao S, Zou H, Chen Y, Li Z, Song B, et al. SHRIMP zircon U–Pb age constraints on Neoproterozoic Quruqtagh diamictites in NW China. Precambrian Research. 2009;**168**(3-4):247-258. DOI: 10.1016/j.precamres.2008.10.008

[3] Xu B, Zou H, Chen Y, He J, Wang Y. The Sugetbrak basalts from northwestern Tarim block of Northwest China: Geochronology, geochemistry and implications for Rodinia breakup and ice age in the late Neoproterozoic. Precambrian Research. 2013;**236**(5):214-226. DOI: 10.1016/j.precamres.2013.07.009

[4] Zhang C, Li Z, Li X, Lu S, Ye H, Li H. Neoproterozoic ultramafic–mafic-carbonatite complex and granitoids in Quruqtagh of northeastern Tarim block, western China: Geochronology, geochemistry and tectonic implications. Precambrian Research. 2007;**152**(3-4):149-169. DOI: 10.1016/j.precamres.2006.11.003

[5] Zhang C, Li Z, Li X, Ye H. Neoproterozoic mafic dyke swarms at the northern margin of the Tarim block, NW China: Age, geochemistry, petrogenesis and tectonic implications. Journal of Asian Earth Sciences. 2009;**35**(2):167-179. DOI: 10.1016/j.jseaes.2009.02.003

[6] Zhang C, Yang D, Wang H, Dong Y, Ye H. Neoproterozoic mafic dykes and basalts in the southern margin of Tarim, Northwest China: Age, geochemistry and geodynamic implications. Acta Geologica Sinica. 2010;**84**(3):549-562

[7] Zhang C, Yang D, Wang H, Yutaka T, Ye H. Neoproterozoic mafic-ultramafic layered intrusion in Quruqtagh of northeastern Tarim block, NW China: Two phases of mafic igneous activity with different mantle sources. Gondwana Research. 2011;**19**(1):177-190. DOI: 10.1016/j.gr.2010.03.012

[8] He J, Zhu W, Ge R. New age constraints on Neoproterozoic diamicites in Kuruktag, NW China and Precambrian crustal evolution of the Tarim Craton. Precambrian Research. 2014;**241**(1):44-60. DOI: 10.1016/j.precamres.2013.11.005

[9] He J, Zhu W, Ge R, Zheng B, Wu H. Detrital zircon U–Pb ages and Hf isotopes of Neoproterozoic strata in the Aksu area, northwestern Tarim Craton: Implications for supercontinent reconstruction and crustal evolution. Precambrian Research. 2014;**254**:194-209. DOI: 10.1016/j.precamres.2014.08.016

[10] Wang C, Wang Y, Liu L, He S, Li R, Li M, et al. The Paleoproterozoic magmatic–metamorphic events and cover sediments of the Tiekelik Belt and their tectonic implications for the southern margin of the Tarim Craton, northwestern China. Precambrian Research. 2014;**254**:210-225. DOI: 10.1016/j.precamres.2014.08.018

[11] Wang C, Liu L, Wang Y, He S, Li R, Li M, et al. Recognition and tectonic implications of an extensive Neoproterozoic volcano-sedimentary rift basin along the southwestern margin of the Tarim Craton, northwestern China. Precambrian Research. 2015;**257**:65-82. DOI: 10.1016/j.precamres.2014.11.022

[12] Wang C, Zhang J, Li M, Li R, Peng Y. Generation of ca. 900-870 Ma bimodal rifting volcanism along the southwestern margin of the Tarim Craton and its implications for the Tarim–North China connection in the early Neoproterozoic. Journal of Asian Earth Sciences. 2015;**113**:610-625. DOI: 10.1016/j.jseaes.2015.08.002

[13] Ge R, Zhu W, Wilde S, He J, Cui X, Wang X, et al. Neoproterozoic to Paleozoic long-lived accretionary orogeny in the northern Tarim Craton. Tectonics. 2014;**33**(3):302-329. DOI: 10.1002/2013tc003501

[14] Jia C. Structural characteristics and oil/gas accumulative regularity in Tarim Basin. Xinjiang Petroleum Geology. 1999;**20**(3):177-183. DOI: 10.3969/j.issn.1001-3873.1999.03.001

[15] Turner SA. Sedimentary record of late Neoproterozoic rifting in the NW Tarim Basin, China. Precambrian Research. 2010;**181**(1-4):85-96. DOI: 10.1016/j.precamres.2010.05.015

[16] Yang Y, Shi K, Liu B, Qing S, Wang J, Zhang X. Tectono-sedimentary evolution of the Sinian in the Northwest Tarim Basin. Chinese Journal of Geology. 2015;**49**(1):19-29. DOI: 10.3969/j.issn.0563-5020.2014.01.002

[17] Feng X, Liu Y, Han C, Yan W, Dong L, He Y. Sinian rift valley development characteristics in Tarim Basin and its guidance on hydrocarbon exploration. Petroleum Geology and Engineering. 2015;**29**(2):5-10

[18] Cui H, Tian L, Zhang N, Liu J. Nanhua-Sinian rift distribution and its relationship with the development of lower Cambrian source rocks in the southwest depression of Tarim Basin. Acta Petrolei Sinica. 2016;**37**(04):430-438. DOI: 10.7623/syxb201604002

[19] Wu L, Guan S, Ren R, Wang X, Yang H, Jin J, et al. The characteristics of Precambrian sedimentary basin and the distribution of deep source rock: A case study of Tarim Basin in Neoproterozoic and source rocks in early Cambrian, Western China. Petroleum Exploration and Development. 2016;**43**(6):988-999. DOI: 10.11698/PED.2016.06.07

[20] Guan S, Wu L, Ren R, Zhu G, Peng C, Zhao W, et al. Distribution and petroleum prospect of Precambrian rifts in the main cratons, China. Acta Petrolei Sinica. 2017;**38**(1):9-22. DOI: 10.7623/syxb201701002

[21] Shi K, Liu B, Tian J, Pan W. Sedimentary characteristics and lithofacies paleogeography of Sinian in Tarim Basin. Acta Petrolei Sinica. 2016;**37**(11):1343-1360. DOI: 10.7623/syxb201611003

[22] Shi K, Liu B, Liu H, Liu J, Pan W. Neoproterozoic tectono-sedimentary evolution in Quruqtagh area NE Tarim basin, Xinjiang, China. Earth Science Frontiers. 2011;**24**(1):297-307. DOI: 10.13745/j.esf.2017.01.020

[23] Shi K, Liu B, Jiang W, Luo Q, Gao X. Nanhua-Sinian tectono-sedimentary framework of Tarim Basin, NW China. Oil & Gas Geology. 2018;**39**(5):862-877. DOI: 10.11743/ogg20180502

[24] Jia C. Tectonic Characteristics and Petroleum Tarim Basin China. Beijing: Petroleum Industry Press; 1997

[25] Zhou X, Li J, Wang H, Li W, Cheng Y. The type of prototypic basin and tectonic setting of Tarim Basin formation from Nanhua to Sinian. Earth Science Frontiers. 2015;**22**(3):290-298. DOI: 10.13745/j.esf.2015.03.025

[26] Zhu G, Chen F, Chen Z, Zhang Y, Xing X, Tao X, et al. Discovery and basic characteristic of the high-quality source rocks of the Cambrian Yuertusi formation in Tarim Basin. Narural Gas Geoscience. 2016;**27**(1):8-21. DOI: 10.11764/j.issn.1672-1926.2016.01.0008

[27] Zhu C, Yan H, Yun L, Han Q, Ma H. Characteristics of Cambrian source rocks in well XH1, Shaya uplift, Tarom Basin. Petroleum Geology & Experiment. 2014;**36**(5):626-632. DOI: 10.11781/sysydz201405626

[28] Gu Y, Zhao Y, Jia C, He G, Luo Y, Wang B, et al. Analysis of hydrocarbon resource potential in Awati depression of Tarim Basin. Petroleum Geology & Experiment. 2012;**34**(3):257-266

[29] Wang Z, Xie H, Chen Y, Qi Y, Zhang K. Discovery and exploration of Cambrian subsalt dolomite original hydrocarbon reservoir at Zhongshen-1 well in Tarim Basin. China Petroleum Exploration. 2014;**19**(2):1-13. DOI: 10.3969/j.issn.1672-7703.2014.02.001

[30] Xu Z, He B, Zhang C, Zhang J, Wang Z, Cai Z. Tectonic framework and crustal evolution of the Precambrian basement of the Tarim block in NW China: New geochronological evidence from deep drilling samples. Precambrian Research. 2013;**235**:150-162. DOI: 10.1016/j.precamres.2013.06.001

[31] Lu S, Li H, Zhang C, Niu G. Geological and geochronological evidence for the Precambrian evolution of the Tarim Craton and surrounding continental fragments. Precambrian Research. 2008;**160**:94-107. DOI: 10.1016/j.precamres.2007.04.025

[32] Long X, Yuan C, Sun M, Zhao G, Xiao W, Wang Y, et al. Archean crustal evolution of the northern Tarim craton, NW China: Zircon U–Pb and Hf isotopic constraints. Precambrian Research. 2010;**180**:272-284. DOI: 10.1016/j.precamres.2010.05.001

[33] Zhang C, Li H, Santosh M, Li Z, Zou H, Wang H, et al. Precambrian evolution and cratonization of the Tarim block, NW China: Petrology, geochemistry, Nd-isotopes and U–Pb zircon geochronology from Archaean gabbro-TTG–potassic granite suite and Paleoproterozoic metamorphic belt. Journal of Asian Earth Sciences. 2012;**47**(1):5-20. DOI: 10.1016/j.jseaes.2011.05.018

[34] Zhang C, Zou H, Santosh M, Ye X, Li H. Is the Precambrian basement of the Tarim Craton in NW China composed of discrete terranes? Precambrian Research. 2014;**254**:226-244. DOI: 10.1016/j.precamres.2014.08.006

[35] Huang B, Xu B, Zhang C, Li Y, Zhu R. Paleomagnetism of the Baiyisi volcanic rocks (ca. 740 Ma) of Tarim, Northwest China: A continental fragment of Neoproterozoic Western Australia? Precambrian Research. 2005;**142**(3-4):83-92. DOI: 10.1016/j.precamres.2005.09.006

[36] Li Z, Bogdanova S, Collins A, Dacidson A, Waele B, Ernst R, et al. Assembly, configuration, and break-up history of Rodinia: A synthesis. Precambrian Research. 2008;**160**(1-2):179-210. DOI: 10.1016/j.precamres.2007.04.021

[37] Li Z, Evans DA, Halverson GP. Neoproterozoic glaciations in a revised global palaeogeography from the breakup of Rodinia to the assembly of Gondwanaland. Sedimentary Geology. 2013;**294**:219-232. DOI: 10.1016/j.sedgeo.2013.05.016

[38] Zhao P, Chen Y, Zhan S, Xu B, Faure M. The apparent polar wander path of the Tarim block (NW China) since the Neoproterozoic and its implications for a long-term Tarim–Australia connection. Precambrian Research. 2014;**242**(1):39-57. DOI: 10.1016/j.precamres.2013.12.009

[39] Bin W, David A, Li Y. Neoproterozoic paleogeography of the Tarim block: An extended or alternative "missing-link" model for Rodinia? Earth and Planetary Science

Letters. 2017;**1**:1-15. DOI: 10.1016/j.epsl.2016.10.030

[40] Ren R, Guan S, Zhang S, Wu L, Zhang H. How did the peripheral subduction drive the Rodinia breakup: Constraints from the Neoproterozoic tectonic process in the northern Tarim Craton. Precambrian Research. 2020;**339**:1-17. DOI: 10.1016/j.precamres.2020.105612

[41] Chen W, Zhu G, Zhang K, Zhang Y, Yan H, Du D, et al. Late Neoproterozoic intracontinental rifting of the Tarim carton, NW China: An integrated geochemical, geochronological and Sr–Nd–Hf isotopic study of siliciclastic rocks and basalts from deep drilling cores. Gondwana Research. 2020;**80**:142-156. DOI: 10.1016/j.gr.2019.10.007

[42] Xiao S, Bao H, Wang H, Alan J, Zhou C, Li G, et al. The Neoproterozoic Quruqtagh group in eastern Chinese Tianshan: Evidence for a post-Marinoan glaciation. Precambrian Research. 2004;**130**:1-26. DOI: 10.1016/j.precamres.2003.10.013

[43] Gao Z, Wang W, Peng W. The Sinian System of Xinjiang. Urumqi: Xinjiang People's Publishing House; 1985

[44] Xia B, Zhang L, Du Z, Xu B. Petrology and age of Precambrian Aksu blueschist, NW China. Precambrian Research. 2019;**326**:295-311. DOI: 10.1016/j.precamres.2017.12.041

[45] Zhang J, Zhang C, Li H, Ye X, Geng J, Zhou Y. Revisit to time and tectonic environment of the Aksu blueschist terrane in northern Tarim, NW China: New evidence from zircon U-Pb age and Hf isotope. Acta Petrologica Sinica. 2014;**30**(11):3357-3365

[46] Tong Q, Wei W, Bei X. Neoproterozoic sedimentary facies and glacial periods in the southwest of Tarim block. Science China: Earth Sciences. 2013;**56**(6):901-912. DOI: 10.1007/s11430-013-4595-4

[47] Zhou X. Deep level structures and paleogeography reconstruction of Tarim Basin, NW China [thesis]. Beijing: Peking University; 2015

[48] Ma Y, Luo J, Tang Y, Li Y, Wu Q, Li W. Geological age of the basement of the eastern segment of the Kuche depression and its geological significances. Chinese Journal of Geology. 2011;**46**(2):475-482

[49] Han Q, Zhu Y, Zhu C, Wang C, Chen Z, Fei J. Petrological characteristics and zircon U-Pb age for magmatic rocks from pre-Sinian basement of the SDQ area of Shaya Rise in Tarim Basin, NW China. Acta Petrologica Sinica. 2016;**32**(5):1493-1504

[50] Yang X, Xu X, Chen Q, Qian Y, Chen Y, Chu C. Palaeotectonics pattern in pre-Cambrian and its control on the deposition of the lower Cambrian source rocks in Tarim Basin, NW China. Natural Gas Geoscience. 2017;**25**(8):1164-1171. DOI: 10.11764/j.issn.1672-1926.2014.08.1164

[51] Wu G, Li H, Xu Y, Su W, Chen Z, Zhang B. The tectonothermal events, architecture and evolution of Tarim craton basement palaeo-uplifts. Acta Petrologica Sinica. 2012;**28**(8):2435-2452

[52] Yang W, Zhang L, Xu Y, Yu C, Yu P, Zhang B, et al. Three dimensional electrical resistivity structure of the Tarim basin. Acta Geologica Sinica. 2015;**89**(12):2203-2212

[53] Ye X. Precambrian tectonic evolution and crust growth of southern Tarim Terrane, Xinjiang, NW China [thesis]. Beijing: Chinese Academy of Geological Sciences; 2016

[54] Zhang C, Ye X, Zou H, Chen X. Neoproterozoic sedimentary basin evolution in southwestern Tarim, NW China: New evidence from field observations, detrital zircon U–Pb ages and Hf isotope compositions. Precambrian Research. 2016;**280**:31-45. DOI: 10.1016/j.precamres.2016.04.011

[55] Zhu W, Zheng B, Shu L, Ma D, Wu H, Li Y, et al. Neoproterozoic tectonic evolution of the Precambrian Aksu blueschist terrane, northwestern Tarim, China: Insights from LA-ICP-MS zircon U–Pb ages and geochemical data. Precambrian Research. 2011;**185**(3-4):215-230. DOI: 10.1016/j.precamres.2011.01.012

[56] Guo Z, Zhang Z, Liu S, Li H. U-Pb geochronological evidence for the early Precambrian complex of the Tarim Craton, NW China. Acta Petrologica Sinica. 2003;**19**(3):537-542

[57] Zheng B, Zhu W, Shu L, Zhang Z, Yu J, Huang W. The protolith of the Aksu Precambrian blueschist and its tectonic setting. Acta Petrologica Sinica. 2008;**24**(12):2839-2848

[58] Wang F, Wang B, Shu L. Continental tholeiitic basalt of the Akesu area (NW China) and its implication for the Neoproterozoic rifting in the northern Tarim. Acta Petrologica Sinica. 2010;**26**(2):547-558

[59] Wang C. Precambrian tectonic of south margin of Tarim Basin, NW China [thesis]. Xian: Northwest University; 2011

[60] Zhang J, Li H, Meng F, Xiang Z, Yu S, Li J. Polyphase tectonothermal events recorded in metamorphic basement from the Altyn Tagh, the southeastern margin of the Tarim basin, western China: Constraint from U-Pb zircon geochronology. Acta Petrologica Sinica. 2011;**27**(1):23-46

[61] Ge R. Precambrian tectono-thermal events and crustal evolution in the Kuruktag Block, Northern Tarim Craton [thesis]. Nanjing: Nanjing University; 2014

[62] Zhang Z, Kang J, Kusky T, Santosh M, Huang H, Zhang D, et al. Geochronology, geochemistry and petrogenesis of Neoproterozoic basalts from Sugetbrak, Northwest Tarim block, China: Implications for the onset of Rodinia supercontinent breakup. Precambrian Research. 2012;**220-221**(8):158-176. DOI: 10.1016/j.precamres.2012.08.002

[63] Gehrels GE, Yin A, Wang X. Detrital-zircon geochronology of the northeastern Tibetan plateau. Geological Society of America Bulletin. 2003;**115**(7):881-896. DOI: 10.1130/00167606(2003)115<0881:dgotnt>2.0.co;2

[64] Wang C, Liu L, Yang W, Zhu X, Cao Y, Chen S, et al. Provenance and ages of the Altyn complex in Altyn Tagh: Implications for the early Neoproterozoic evolution of northwestern China. Precambrian Research. 2013;**230**:193-208. DOI: 10.1016/j.precamres.2013.02.003

[65] Wang L, Zhang W, Duan X, Long X, Ma Z, Song Z, et al. Isotopic age and genesis of the mozogranitic gneiss at the Huanxingshan in middle Altyn Tagh. Acta Petrologica Sinica. 2015;**31**(1):119-132

[66] Yin D, Zheng Y, Wu H. Study on the tectonic setting and geologic feature of Sailajiazitage group in western Kunlun. Xinjiang Geology. 2014;**32**(3):295-300. DOI: 10.3969/j.issn.1000-8845.2014.03.003

[67] Yang Z. Early palaeozoic tectonic evolution in Hongliugou, Altyn, Xinjiang [thesis]. Beijing: Chinese Academy of Geological Sciences; 2012

[68] Liu S, Sun W, Song J, Deng B, Zhong Y, Luo C, et al. Tectonics-controlled distribution of marine petroleum accumulations in the Sichuan Basin, China. Earth Science Frontiers. 2015;**22**(3):146-160. DOI: 10.13745/j.esf.2015.03.013

[69] Liu S, Wang Y, Sun W, Zhong Q, Hong H, Deng B, et al. Control of intracratonic sags on the hydrocarbon accumulations in the marin strata across the Sichuan Basin, China. Journal of Chengdu University of Technology. 2016;**43**(1):1-23. DOI: 10.3969/j.issn.1671-9727.2016.01.01

[70] Du J, Wang Z, Zou C, Xu C, Shen P, Zhang B, et al. Discovery of intra-cratonic rift in the Upper Yangtze and its control effect on the formation of Anyue giant gas field. Acta Petrolei Sinica. 2016;**37**(1):1-16. DOI: 10.7623/syxb201601001

Chapter 12

The Onshore Southern Carnarvon Basin in Coastal Western Australia during the Quaternary: Tectonic Setting and Facies-Complicated Heterogeneous Stratigraphic Patterns

Vic Semeniuk and Margaret Brocx

Abstract

The onshore southern Carnarvon Basin in Western Australia, in existence since the early Palaeozoic, has a history during the Palaeozoic and Tertiary of relatively uniform sedimentary styles with thick laterally-extensive sequences of sediment. Its sedimentary history became more complicated in the Quaternary period with complex tectonics and arrays of sedimentary facies and packages and basin complexity over relatively short distances, with several regions that are sedimentologically and stratigraphically distinct related to the factors of physiographic and geological setting, riverine input, arid climate, migrating climate, tectonism, and degree of protection from open ocean. For the Pleistocene and Holocene epoch, there are distinct north-trending stratigraphic packets, each with their environmentally distinctive shoaling facies sharply juxtaposed against each other or separated by Pleistocene non-marine sediments; in geographic order, from south to north, these are: a limestone aeolianite barrier along western Shark Bay; pocket seagrass bank carbonate complexes of central western Shark Bay that are nestled in the northerly-oriented inter-dune depressions developed as swales of the north-trending parabolic dunes deriving from the limestone aeolianite barrier; an aeolian red sand shoestring of the north-trending Peron Peninsula longitudinally bisecting central Shark Bay; metahaline to hypersaline shoaling carbonate sedimentary packages of south-eastern Shark Bay that fringe Hamelin Pool; the Wooramel delta, a wave-dominated delta composed of quartz sand and locally-generated carbonate sediment; the Wooramel seagrass bank (an extensive shore-parallel wedge of seagrass bank carbonate sequence along the eastern coast, central to northern Shark Bay); metahaline carbonate and quartz sand platforms fringing both sides of the red-sand Peron Peninsula; metahaline to hypersaline carbonate sediments that underlie the deeper-water axially-oriented embayments of Shark Bay; the Boodalia Pleistocene reddened (quartzose) deltaic sediment sequence; the Gascoyne Delta and laterally equivalent beach-ridge complex, the former comprising subtidal quartz-dominated sand capped by tidal sand-and-mud sequences, and the latter comprising subtidal quartz-dominated sand capped by beach-to-beach-ridge deposits; the Lake MacLeod evaporite basin filled with a shoaling sequence of carbonate sediments, halite, and gypsite; Tertiary limestone and Pleistocene aeolian sediments acting as a barrier to Lake McLeod; and the uplifted Tertiary limestone barrier of Cape Range that is fringed by Holocene coral complexes of the Ningaloo Reef. The coastal and onshore near-coastal southern Carnarvon Basin is an example of a complex sedimentary basin, where sedimentary packages can be markedly different over short distances, and illustrates the complexities a geologist would face if analyzing such a basin in the stratigraphic column. This feature of extreme diversity of sedimentary facies and packages within and between separate contemporaneous 'sedimentary basins' is the theme of this contribution.

Keywords: Carnarvon Basin, Western Australia, Quaternary, Shark Bay, Lake MacLeod, Gascoyne Delta, Quaternary tectonics, facies-complicated sedimentary basins

1. Introduction

The onshore epicratonic Carnarvon Basin in the mid western part of Western Australian has existed since the early Palaeozoic and its Palaeozoic and Tertiary history was generally one of relatively uniform sedimentary styles with thick laterally-extensive sequences of sediment [1, 2]. In this context, it conforms with many sedimentary basins throughout the World (*e.g.*, the Sydney Basin [3], the Kimberley Basin [4], the Eucla Basin [5], the Paris Basin [6, 7], the Paradox Basin [8, 9], the Tindouf Basin [10]; the Gourara Basin [11], the Karoo Basins [12], amongst many others). However, the sedimentary history of the Carnarvon Basin became more complicated in its southern part in the Quaternary period with complex tectonics and arrays of distinct but separated sedimentary facies and sedimentary packages.

During the Pleistocene-Holocene in the southern part of the Carnarvon Basin, the coastal and near-coastal onshore zone has basin complexity over its relatively short latitudinal distance of 500 km, with several separate regions distinct from both a sedimentologic and stratigraphic point of view, as related to several control factors, including the physiographic and geological setting, the river input, the arid climate, the tectonic setting, and the degree of protection from the open ocean. As a consequence of these controls, distinct south-north trending sedimentary packages have formed, each one extending tens of kilometres in length and several kilometres in width. These distinct sedimentary packages are completely different one from each other and are either generally sharply latitudinally and longitudinally juxtaposed one against each other or separated by Pleistocene uplands or fluvial non-marine sediments. Each one of the south-north trending stratigraphic sequences has environmentally distinctive shoaling facies and, being separated over short distances by uplands, gives the impression of a series of closely juxtaposed 'sub-basins'. In fact, particularly for the Holocene, the stratigraphic packages in the southern Carnarvon Basin illustrates the diversity and the complexity of laterally equivalent contemporaneous units that would provide a geologist working in the stratigraphic record a difficulty and a dilemma on how to interpret and to correlate various facies and intervening uplands.

2. The Carnarvon Basin: its geological and tectonic setting

The Carnarvon Basin is an epicratonic, faulted and folded basin some 535,000 km^2 in the offshore portion and some 115,000 km^2 in the onshore area.

Existing since the Palaeozoic Era, it is a basin elongated northeast-southwest, and is bordered by Precambrian rocks of the Pilbara Craton and Yilgarn Craton [13], and is adjoined by the Perth Basin to the south and the Canning Basin to the north (**Figure 1**). It has been subdivided into a range of sub-basins and tectonic ridges, the most important of which to this paper is the large-scale division into Northern Carnarvon Basin and Southern Carnarvon Basin (**Figure 2**). But while there is crustal sagging and crustal faulting on the regional scale in the general Carnarvon Basin [1], tectonism did not seem to play a major part at smaller scales during sedimentation of formational units as it has during the Quaternary epoch. The Palaeozoic, Mesozoic, and lower to middle Cainozoic sedimentary sequences of the

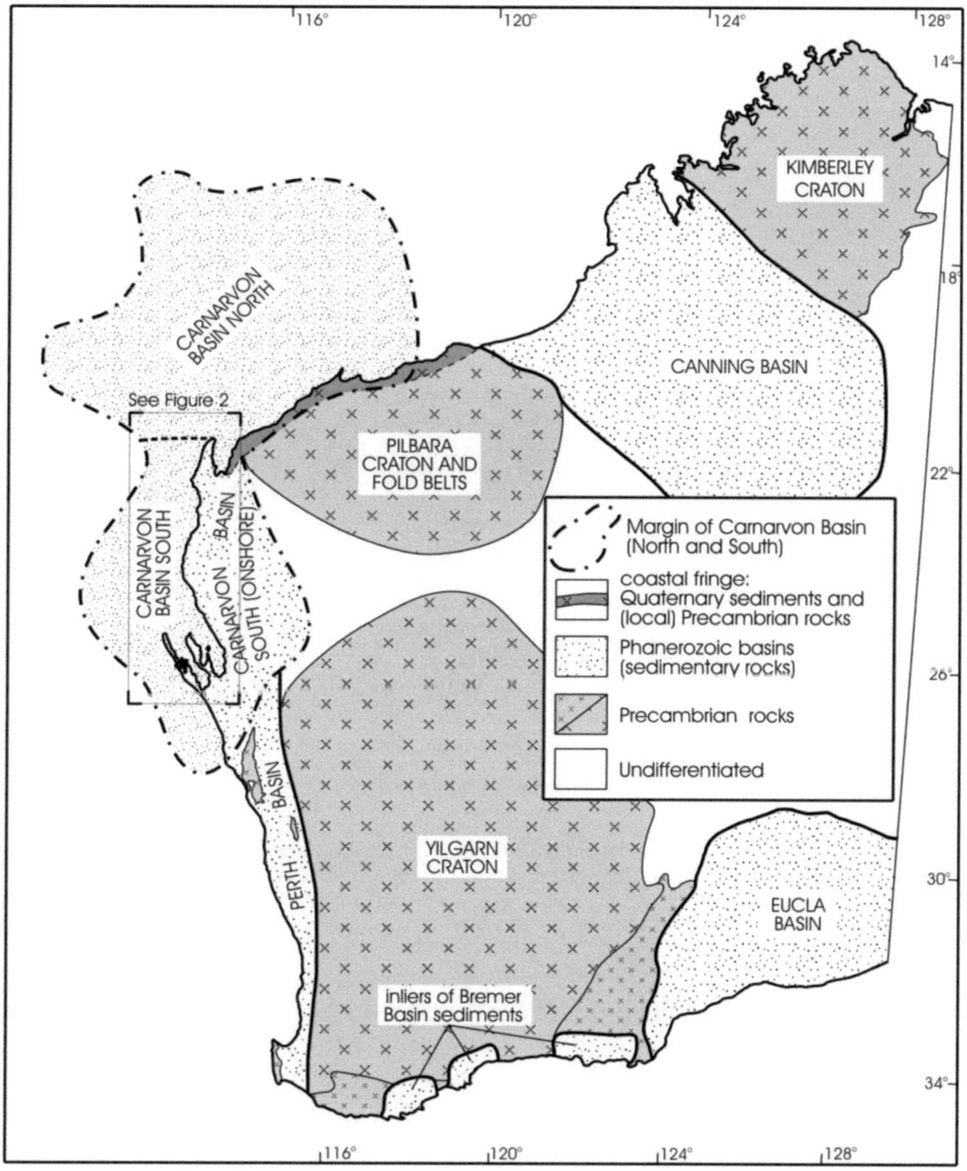

Figure 1.
Geological map of Western Australia (from Brocx and Semeniuk [14]) showing cratons and basins, the location of the Carnarvon Basin, and subdivision of the Carnarvon Basin into northern and southern basins.

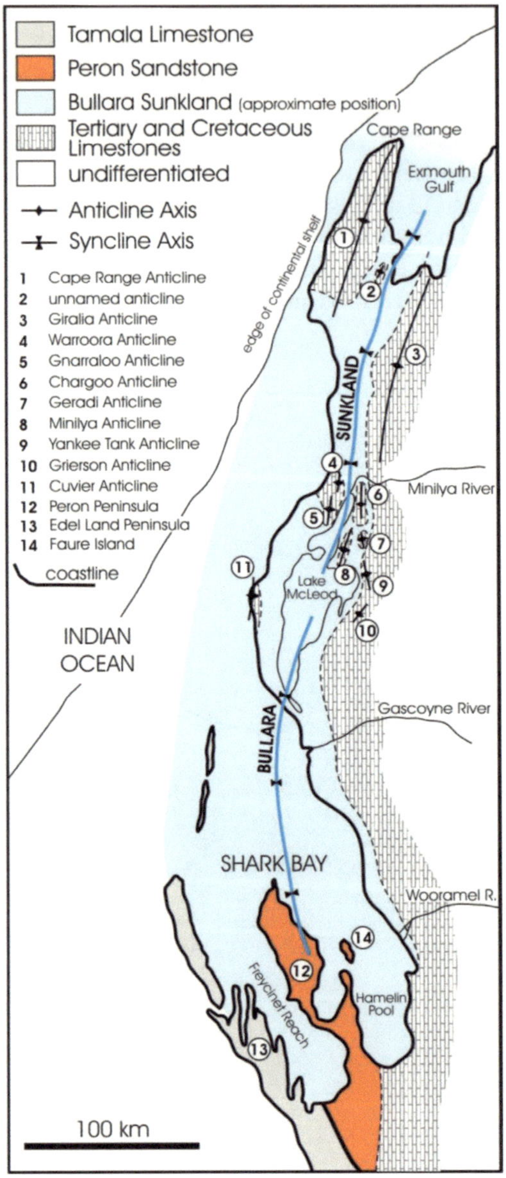

Figure 2.
Geological map of Southern Carnarvon Basin Western Australia (modified from [15]) showing the tectonic framework (with the main emerging tectonic structures as anticlines) and the major depositional depression, viz., the Bullara Sunkland. Logan [16] subdivided the Bullara Sunkland into a southern part termed the 'Shark Bay Depression', a central part termed the 'MacLeod Graben', and a northern part termed the 'Dingo Syncline'. Main geographic names are also shown.

southern Carnarvon Basin comprise formational sheets that are laterally and basin-wide extensive [1, 2] (**Figure 2**). There is an extensive literature on the Northern Carnarvon Basin and the inland Southern Carnarvon Basin [17–22] but information on their geology, sub-basins, and stratigraphy is not relevant to the coastal and near-coastal onshore zone of the Southern Carnarvon Basin as presented in this Chapter as the latter in the Quaternary had a change in basin dynamics (with development of smaller isolated sedimentary packages formed in environmentally discrete areas), and is therefore distinct from the general geological setting of the Carnarvon Basin.

3. The southern Carnarvon Basin in the Quaternary: a mosaic of facies-complicated sedimentary basins

The Quaternary southern Carnarvon Basin, arrayed along some 500 km of the Western Australian coastal and near-coastal zone, shows tectonic and palaeogeographic complication and inter-basinal heterogeneity. There are a multitude of smaller-scale depositional sites (or 'depositional basins'), each one limited in area extent and each one sharply bordered by uplands or by a contrasting sedimentation style. There are thirteen main Quaternary (Pleistocene and Holocene) stratigraphic packages along the southern Carnarvon Basin; ordered geographically from south to north, these are (**Figure 3**): a limestone aeolianite barrier along western Shark Bay; pocket seagrass bank carbonate complexes of central western Shark Bay that are nestled in the northerly-oriented inter-dune depressions developed as swales of the north-trending parabolic dunes deriving from the limestone aeolianite barrier; an aeolian red sand shoestring of the north-trending Peron Peninsula longitudinally bisecting central Shark Bay; metahaline to hypersaline shoaling carbonate sedimentary packages of south-eastern Shark Bay that fringe Hamelin Pool; the Wooramel delta, a wave-dominated delta composed of quartz sand and locally-generated carbonate sediment; the Wooramel seagrass bank (an extensive shore-parallel wedge of seagrass bank carbonate sequence along the eastern coast, central to northern Shark Bay); metahaline carbonate and quartz sand platforms fringing both sides of the red-sand Peron Peninsula; metahaline to hypersaline deep basin carbonate sediments that underlie the deeper-water axially-oriented embayments of Shark Bay;

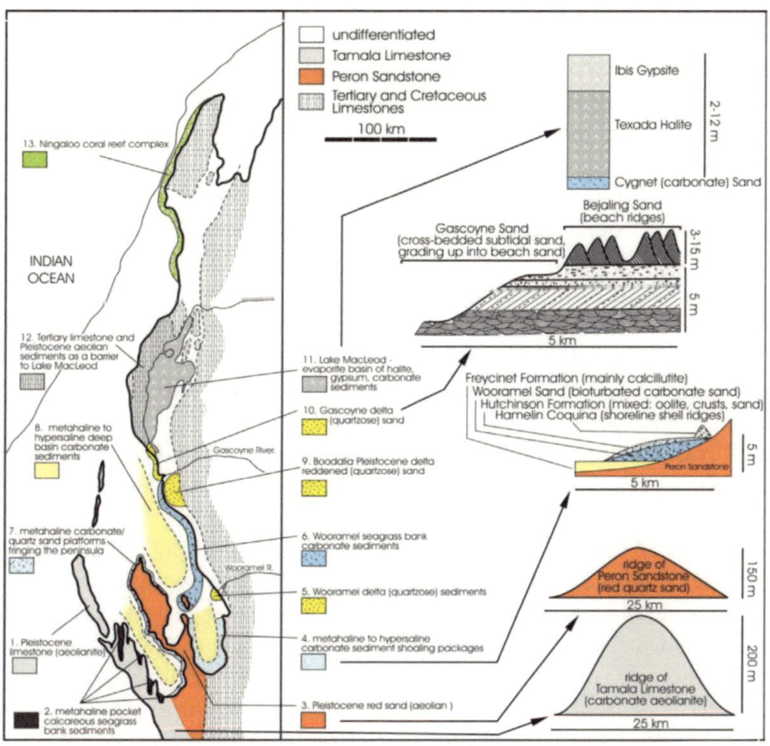

Figure 3.
Distribution of the 13 main Quaternary stratigraphic packages in the Southern Carnarvon Basin. Most of the stratigraphic packages are Holocene in age, with the Pleistocene and Tertiary rocks (and some Cretaceous rocks) noted in the Legend. Of special note is the south-to-north elongation of the various Holocene stratigraphic packages and the older stratigraphic units.

The Boodalia—a Pleistocene reddened (quartzose) deltaic sediment sequence; the Gascoyne Delta and laterally equivalent beach-ridge complex, the former comprising subtidal quartz-dominated sand and tidal sand-and-mud sequences and the latter comprising subtidal quartz-dominated sand capped by beach to beach-ridge deposits; the Lake MacLeod evaporite basin filled with a shoaling sequence of carbonate sediments, halite, and gypsite; Tertiary limestone and Pleistocene aeolian sediments acting as a barrier to Lake McLeod; and uplifted Tertiary limestone barrier of Cape Range that is fringed by Holocene coral-reef complexes of the Ningaloo Reef. Selected stratigraphic profiles of the thirteen packages are shown in **Figure 3**.

The stratigraphic packages are separated by 1. tectonically-uplifted Tertiary limestone, 2. emerging Quaternary anticlines form emergent ridges, or 3. aeolian-emplaced Quaternary sediments, with the packages occurring in relatively small and distinct self-contained well-defined depositional basins (such as the bowls of parabolic dunes; terms from Semeniuk et al. [23]). The stratigraphic sequences of the thirteen packages are described in **Table 1**, and their distribution is illustrated in **Figure 4**. Detailed stratigraphy and lithologies of the Holocene-age stratigraphic sequences are described in various publications [15, 16, 24–28, 30, 31], and summarised in **Table 2**. **Figure 4** illustrates the location of the distinct and discrete sedimentary packages and their inter-relationship as arrayed along the length of the southern Carnarvon Basin.

There are a number of reasons why these stratigraphic packages have developed as separate systems and separated in these sedimentologically distinct geographic regions; they are: physiographic and geological setting, riverine input, arid climate, migrating climate, tectonism, and degree of protection from open ocean.

The physiography and geology of this part of the coastal and near-coastal Carnarvon basin comprises 1. tectonically-emerging ridges (that Logan et al. [15] term 'anticlines'), 2. aeolian-emplaced coastal calcareous sand bodies (that when cemented form resistant upstanding ridges and, when mobilized into parabolic dunes and when later cemented form a limestone-rocky invaginated coast), and 3. aeolian-emplaced coastal quartz sand ridge. These large-scale and smaller-scale ridges and uplands form the cradle within which Holocene sedimentation was localised. As such, the Holocene sedimentation was/is occurring in lowlands, or along the open coast. The Pleistocene stratigraphic units generally form the uplands to the Holocene sediments as limestone aeolianites or red sand dune deposits [15].

Locally, riverine input has developed deltoid sedimentary accumulations along the coast, *viz.*, the Wooramel Delta and the Gascoyne Delta. The Gascoyne Delta, since it is partly exposed to open Indian Ocean conditions and strong southerly winds, is an extremely asymmetric delta, with northerly-propelled deltaic sand forming a ribbon of shore-parallel beach ridges north of the river mouth.

The arid climate is a major factor determining sedimentation patterns and diagenesis of the Holocene sediments and, during the Holocene, with Earth axis precession the Tropic of Capricorn (separating the Tropical climate from Subtropical climate) has migrated northwards-centred on Shark Bay several thousand years ago it is now located near Exmouth Gulf [30]. This was one of the factors that changed the sedimentation styles in Shark Bay and along the coast of Cape Range.

In recent times, Shark Bay is partitioned into hydrochemical fields from its oceanic exterior to its protected interior (*e.g.*, Hamelin Pool, **Figure 2**), *viz.*, the salinity changes from oceanic in the north to metahaline and then hypersaline in southern parts [31]. This is partly due to its south-eastern interior portions being isolated from oceanic influences and, as such, there is more restricted circulation from ocean to interior, but also is due to the extreme evaporation acting on the south-easterly restricted parts of the Bay and to the evolution of a circulation-restricting

Stratigraphic package (south to north)	Height, thickness, width, and length of the stratigraphic package[2]	Main stratigraphic units[3]	Relation to adjoining uplands and to adjoining stratigraphic packages
1. Limestone aeolianite barrier along western Shark Bay [15]	10–40 m above MSL. 10–20 km wide, 180 km long, up to 150 m thick	Tamala limestone: a Pleistocene aeolianite comprising carbonate and quartz calcarenite	Forms a barrier to Shark Bay
2. Pocket seagrass bank carbonate complexes of central western Shark Bay [24, 25]	In small packets within the inlets of eastern shore Edel Land 8 km × 3 km, up to 30 km × 6 km; 3–5 m thick seaward-thickening wedge	Boat Haven Sand[3], a stratigraphic unit of bioturbated sand formed under seagrass bank conditions	Packets of seagrass-bank-dominated calcareous sediment cradled between uplands in a limestone rocky shore invaginated coastline
3. Aeolian red sand shoestring of the north-trending Peron Peninsula longitudinally bisecting central Shark Bay [15]	120 km long × 20 km wide, with heights 20–70 m above MSL	Peron Sandstone: a Pleistocene red sand broadly elongate dune complex with cuspate, straight, curved, and irregular margins (former coastal barrier); stratigraphically, composed of several episodes of dune building separated by soils	Forms a peninsula partition within Shark Bay and the source of quartz sand within the Shark Bay depositional system
4. Metahaline to hypersaline shoaling sedimentary package of south-eastern Shark Bay [25]	3–10 km wide, 100 km overall length as measured along the Hamelin Basin shore	Wooramel Sand[3] (earlier Holocene metahaline seagrass bank sediments) overlain by Hutchinson Formation and (onshore) the shoreline ribbon of Hamelin Coquina	Ribbon-like (metahaline) calcareous seagrass bank sediments shore-parallel and flanking the south-eastern shore of Shark Bay lithologically passing up into (hypersaline) oolite sediment and tidally into stromatolites
5. Wooramel Delta, central-eastern Shark Bay [15]	Triangular wedge 15 km × 15 km of sand, 1–2 m thick, with local spits and cheniers	Un-named	Deltoid accumulation at the mouth of the Wooramel River
6. Wooramel seagrass bank deposits: extensive shore-parallel wedge of seagrass bank carbonate sequence eastern coast, central to northern Shark Bay [24–26]	5 km × 90 km, wedging up to 3 m thick	Wooramel Sand[3]	Wedge-like calcareous seagrass bank sediments flanking the eastern shore of Shark Bay

Stratigraphic package (south to north)	Height, thickness, width, and length of the stratigraphic package[2]	Main stratigraphic units[3]	Relation to adjoining uplands and to adjoining stratigraphic packages
7. Mainly wave-built platforms and some seagrass bank deposits: extensive shore-parallel wedge of sediment western and eastern coast of the Peron Peninsula, central Shark Bay [25]	30 km long, 5–6 km wide; subtidal units up to 3 m thick, tidal units up to 1 m thick, beach-ridge units 3–15 m thick	Denham Sand[3]—metahaline wave-built sand platform and seagrass bank sediments	Sharply adjoins the Wooramel (seagrass bank) Sand, and onlaps the red sand hinterland
8. Deep water marine embayment plains (10–15 m deep) underlain by calcareous mud, shelly mud, and sand [25]	100 km long, 20–25 km wide, 5–6 km wide, 0.5–1.5 m thick	Freycinet formation[3]	Sharply adjoins the shallow water seagrass bank sediments and wave-built platform sediments
9. The Boodalia, a Pleistocene reddened (quartzose) deltaic deposit, as a subset of the Gascoyne Delta, and its laterally equivalent beach-ridge complex north of Shark Bay [15]	25 km along coast, 25 km wide, up to 6 m thick [26], triangular seaward-lobed sedimentary deposit	Boodalia Formation[3]—reddened deltaic sediment complex	Sharply and laterally adjoins the Gascoyne Sand and Babbage Island Formation, and underlies and laterally adjoins the Wooramel (seagrass bank) Sand
10. Gascoyne Delta and laterally equivalent beach-ridge complex (Bejaling Beach Ridges) north of Shark Bay [27]	30 km long, 5–6 km wide; subtidal units up to 3 m thick, tidal units up to 1 m thick, beach-ridge units 3–15 m thick	Babbage Island Formation[3]—beach-ridge and tidal lagoon complexes at the river mouth; Gascoyne Sand[3]—the offshore and near-shore cross-bedded sand that forms the delta slope of the Gascoyne Delta; Bejaling Sand[3]—beach ridge complex to the north of the river mouth	Sharply and laterally adjoins the Wooramel (seagrass bank) Sand, and onlaps the red sand hinterland
11. Tertiary-limestone-barred evaporite sequence of Lake McLeod [16]	Evaporites 130 km long, 20–40 km wide, 8 m thick	MacLeod Evaporite comprised of the members Texada Halite, Ibis Gypsite, and Cygnet Carbonate Member [16]	Evaporite sequence cradled in a graben
12. Barrier of Tertiary limestone and cemented Pleistocene aeolian sediments as a barrier to Lake McLeod [16]	Tertiary and Pleistocene limestone barrier 20–50 m above MSL 130 km long	Thickness depending on height (*i.e.*, 20–50 m thick)	Sharply bordered by Holocene sediments

Stratigraphic package (south to north)	Height, thickness, width, and length of the stratigraphic package[2]	Main stratigraphic units[3]	Relation to adjoining uplands and to adjoining stratigraphic packages
13. Holocene Ningaloo coral-reef complex [28] fringing the uplifted Tertiary limestone barrier of Cape Range	Coral reef 150 km long, 2–5 km wide, up to 5 m thick	Ningaloo Limestone[3]	Plastered on and flanking rocky shore cut into Tertiary limestone

[1]*Location of stratigraphic packages shown in* ***Figure 3****.*
[2]*These geomorphic and stratigraphic systems are variable and the figures for length, width, and height, while providing an estimate of their dimensions, are indicative only.*
[3]*Some of the stratigraphic units were not properly defined in the various original publications on this region [15, 16, 27, 28], and are in the process of being formally named [29].*

Table 1.
Description and settings of the 13 major stratigraphic packages in generalised sequences.[1]

shallow-water submarine seagrass bank that evolved in the Faure Island area [31]. Seagrass bank sedimentation occurs in those parts of Shark Bay where there is oceanic and metahaline salinity, and is eliminated when salinity becomes hypersaline.

Much of Shark Bay is protected from the open ocean and, as such, many of the facies therein reflect this. Where the coast is exposed to the Indian Ocean, it is subject to prevailing swell, wind waves, and strong sea breezes, and wave-dominated conditions prevail, *e.g.*, the wave-dominated and asymmetric delta of the Gascoyne River. Further, where there is a rocky coast (such as the tectonically-uplifted Cape Range), with open ocean conditions and a tropical climate, coral reefs are developed (*e.g.*, the Ningaloo Reef).

4. Discussion and conclusions

The Quaternary Southern Carnarvon Basin in its modern sedimentology, palaeosedimentology, and palaeogeography is an array of facies-complicated patterns determined by megascale geomorphic architecture, tectonics, oceanography, and climate. It contrasts with other sedimentary basins mentioned above (*e.g.*, the Paris Basin, the Paradox Basin, the Sydney Basin, amongst others) that have more laterally-extensive formational sheets. Indeed, the Quaternary Southern Carnarvon Basin *does not have latitudinally and longitudinally extensive formational sheets*, and the individual sedimentary packages of significant thicknesses are discrete and cannot (axiomatically) be correlated from depositional basin to depositional basin. Each of these sedimentary packages would appear in the geological record as discrete isolated lenses (each with its own internal sedimentary signatures and sedimentary sequences indicative of an internal relationship of lithofacies within the package), and would be assigned to different formations.

To highlight and contrast this facies-complicated array within the Southern Carnarvon Basin, we use the coastal zone of the Perth Basin in the southwest of Western Australia, the Canning Coast in northwest of Western Australia, and selected basins from around the World. In this comparison, we have direct field experience with the Perth Basin, Canning Basin, Kimberley Basin, Eucla Basin, the Paris Basin, the Tindouf Basin, and southern parts of the Karoo Basins. The remaining information has been obtained from the literature.

The coastal zone of the Perth Basin extends for some 400 km and (apart from the occasional shore-normal narrow estuary) portrays a fairly consistent sequence

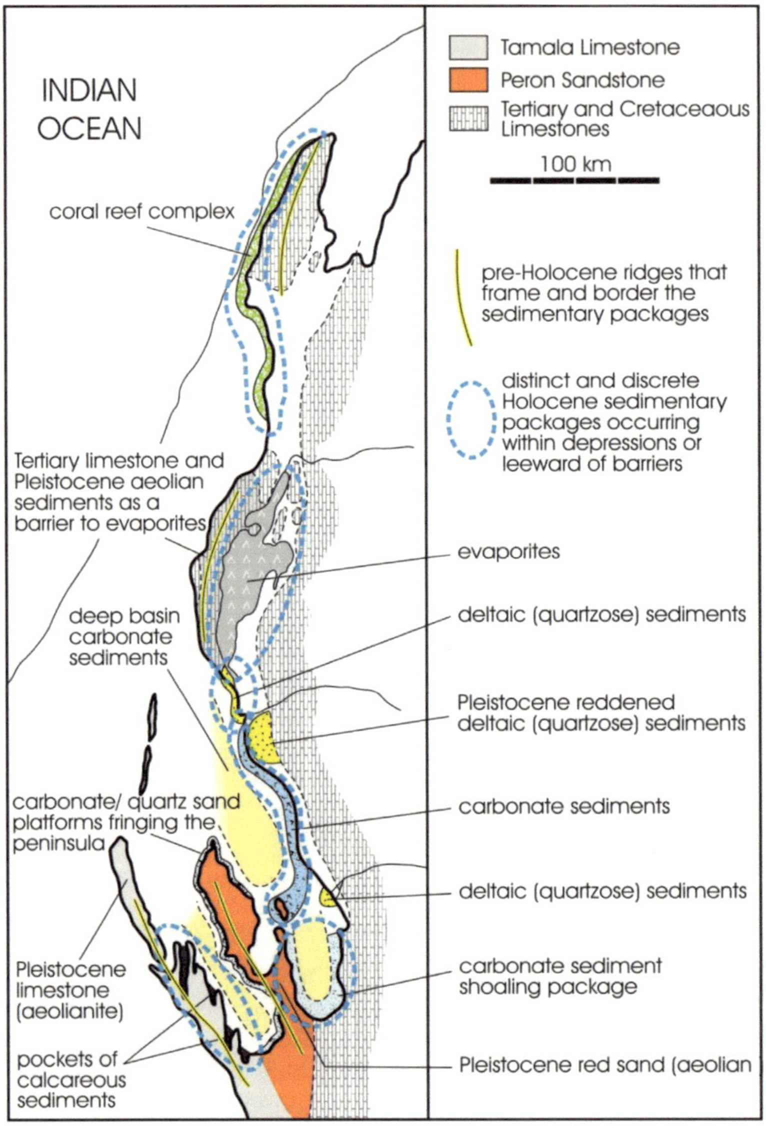

Figure 4.
The disposition of the lithologically distinct and spatially discrete sedimentary packages arrayed along the length of the southern Carnarvon Basin. Also shown are the south-to-north uplands of Pleistocene aeolian and aeolianite bodies and the uplifted Tertiary and Cretaceous limestone formations that either act as barriers to or cradle the Holocene depositional sites.

of beaches, beach ridges, and coastal dunes [23, 32, 33], that while largely shore-parallel and close to shore can extend inland for kilometres and, locally, have prograded seawards for over 10 km [33]. These coastal formations are composed of lithologically fairly uniform quartzose calcareous medium sand. Similarly, along 600 km of the Canning Coast [34], the coastal zone of the Canning Basin is relatively consistent comprising longitudinally-extensive beaches, beach ridges, and (in small embayments) limestone barriers sheltering calcilutite basins; the coastal sands are quartzose calcareous medium and coarse sand (*viz.*, Shoonta Hill Sand and Cable Beach Sand of Semeniuk, respectively [34]).

However, prior to comparing the sedimentology and stratigraphy of the Southern Carnarvon Basin with other sedimentary basins globally, there are two

Stratigraphic package (south to north) (numbering follows that in Table 1)	Fine scale stratigraphy
2. Pocket seagrass bank carbonate complexes of central western Shark Bay	Boat Haven Sand[1]—bioturbated shelly calcareous quartzose and underlain by thin calcilutite, and overlain by laminated sheet laminated sand, laminated beach sand, and beach-ridge sand; inland, where hypersaline, there is development of carbonate crusts [24, 25]
4. Shoaling sedimentary packages of south-eastern Shark Bay	Bioturbated Wooramel Sand[1] overlain by laminated and cemented Hutchinson Formation, stromatoliths and Hamelin Coquina
5. Deltoid deposit of river sediments at the mouth of the Wooramel River central-western Shark Bay	Bioturbated Wooramel Sand[1] overlain by laminated and cemented Hutchinson Formation, stromatoliths and Hamelin Coquina
6. Extensive shore-parallel wedge of seagrass bank sequence eastern coast, northern Shark Bay	Wooramel Sand[1]—bioturbated seagrass bank sand overlying a thin calcilutite and overlain by laminated sand of beach ridges and spits
7. Mainly wave-built platforms and some seagrass bank deposits: extensive shore-parallel wedge of sediment western and eastern coast of the Peron Peninsula, central Shark Bay [25]	30 km long, 5–6 km wide; subtidal units up to 3 m thick, tidal units up to 1 m thick, beach-ridge units 3–15 m thick Denham Sand[1]—metahaline wave-built sand platform and seagrass bank sediments; sharply adjoins the Wooramel (seagrass bank) sand and the Boodalia Formation[1], and onlaps the red sand hinterland of Peron Sandstone
8. Deep water marine embayment plains underlain by sheets of calcareous mud, shelly mud, and sand	100 km long, 20–25 km wide, 5–6 km wide, 0.5–1.5 m thick Freycinet Formation[1] sharply adjoins the shallow water seagrass bank sediments and wave-built platform sediments
10. Gascoyne Delta and laterally equivalent beach-ridge complex north of Shark Bay	Cross-bedded Gascoyne Sand[1] overlain in the south by laminated to bioturbated (root-structured) beach-ridge sand and ribbons of tidal mud of the Baggage Island Formation, and overlain in the north by laminated to bioturbated (root-structured) beach-ridge sand referred to the Bejaling Sand
11. Evaporite sequence of Lake McLeod	Thin unit of the Cygnet Carbonate Member overlain by the crystalline 5–6 m thick Texada Halite and, in turn, by the Ibis Gypsite (*in situ* and reworked gypsum crystals)
13. Coral-reef complex of the Ningaloo Reef fringing the uplifted Tertiary limestone barrier	Coral framework (domal, arborescent, mixed, tabulate and encrusting corals), coral rubble, skeletal sand, and intercalated sheets of alluvial-fan sediment = Ningaloo Limestone

Stratigraphic nomenclature in preparation [29].

Table 2.
More detailed description of fine-scale stratigraphy of the Holocene stratigraphic packages.

inter-related factors that need discussion in order to establish a framework for these comparisons; these are: 1. systems can be inherently internally complex in terms of sedimentary dynamics and facies resulting in small-scale complex stratigraphy or, alternatively, relatively simple (even if reflecting an upward shoaling energy-related system) [35],and 2. the concept of a formation and its use within the comparative **Table 3**.

In the first instance, sedimentary systems can be inherently stratigraphically complex laterally and vertically. For example, floodplain sedimentology and the resulting stratigraphy, and the stratigraphic complexity associated with channel switching in fluvial settings are examples (*e.g.*, Figures 404, 409 & 410 in Reineck & Singh [37]) where there can be marked lateral and vertical lithological variation. So too for deltas –for instance, while a wave-dominated, sand-dominated delta

Sedimentary basin	Location, area, longest length	Key laterally-extensive, relatively thick basin-wide formations, their lithologies and thicknesses
Sydney Basin [3]	New South Wales, Australia; 64,000 km^2; 450 km	Basin-extensive Wianamatta Shale (300 m thick), Hawkesbury Sandstone (230–290 m thick), the Newcastle and Illawarra Coal Measures of coal seams, sandstone sheets, and shale (150–500 m thick)
Kimberley Basin [4]	Western Australia, Australia; 400,000 km^2; 500 km	Five major basin-extensive lithological sequences of Proterozoic sedimentary and volcanic rocks, accounting for a stratigraphic thickness of some 5000 m and assigned to formational and/or group level; sandstone dominates the sequences; lithologies are quartzose sandstones, felspathic sandstones, minor siltstones, and minor volcanic rocks
Eucla Basin [5]	Western Australia to South Australia; 1,141,000 km^2; 1200 km	Basin-extensive sheets of skeletal limestone (bryozoa, foraminifera, molluscs, calcareous algae forming calcarenites, rudstones, and calcareous muddy limestones) and some sandstone; lithologically fairly consistent and extensive across the basin (*e.g.*, Abrakurrie Limestone 120 m thick; Colville sandstone 25 m thick; Nullarbor Limestone 45 m thick; Wilson Bluff Limestone 335 m thick)
Paris Basin [6, 7]	Northern to central France; 140,000 km^2; 600 km	Basin-extensive sheets and lenses of limestone, sandstone, chalk, shale/marl, dolomite, oolite, and evaporites, each lithologic unit of relatively regular thickness, each either 300 m, 400 m, 500 m, or 600 m thick
Paradox Basin [8, 9]	United States of America: Utah, SW Colorado, extending into NE Arizona and NW New Mexico; 85,470 km^2; 280 km	Asymmetric basin with western basin dominated by thick salt (up to 2500 m thick) and eastern basin dominated by limestone with lesser dolomite, bioherms, shale (1400 m thick)
Tindouf Basin [10]	Anti-Atlas, Morocco, and Algeria; 100,000 km^2; 700 km	Basin-extensive sheets of sandstone, shale, limestone, marls, coral beds, conglomerate; individual formations comprise sandstone (500 m thick), interbedded limestone and shale (500 m thick), marls and marly limestone (700 m thick), and evaporite (up to 100 m thick)
Gourara Basin [11]	Algeria; 260,000 km^2; 320 km	Basin-extensive sheets of sandstone and siltstone, total thickness some 4000 m
Karoo Basins [12]	Scattered series of sub-basins over northern to southern Africa and Madagascar, the largest of which are the main Karoo Basin 700,000 km^2, 1500 km long, and the Kalahari Basin 2,500,000 km^2, 1500 km long	The main Karoo Basin: asymmetric, with basin-extensive sheets of mudstone and sandstone, totaling 7000 m in thickness for the uppermost unit, *viz.*, the Beaufort Group of formations, sheets of shale, sandstones, wackestones, chert, dolomite and coal, totaling 3200 m for the middle unit, *viz.*, the Ecca Group of formations, and diamictite, tillite, totaling 800 m for the lowermost unit, *viz.*, the Dwyka Group of formations

Table 3.
Global perspective: generalized and simplified description of selected sedimentary basins and their stratigraphy (for Australian stratigraphy, refer to Geoscience Australia [36]).

may produce a relatively even and repetitive sequence of subtidal to tidal to beach-ridge facies, *cf.*, Allen [38], a mixed sand-and-mud river-dominated system can result in a plethora of facies which are randomly arrayed and which become more complex where there has been major channel switching [27, 37]. In contrast, coastal beach-to-dune sequences and marine shelf sequences result in relatively predictable shore-parallel sheets of stratigraphic sequences reflecting gradients in wave energy, tides, and aeolian effects for the former [35], or vertically stacked lithologies of energy-related sequences for the latter [37].

In the second instance, in literature reviews of ancient sedimentary sequences globally that are to be used for comparative analyses in this Chapter, it is important to address the concept and definition of a formation as used by a given author, *i.e.*, whether an author in defining a formation was focused on small-scale features, or used a broad-scale approach. Given that sedimentary systems can be inherently stratigraphically complex (*e.g.*, fluvial systems as noted above, or deltaic systems, *cf.* Figures 452, 453, & 456 in Reineck and Singh [37]) or, conversely, stratigraphically relatively simple (*e.g.*, shelf systems), in our selection for comparative purposes of the ancient stratigraphic case studies, we focused on those examples where there was enough lithological detail to ensure that the stratigraphic sequences were adequately described and could be validly compared between the basins listed in **Table 3**. Following the criteria set out in the International Codes and Standards (*e.g.*, the International Commission on Stratigraphy; http://www.stratigraphy.org/index.php/ics-stratigraphicguide) for assigning formational status to rock bodies, in our review of the literature we assessed a rock body could be defined as a 'formation' if it had one of the following characteristics: 1. dominantly of one lithology (*e.g.*, the Hawkesbury Sandstone in the Sydney Basin [3]); 2. mainly of two or three recurring interbedded lithologies mappable as a suite; or 3. a mixture of interbedded lithologies complexly inter-related but the whole suite being clearly recognised as a heterogeneous unit against underlying and overlying rock strata. Our conclusions were that is a formation was valid, and if it was mapped and correlated as being basin-extensive then its lithologically defining characteristics were present from one end of the basin to the other.

As noted in Section 1, globally, there are numerous sedimentary basins that are filled with laterally and longitudinally extensive formations. For comparison with the southern Carnarvon Basin, we have selected the Sydney Basin [3], the Kimberley Basin [6], the Eucla Basin [7], the Paris Basin [4, 5], the Paradox Basin [8, 9], the Tindouf Basin [11], the Karoo Basins [12], and the Gourara Basin [12]. Most of these basins have experienced some degree of tectonism resulting in normal faults and low-amplitude folds, but the sedimentary formations therein are easily correlated across large tracts of their respective basins. Information on the size of these basins and the composition of the laterally-extensive formations is presented in **Table 3**. As is evident in **Table 3**, most of the basins selected for comparison with the southern Carnarvon Basin have basin-wide and extensive formations and *contrast markedly* with the Quaternary formations of the southern Carnarvon Basin of this Chapter in that they are thick, not markedly discontinuous, and can be correlated over large distances.

In this context, comparing the Quaternary Southern Carnarvon Basin with other Western Australia sedimentary basins and with other selected basins Worldwide, it is clear that the Southern Carnarvon Basin stands as an important model globally of basin development – there is interplay of megascale geomorphic architecture, tectonics, oceanography, and climate resulting in a series of disparate sedimentary packages that are in relative close proximity to each other.

In summary, the onshore Carnarvon Basin is an example of a complex sedimentary array of facies-complicated sedimentary small basins where distinct but separated sedimentary facies and sedimentary packages have formed over a relatively small scale. It is an example where sedimentary packages can be markedly diverse over short distances. As such, this coastal and near-coastal part of the southern Carnarvon Basin illustrates the complexities a geologist would face if analyzing a basin in the stratigraphic column.

Conflict of interest

We have no conflict of interest.

Author details

Vic Semeniuk[1,2,3]* and Margaret Brocx[3]

1 V & C Semeniuk Research Group, Warwick, Western Australia

2 School of Arts and Sciences, Notre Dame University, Fremantle, Western Australia

3 Environmental and Conservation Sciences, Murdoch University, Perth, Western Australia

*Address all correspondence to: vcsrg@iinet.net.au

References

[1] Hocking RM, Moors HT, van de Graaf WJE. Geology of the Carnarvon Basin, Western Australia. Geological Survey of Western Australia Bulletin. 1987;**133**:288

[2] Hocking RM. Carnarvon Basin. Geological Survey Memoir. 1990;**3**:457-495

[3] Packham GH. The geology of New South Wales. Journal of the Geological Society of Australia. 1969;**16**(1):654

[4] Griffin TJ, Grey K. Kimberley Basin. Western Australia Geological Survey Memoir. 1990;**3**:293-304

[5] Hocking RM. Eucla Basin. In: Western Australia Geological Survey Memoir 3. 1990. pp. 548-561

[6] Claude M, Mégnien F, editors. Synthèse Géologique Du Bassin De Paris: Volume I. Orléans: du BRGM [Bureau de Recherches Géologiques et Minières]; 1980. p. 101

[7] Duval BC. Villeperdue Field. In: Halbouty MT, editor. Giant Oil and Gas Fields of the Decade, 1978-1988. Vol. 54. American Association of Petroleum Geologists Memoir; 1992. p. 526. ISBN: 0891813330

[8] Huffman AC Jr. Evolution of Sedimentary Basins: Paradox Basin. Washington, USA: United States Geological Survey Bulletin; 1993

[9] Nuccio VF, Condon SM. Burial and Thermal History of the Paradox Basin, Utah and Colorado, and Petroleum Potential of the Middle Pennsylvanian Paradox Formation. Washington, USA: United States Geological Survey Bulletin. 1996

[10] Selley RC. African Basins. Vol. 3. Amsterdam, The Netherlands: Elsevier Science; 1997. p. 391

[11] Baouche R, Nedjari A, El Adj S. A sedimentological approach to refining reservoir architecture using the well log data and core analysis in the Saharan Platform of Algeria. WSEAS Transactions on Environment and Development. 2009;**5**:519-534

[12] Catuneanu O, Wopfner H, Eriksson PG, Cairncross B, Rubidge BS, Smith RMH, et al. The Karoo basins of south-central Africa. Journal of African Earth Sciences. 2005;**43**:211-253

[13] Playford PE, Cope RN, Cockbain AE, Low GH, Lowry DC. Carnarvon Basin. In: Geology of Western Australia; Western Australia Geological Survey. Memoir 2. 1975. pp. 269-318

[14] Brocx M, Semeniuk V. Coastal geoheritage: A hierarchical approach to classifying coastal types as a basis for identifying diversity and sites of significance in Western Australia. Journal of the Royal Society of Western Australia. 2010;**93**:81-113

[15] Logan BW, Read JF, Davies GR. History of carbonate sedimentation, Quaternary Epoch, Shark Bay, Western Australia. In: Logan BW, editor. Sedimentary environments of Shark Bay Western Australia. Vol. 13. Tulsa, Oklahoma: American Association of Petroleum Geologists Memoir; 1970. pp. 38-84

[16] Logan BW. The MacLeod evaporite basin, Western Australia: Holocene environments, sediments and geological evolution. American Association of Petroleum Geologists Memoir. 1987;**44**:140

[17] Barber PM. The Exmouth Plateau deep water frontier: A case history. The North West Shelf, Australia. In: Purcell PG, Purcell RR, editors. Proceedings of the North West Shelf Symposium, Perth, Western Australia,

10-12 August. Perth: PESA; 1988. pp. 173-187

[18] Purcell PG, Purcell RR. The Sedimentary Basins of Western Australia. In: Proceedings of Petroleum Exploration Society of Australia Symposium, Perth. 1994. p. 864

[19] Ghori KAR. Petroleum generating potential and thermal history of the Palaeozoic, Carnarvon Basin, Western Australia. In: Purcell PG, Purcell RR, editors. The Sedimentary Basins of Western Australia 2. Proceedings of the Petroleum Exploration Society of Australia Symposium, Perth. 1998. pp. 553-567

[20] Iasky RP, Mory AJ, Shevchenko SI. A structural interpretation of the Gascoyne Platform, southern Carnarvon Basin, WA. In: Purcell PG, Purcell RR, editors. The Sedimentary Basins of Western Australia 2. Proceedings of the Petroleum Exploration Society of Australia, Perth. 1998. pp. 589-598

[21] Tindale K, Newell N, Keall J, Smith N. Structural evolution and charge history of the Exmouth Sub-basin, northern Carnarvon Basin, Western Australia. In: Purcell PG, Purcell RR, editors. The Sedimentary Basins of Western Australia 2. Proceedings of Petroleum Exploration Society of Australia Symposium, Perth. 1998. pp. 447-472

[22] Karner GD, Driscoll NW. Style, timing and distribution of tectonic deformation across the Exmouth Plateau, northwest Australia, determined from stratal architecture and quantitative basin modelling. In: MacNiocaill C, Ryan PD, editors. Continental Tectonics. The Geological Society of London; 1999. pp. 271-311

[23] Semeniuk V, Cresswell ID, Wurm PAS. The Quindalup Dunes: The regional system, physical framework and vegetation habitats. Journal of the Royal Society of Western Australia. 1989;**71**:23-47

[24] Read JF. Carbonate bank and wave-built platform sedimentation, Edel Province, Shark Bay, Western Australia. In: Logan BW, editor. Evolution and Diagenesis of Quaternary Carbonate Sequences, Shark Bay, Western Australia. Vol. 22. American Association of Petroleum Geologists Memoir; 1974. pp. 1-60

[25] Hagan GM, Logan BW. History of Hutchinson Embayment tidal flat, Shark Bay, Western Australia. In: Logan BW, editor. Evolution and Diagenesis of Quaternary Carbonate Sequences, Shark Bay, Western Australia. Vol. 22. American Association of Petroleum Geologists Memoir; 1974. pp. 283-315

[26] Davies GR. Carbonate bank sedimentation, eastern Shark Bay, Western Australia. In: Logan BW, editor. Carbonate Sedimentation and Environments Shark Bay Western Australia. Vol. 13. American Association of Petroleum Geologists Memoir; 1970. pp. 85-168

[27] Johnson DP. Sedimentary facies in an arid zone delta: Gascoyne delta, Western Australia. Journal of Sedimentary Petrology. 1982;**52**:547-563

[28] Collins LB, Zhu ZR, Wyrwoll KH, Eisenhauer A. Late Quaternary structure and development of the northern Ningaloo Reef, Australia. Sedimentary Geology. 2003;**159**:81-94

[29] Semeniuk V. Rationalization of stratigraphic nomenclature in the Southern Carnarvon Basin. 2020. In preparation for episodes

[30] Semeniuk V. Predicted response of coastal wetlands to climate changes—A Western Australian model. Hydrobiologia. 2013;**708**:23-43. DOI: 10.1007/s10750-012-1159-0

[31] Logan BW, Cebulski DE. Sedimentary environments of Shark Bay Western Australia. In: Logan BW, editor. Sedimentary Environments of Shark Bay Western Australia. Vol. 13. Tulsa, Oklahoma: American Association of Petroleum Geologists Memoir; 1970. pp. 1-37

[32] Searle DJ, Semeniuk V. The natural sectors of the Rottnest Shelf Coast adjoining the Swan Coastal Plain. Journal of the Royal Society of Western Australia. 1985;**67**:116-136

[33] Searle DJ, Semeniuk V, Woods PJ. The geomorphology, stratigraphy and Holocene history of the Rockingham-Becher plain. Journal of the Royal Society of Western Australia. 1988;**70**:89-109

[34] Semeniuk V. Sedimentation, stratigraphy, biostratigraphy, and Holocene history of the Canning Coast, north-western Australia. Journal of the Royal Society of Western Australia. 2008;**91**:53-148

[35] Semeniuk V. Pleistocene coastal palaeogeography in southwestern Australia—Carbonate and quartz sand sedimentation in cuspate forelands, barriers and ribbon shoreline deposits. Journal of Coastal Research. 1997;**13**:468-489

[36] Australian Stratigraphic Units Database 2020. Available from: https://www.ga.gov.au/data-pubs/datastandards/stratigraphic-units

[37] Reineck HE, Singh IB. Depositional Sedimentary Environments. 2nd ed. Berlin: Springer-Verlag; 1980

[38] Allen JRL. Sediments of the modern Niger Delta: A summary and review. In: Morgan JP, editor. Deltaic Sedimentation—Modern and Ancient. Vol. 15. Tulsa, Oklahoma: Society of Economic Palaeontologists and Mineralogists Special Publication; 1970. pp. 138-151

Chapter 13

Middle Miocene Evaporites from Northern Iraq: Petrography, Geochemistry, and Cap Rock Efficiency

Ali I. Al-Juboury, Rana A. Mahmood and Abulaziz M. Al-Hamdani

Abstract

Evaporites (gypsum and anhydrite) of the middle Miocene age (Fat'ha Formation) form one of the main sulfate cap rocks in the Middle East oilfields. Detailed petrographic and diagenetic investigations accompanied with geochemical analysis of these evaporite rocks in Mosul and Kirkuk areas of northern Iraq have revealed that nodular gypsum is the dominant type, whereas laminated, structureless, and secondary (selenite and satin spar) also are present. Nodular gypsum was deposited in a very shallow, arid, and semi-restricted lagoonal environment which has undergone influx and reflux processes, while laminated gypsum may represent pulses of freshwater into the lagoonal basin of Fat'ha Formation. Low strontium values of the secondary and laminated gypsum may attribute to their secondary origin by hydration processes from the original anhydrite. Based on petrographic, diagenetic, and petrophysical (porosity and permeability) properties, it appears that the efficiency of the Fat'ha sulfates as petroleum cap rocks increases with increasing nodular growth and compaction degree. The occasional presence of bitumen inclusions with both nodular gypsum and host materials relates to early leakage of the hydrocarbons which were being halt due to the growing and packing of nodules and host materials.

Keywords: evaporites, petrography, geochemistry, cap rock potential, miocene, Iraq

1. Introduction

More than 70% of the world's giant oilfields in carbonate rocks bear a relationship to evaporites [1]. The association among evaporates, carbonates, and hydrocarbons is more than fortuitous as evaporates constitute less than 2% of the world's platform sediments [2].

Evaporites form about 50% of the total thickness of the middle Miocene Fat'ha Formation in Iraq [3]. Gypsum ($CaSO_4.2H_2O$) is the most common type in surface (outcrop) sections, while in subsurface sections, anhydrite ($CaSO_4$) and halite ($NaCl$) are the most common evaporites.

The Fat'ha Formation was deposited in a NW-SE-oriented basin which extended from NE Syria through N and NE Iraq into SW Iran (**Figure 1**). This basin is called the "Mesopotamian Basin" which is a foreland basin situated on the leading edge of the Arabian Plate attached to the African Plate [3].

The basin-center model is also manifested by the concentric arrangement of evaporite beds interbedded with limestone and marly limestone with gypsum and anhydrite along the margins to soluble halite in the depocenter. During high-frequency sea-level lowstands, intra-basinal and regional structural barriers may have isolated the hypersaline basin from the open sea, such that evaporation exceeded

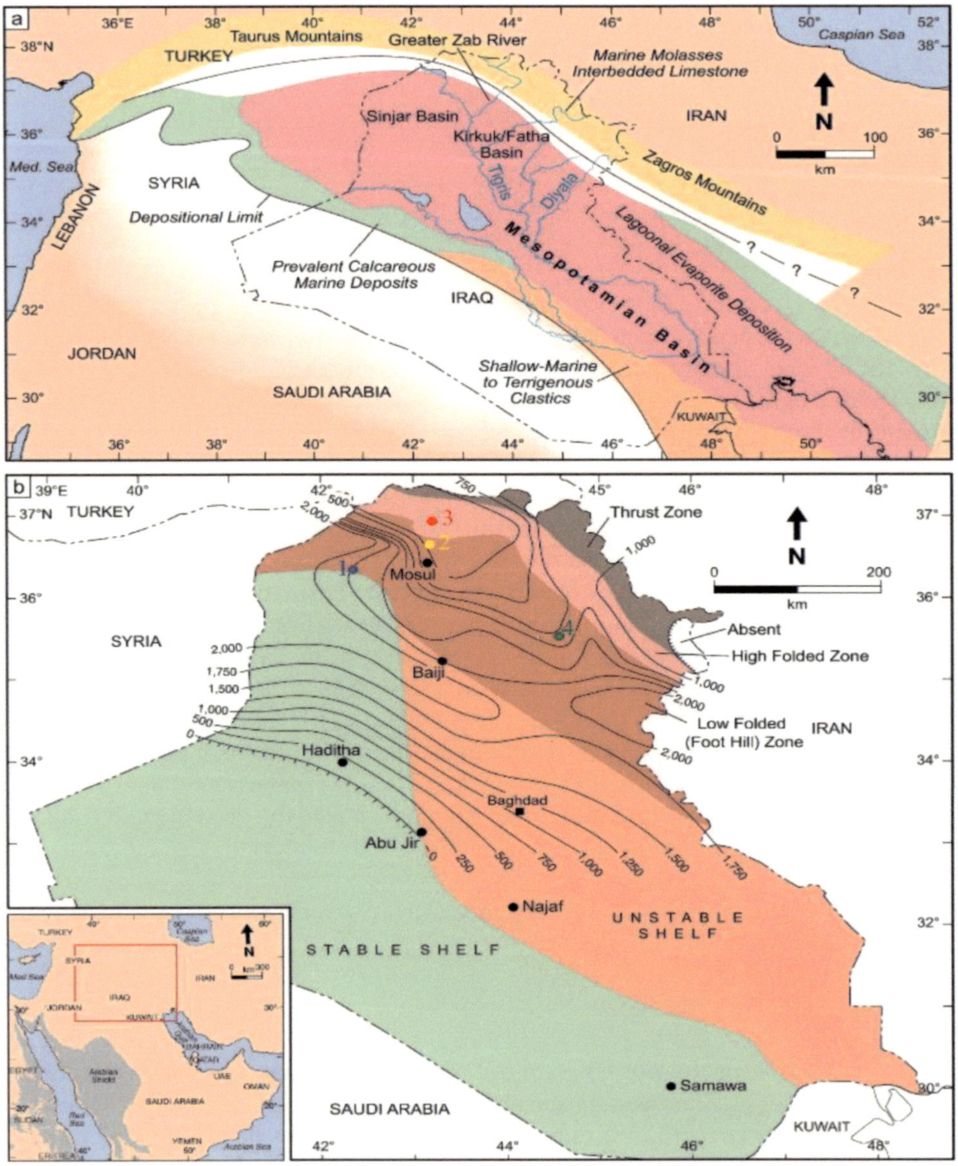

Figure 1.
(a) Location map showing the Mesopotamian Basin in Iraq and Fat'ha/Kirkuk and Sinjar sub-basins. Simplified middle Miocene lithofacies distribution map (after [4–7]). (b) Locations of the four studied sections: (1) Sheikh Ibrahim, (2) Telkif, (3) Batnaya, and (4) well in Kirkuk. Also shown is an isopach map of the Fat'ha Formation (after [4]) and tectonic provinces of Iraq (modified from [3] after permission from GeoArabia).

the ingress of water in an arid climate [2, 3]. The Fat'ha Formation is one of the most extensive and economically important formations in the entire Middle East region [8].

The formation covers a large area (approximately 1500 km x 300 km) and extends northwestward into Syria (there termed Lower Fars Formation) and southeastward into Iran (there termed upper part Gachsaran Formation) [9] (**Figure 1**). The Fat'ha Formation is a seal to numerous oil reservoirs in Iraq and Iran and, in certain areas, is a reservoir in its own right (e.g., Kirkuk, northern Iraq, [4, 10]).

In the present study, lithofacies analyses of various gypsum and anhydrite successions from both surface and subsurface sections (**Figure 1**) are studied accompanied by petrographic investigation using traditional petrographic microscope supported by scanning electron microscopy (SEM) for better determination of their petrographic, textural, and diagenetic features. The study also includes mineralogical determination using X-ray diffraction (XRD), geochemical, X-ray fluorescence (XRF), and petrophysical (porosity and permeability) measurements for selected samples from both surface and subsurface sections.

The aim of the study is to elucidate the lithofacies and related petrographic, textural, and diagenetic and geochemical characteristics of the gypsum and anhydrites of the Fat'ha Formation and to determine their ability as seal or cap rocks.

2. Geologic setting

The Neo-Tethys Ocean began to close in the late Cretaceous as evidenced by the obduction of ophiolites in Oman and elsewhere along the margin of the Arabian Plate [11, 12]. In the late Miocene and early Pliocene, the Neo-Tethys Ocean was closed by the collision of the Arabian and Eurasian plates (Central Iran and Turkey), and the Zagros and Taurus Mountain belts started to be uplifted [13, 14]. Between these two tectonic events, starting in the late Eocene and continuing through the middle Miocene, crustal loading and flexure of the eastern Arabian Plate formed the broad and shallow Mesopotamian Basin as a NW-oriented foreland basin [15, 16]. This 2000-km-long basin extended from Bandar Abbas, in Iran, across Iraq and Syria to the Mediterranean Sea, and it was located southwest of the Zagros and Taurus Mountains (**Figure 1**).

The Fat'ha Formation is largely an evaporatic sequence. It consists of numerous shallowing-upward cycles of alternating mudrocks, limestones, gypsum, anhydrite, and halite which are present in the basin center. The rich sulfur deposits are found in evaporite beds consisting mainly of gypsum and anhydrite, limestone, marl, and claystone [17]. The formation comprises a cyclic succession deposited in shallow marine, supra-tidal, and continental environments [5, 18]. The formation of the Zagros-Taurus mountain range led to the development of the Mesopotamian Basin as a result of crustal loading and flexure. Major orogeny also occurred in the late Miocene–Pliocene as a result of regional changes in the rates of plate motion, which produced a preferential northward movement of the Arabian Plate relative to the Iranian-Turkish plates, and the collision of the Turkish-Iranian plates with the Eurasian plate to the north.

3. Materials and methodology

Forty five samples from the middle Miocene evaporate succession were selected for the present work. Lithofacies analysis is conducted in the field based on systematic classification of gypsum/anhydrite by Holliday (1971) [19] and comparison with classifications of [20, 21].

Petrographic investigation using traditional petrographic microscopy are achieved at the Geology Department of Mosul University, Iraq. Furthermore, a deeply focusing of textural and diagenetic identification using scanning electron microscopy (SEM) was conducted on selected samples using Camscan MV 2300 SEM at Steinmann Institute, Bonn University, Germany. Mineralogical XRD analysis using D8 ADVANCE [Bruker AXS] with Cu-∞radiation and geochemical analysis using Siemens SRS 303 XRF also are conducted at Steinmann Institute, Bonn University, Germany, whereas porosity and permeability measurements were conducted at the Geology Department of University of Mosul, Iraq, using dimension measurement and wax method using Soxlith instrument after bitumen extraction for porosity and the pipette method for permeability measurement, respectively.

4. Results

4.1 Lithofacies

Several lithofacies have been recognized through the field study of the evaporitic successions of the Fat'ha Formation; these include the following:

1. Nodular and structureless gypsum/anhydrite lithofacies

This form is the common lithofacies in the studied successions. They are commonly bedded with thickness varying between 0.1 and 50 meters. Nodules are white sucrose or of other colors depending on the included impurities. These nodules are surrounded by different colors of clayey or carbonate stripes. Nodules are finger-shaped or cylindrical in the lower parts of the beds to condensed circle in shape in the upper parts (**Figure 2A**) or as compound nodular texture (**Figure 2B**).

Based on the nature of the nodules and their interstitial materials, compaction and growth nature of these nodules, deformation features, and nature of bedding, several sublithofacies could be recognized, and these include nodular, nodular mosaic, mosaic, wispy, and massive (structureless) gypsum/anhydrite sublithofacies (see **Figures 2–4**). Laminated and enterolithic structures (as a result of anhydrite to gypsum transformation) are common in the mosaic secondary sublithofacies. This lithofacies could be correlated with the Miocene sulfate facies of Seven River Formation of southeast Mexico [22], Codo Formation evaporate of northern Brazil [23], and middle Miocene gypsum unit (Ninyerola) near Valencia, Italy [24].

2. Laminated gypsum lithofacies

This lithofacies is less dominated than the previous one and characterized by thin lamination with lamina of less than 2 mm thick and interlaminated with other marly, limy, or secondary satin spar or selenite laminas (see **Figures 2F** and **3F**). This interlamination may reflect cyclic dynamic changes of the sedimentary basin where the thickness of lamina reflects the stability period of the basin [25]. The gypsum laminas are formed of fine white sucrose (alabastrine type) of gypsum, whereas other laminas are of pale to greenish-gray in color. This color variation may reflect the seasonal changes in temperature and water chemistry of the basin [25].

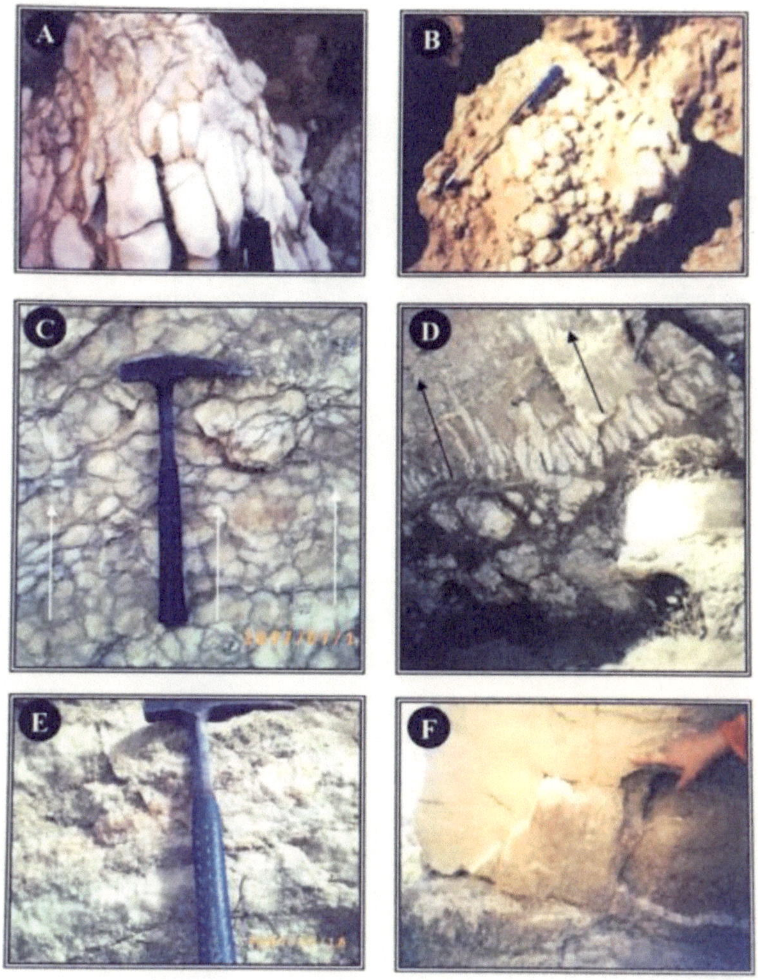

Figure 2.
(A) Elongated nodules in the lower part of the gypsum beds. (B) Compound spherical gypsum nodules as representative for the compound nodular lithofacies. (C) Gradual change of mosaic to wispy gypsum upward. (D) Gradual change of singular to compound nodular to mosaic then to wispy and massive structureless gypsum. (E) Gypsum bed composed of alternative mosaic nodules. (F) Satin spar laminas fill bedding planes in gypsum bed, Sheikh Ibrahim section.

3. Satin spar and selenite gypsum lithofacies

This lithofacies is dominated in the evaporate successions of the Fat'ha Formation and in their interlaminated marly, clayey, and limestone beds as veins, lenses, and fibrous nodules along bedding planes or within joints, cracks, and cavities and commonly is dominated in the upper parts of the formation. Two sublithofacies are recognized in the present study, satin spar and selenite (**Figure 5**).

4.2 Petrography and diagenesis

4.2.1 Petrographic investigation

Detailed petrographic analysis of the studied evaporitic succession by the means of polarized microscopy supported by scanning electron microscopic study has revealed that nodular gypsum is the dominant gypsum type, although laminated

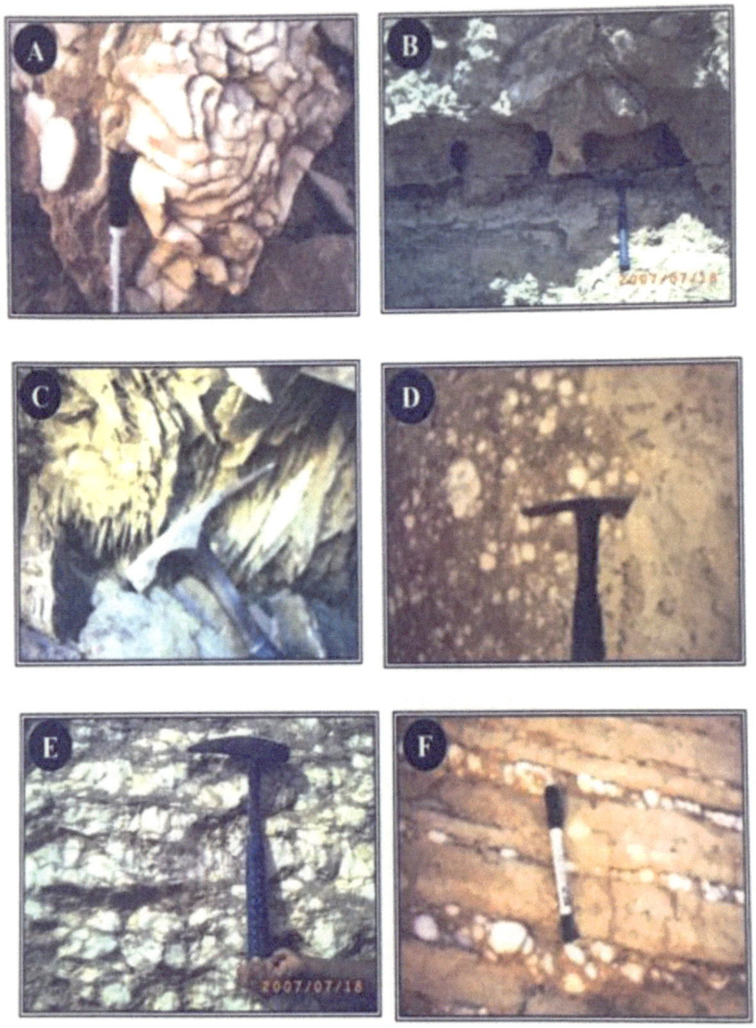

Figure 3.
(A) Enterolithic structure. (B) Erosional caving in the lower part of the gypsum bed within marl hosting nodular gypsum. (C) Desert rose feature in secondary gypsum forming large twinned laminas of fibrous gypsum. (D) Lime interstitial materials in between spherical gypsum nodules. (E) Sutured marly materials in between gypsum nodules in thinly laminated gypsum unit. (F) Laminated nodular gypsum beds with marl (A, C, D, F—Sheikh Ibrahim section; B—Batnaya; and E—Telkif section).

and thick-bedded gypsum are also present. Nodular gypsum passes gradually and vertically into thick to very thick-bedded gypsum. Secondary gypsum (selenite and satin spar) also occurs. Gypsum is white and sugary or creamy in color, but red pink and greenish white varieties also are present. The greenish white color is usually related to secondary coloration as result of enveloping cover of green marl in the succession of the Fat'ha Formation.

In the current study, several textures for gypsum and anhydrite are recognized.

Gypsum textures: Four principal textures are distinguished, some are subdivided into secondary types based on the form, size, and relationships between gypsum crystals, and these include:

1. Alabaster texture, which is characterized by fine-grained and oriented nature due to recrystallization and reorientation from their primary rocks as a result

of direct hydration to gypsum [26, 27]. According to textural stages of Holliday (1971) [19], this texture has three stages as follows:

- Stage 1: feathery texture which is common in the lower parts of the Fat'ha Formation as anhedral and sutured crystals of up to 50 micron in size and commonly includes mineral inclusions (**Figure 6A1** and **A2**); it is represented by nodular gypsum lithofacies.
- Stage 2: grained texture, up to 200 micron in size, more clear crystals than the feathery texture with rare inclusions and curved crystal contacts (**Figure 6B1** and **B2**), represented by wispy gypsum lithofacies.
- Stage 3: a developed texture from either stage 1 or stage 2, up to 400 micron in size, subhedral to euhedral crystals with no inclusions, and clear crystal contacts (**Figure 6C1** and **C2**) represented by massive (structureless), compound mosaic and laminated lithofacies.

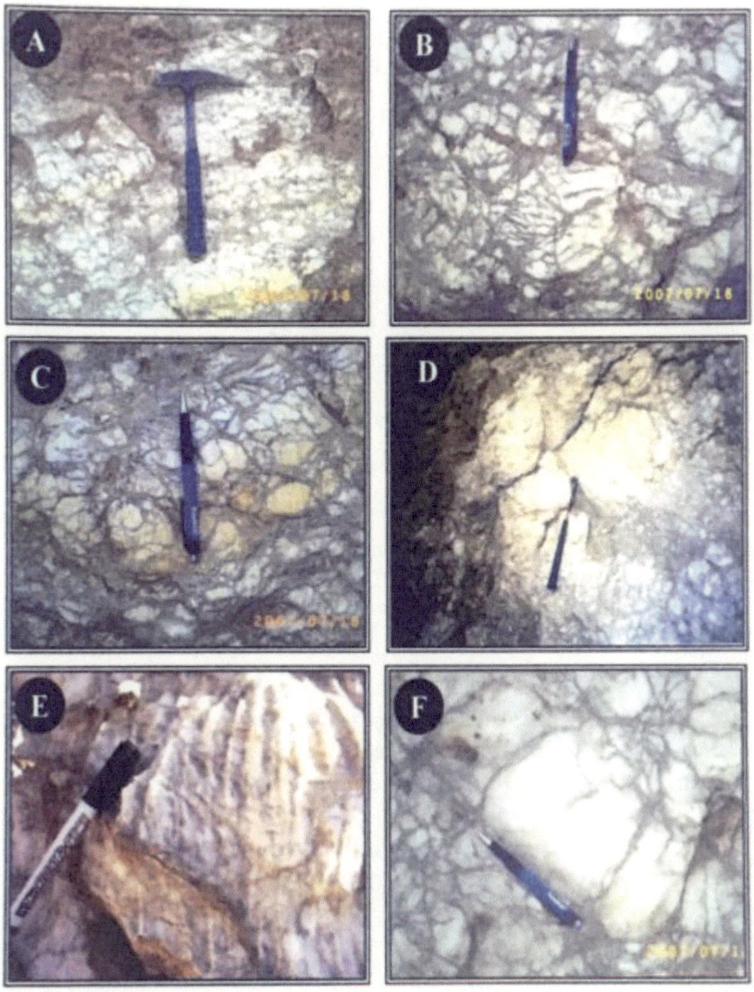

Figure 4.
(A) Mosaic compound gypsum lithofacies, Telkif section; (B) wispy compound gypsum lithofacies, Telkif section; (C) mosaic gypsum lithofacies, Telkif section; (D) graded size in gypsum nodules bed, Sheikh Ibrahim section; (E) wispy gypsum lithofacies, note erosional starching, Sheikh Ibrahim section; (F) massive gypsum lithofacies surrounded by wispy and mosaic lithofacies, Telkif section.

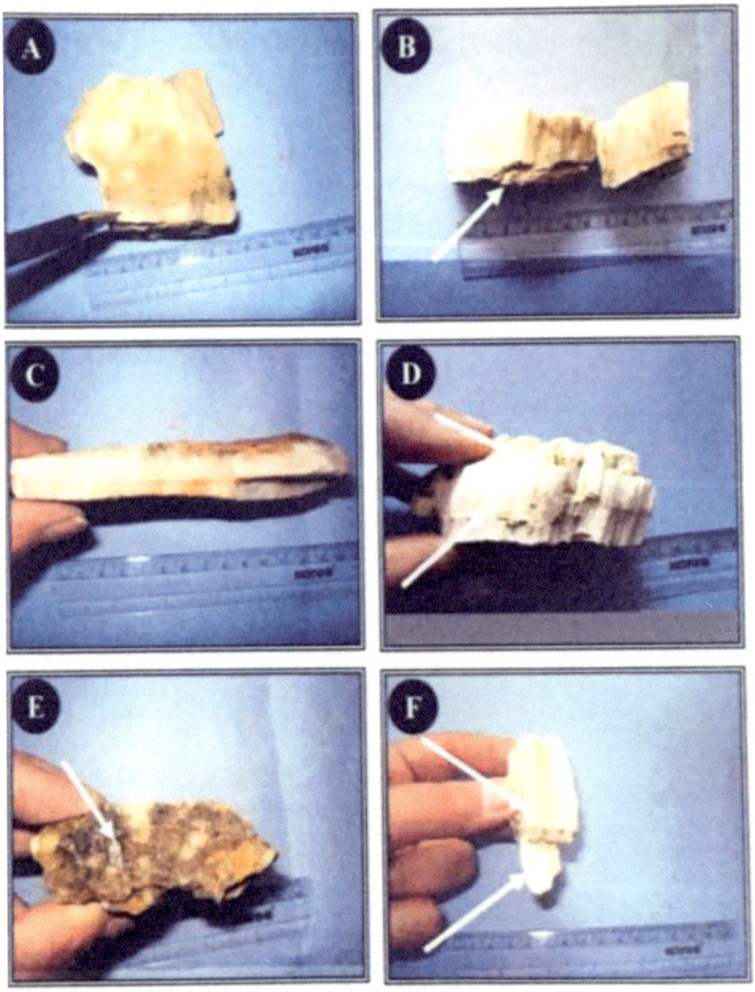

Figure 5.
(A) Gypsum nodule surrounded by fibrous gypsum, Sheikh Ibrahim section; (B) two cross sections of satin spar showing thicker upper lobes, Sheikh Ibrahim section; (C) satin spar veins showing parting in the medium part, Sheikh Ibrahim section; (D) brown mud inclusions in veins of fibrous gypsum, Batnaya section; (E) laminar selenite crystals below satin spar veins, Sheikh Ibrahim section; (F) thick vein of satin spar with curved fibers, Sheikh Ibrahim section.

2. Porphyroblastic texture, which is recognized as large platy crystals with more than 1 cm length which may reflect slow growth of crystals and nuclei [19]. Most of these crystals are embedded in fine alabaster groundmass as a result of anhydrite dissolution and re-precipitation as secondary gypsum (**Figure 7A1** and **A2**). Porphyroblastic texture accompanied also with alabastrine gypsum representing the first growth stage of anhydrite to gypsum (**Figure 7B**). In the field it is represented by mosaic nodular or laminated gypsum lithofacies.

3. Satin spar texture, which commonly are parallel longitudinal fibrous crystals, twinned and oriented with different colors, white, gray and yellow, up to 50 mm long. It is found in either fine (0.11 mm long) (**Figure 8A1** and **A2**) or coarse (0.37 mm) crystals long (**Figure 7C1** and **C2**) represented by fibrous and satin spar lithofacies.

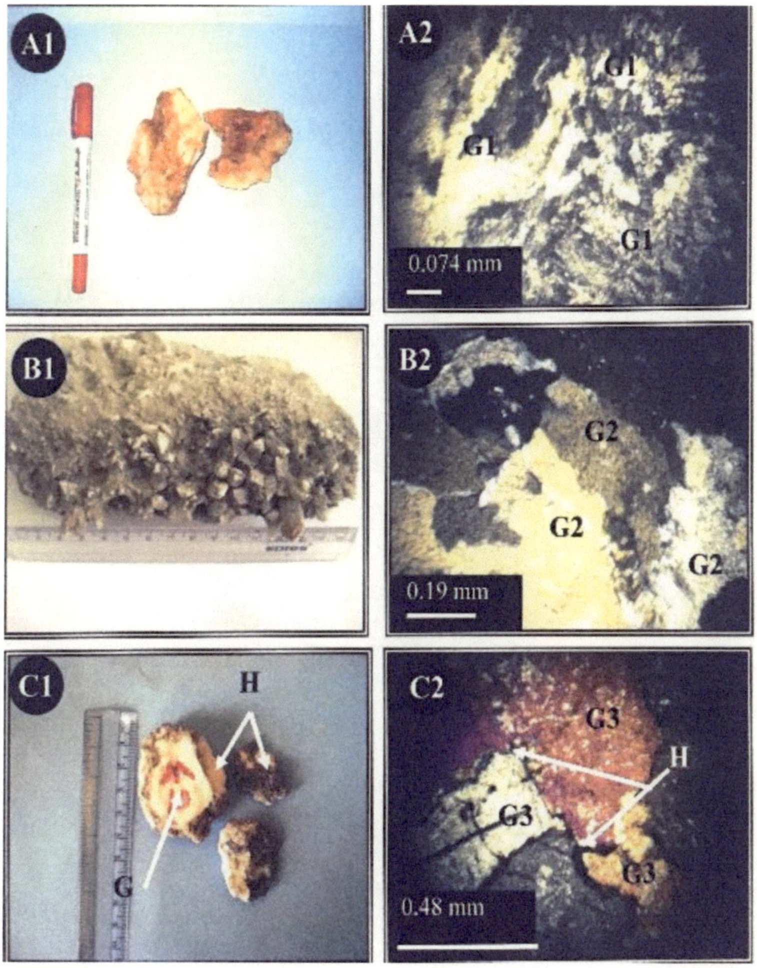

Figure 6.
(A1) Gypsum nodule with brown inclusions, Telkif section; (A2) thin section of the same sample in A1, showing G1 stage of growth of alabastrine gypsum; (B1) gypsum nodules finely crystalline, Sheikh Ibrahim section; (B2) thin section for the same sample in B1 showing G2 growth of alabastrine gypsum; (C1) single gypsum grain (G) surrounded by host carbonate rich in hydrocarbon materials (H), Sheikh Ibrahim section; and (C2) thin section of the same sample in C1 showing G3 growth of alabastrine gypsum and hosting carbonates.

4. Granular texture, which is medium to coarse grained. It exists in two forms as follows:

    - Integrated granular: interconnected crystals of 0.1–0.55 mm in size, represented by nodular gypsum lithofacies (**Figure 8B1** and **B2**).

    - Unintegrated granular: 0.11–0.29 mm size grains of angular edges and also represented by the nodular gypsum lithofacies (**Figure 8C1** and **C2**).

Anhydrite textures: These textures are distinguished in the subsurface sections of the Fat'ha Formation. Based on the crystal shape and size of the anhydrite, six textures are distinguished, these are as follows:

1. Felty texture: crystals in the form of plates of 0.5 mm long with random distribution of crystals which form the advanced stage of recrystallization of finely

Figure 7.
(A1) Fine mosaic gypsum nodule, Sheikh Ibrahim section; (A2) thin section of the same sample A1 showing porphyroblastic gypsum crystals; (B) thin section of mosaic gypsum showing porphyroblastic texture (Gp) accompanied with alabastrine texture that represent G1 stage of gypsum growth, Telkif section; (C1) satin spar with coarse acicular crystals, Sheikh Ibrahim section; (C2) thin section of the same sample C1 showing coarse fibrous gypsum fibers.

crystalized textures. Hydrocarbon materials are concentrated between crystals (**Figure 9A1** and **A2**).

2. Lath texture: long euhedral plates. They are arranged subparallel to radial forms. Commonly they are distributed in groundmass of felty texture anhydrite (**Figure 9B1** and **B2**).

3. Gneissoid texture: oriented parallel plates presented in curved (v) shape "Chevron" folded shape, which may be formed due to gypsum to anhydrite under high-pressure conditions [28] (**Figure 10A1** and **A2**).

4. Microcrystalline texture: fine crystalline below 0.06 mm in size and equidimensional, accompanied with sub-felty textures (**Figure 10B1** and **B2**).

5. Bacillar texture: fine bladed to prismatic in shape with hydrocarbon materials within this texture (**Figure 11A1–A3**)

6. Porphyroblastic texture: medium-sized (0.2–0.3 mm) anhedral crystals that may reflect advanced stage in anhydrite growth (**Figure 10C1** and **C2**), hydrocarbon also present, (**Figure 11B1** and **B2**).

4.2.2 Diagenetic processes

Due to high solubility of evaporates and their rapid susceptibility to deformation and destruction, most evaporitic succession commonly are changed or deformed after deposition and burial; therefore, it is seldom to find evaporates of primary origin in the geologic record of age earlier than 25my [29].

Facies analysis and petrographic description of the studied evaporates revealed that several diagenetic processes have affected on the studied rocks; these include dehydration (e.g., presence of fine pseudo-gypsum plates with anhydrite), cementation (e.g., either presence of calcareous gypsum plates filling cavities or calcite cementing materials around gypsum nodules), compaction (e.g., continuous growth and suturing of gypsum nodules), hydration (or gypsification, e.g., various

Figure 8.
(A1) Fine secondary satin spar gypsum with fine acicular crystals, Batnaya section; (A2) thin section of the same sample in A1, showing twinning in fine fibrous gypsum; (B1) compound mosaic gypsum nodule, Telkif section; (B2) thin section for the same sample in B1 showing growth in granular gypsum (Gi) with calcite crystals (C) colored red by alizarin red stain; (C1) carbonate grain including very fine gypsum nodule (G), Telkif section; (C2) thin section of the same sample C1 showing granular gypsum with no growth (Gu) surrounded by calcite stained red crystal (C).

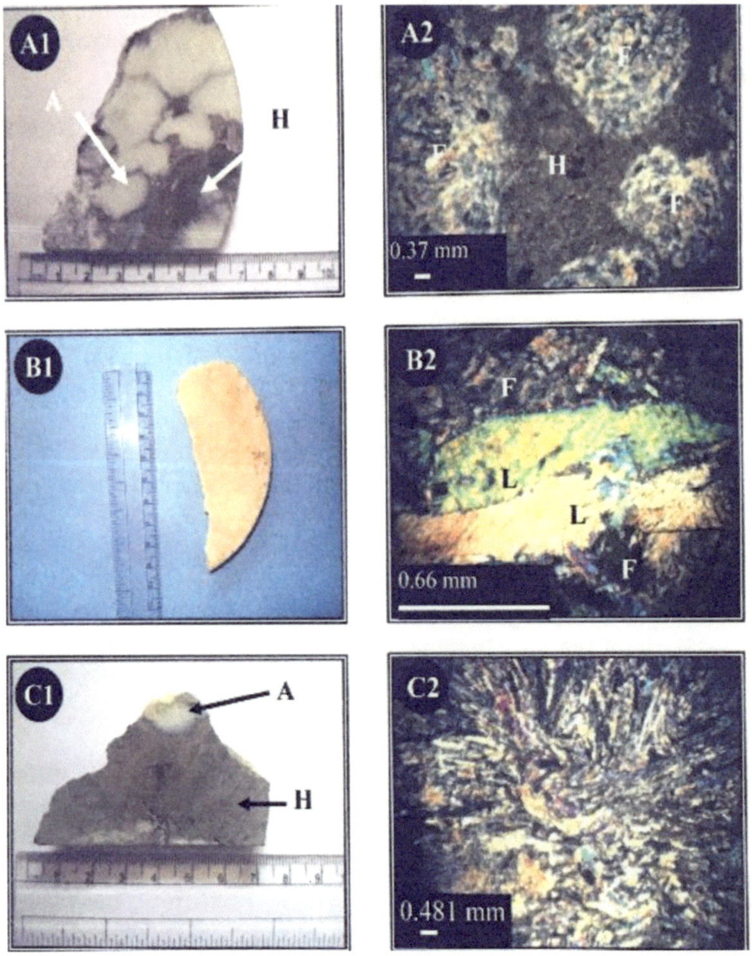

Figure 9.
(A1) Nodular anhydrite (A) with interstitial carbonates rich in hydrocarbons (H), Kirkuk well; (A2) thin section of the same sample in A1, showing felty texture (F), note hydrocarbon existing only on the interstitial materials; (B1) part of pure anhydrite nodule, Kirkuk well; (B2) thin section for the same sample in B1 showing the platy anhydrite in the felty texture; (C1) carbonate including fine anhydrite nodules (A) showing hydrocarbon (H) disseminated in host ground mass not in the anhydrite nodules, Kirkuk well; (C2) thin section of the same sample in C1 showing Chevron folding in anhydrite nodule.

secondary textures such as alabastrine, porphyroblastic, and common satin spar veins), replacement (e.g., calcite replacing gypsum and vice versa), and recrystallization (commonly in the subsurface anhydritic samples, e.g., presence of chevron folding and flow structures). These characteristic features of diagenesis are shown in the previous section and the **Figures 2–11**.

Scanning electron microscopic investigation shows deep focusing various gypsum structures such as coarse crystalline associated with calcite bands (**Figure 12A**) and alternated bands of dark and white folias in the selentic (fibrous) gypsum (**Figure 12B–D**) with carbonate inclusions.

XRD analysis revealed that gypsum is the common mineralogical phase in all the studied samples (**Figure 13**) in addition to rare calcite and/dolomite.

4.3 Geochemistry

Major and trace elements geochemical data for selected gypsum samples are illustrated in **Table 1**. In general, the low content of silica and alumina reflects the

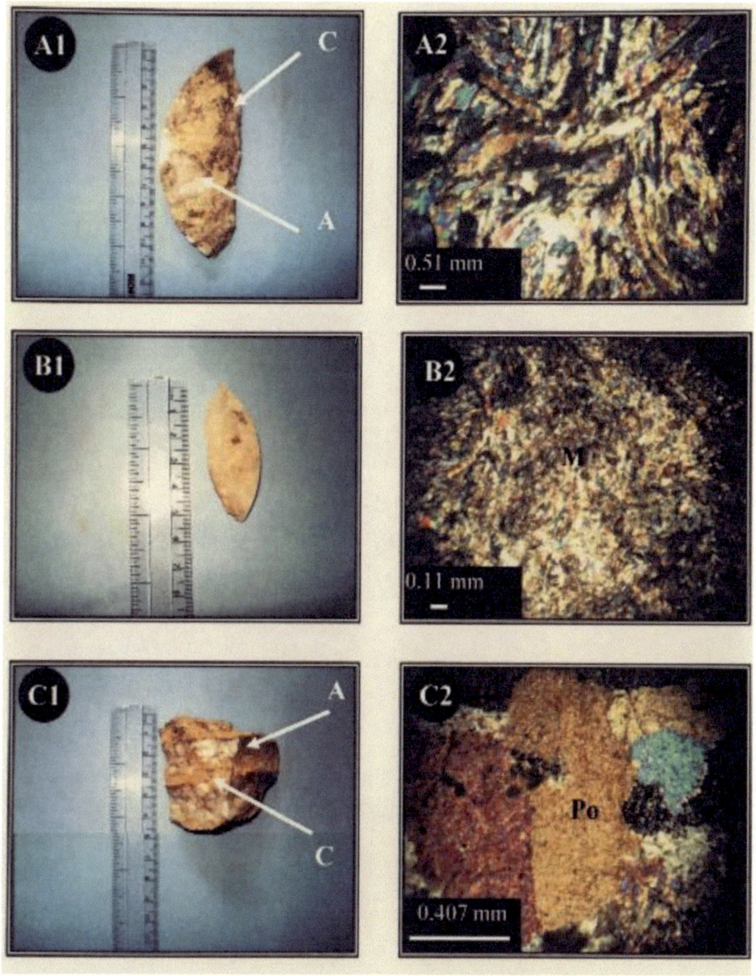

Figure 10.
(A1) Fine anhydrite nodule (A) in carbonate specimen (C), Kirkuk well; (A2) thin section of the same sample in A1, showing flow structures in anhydrite; (B1) finely crystalline anhydrite nodule, Kirkuk well; (B2) thin section for the same sample in B1 showing microcrystalline texture; (C1) finely intercalations of anhydrite nodules (A) and carbonates (C), Kirkuk well; (C2) thin section of the same sample C1 showing porphyroblastic texture (Po).

presence of fine clayey materials in the studied evaporates as brown and gray inclusions. Calcium and magnesium content reflects the accompanied carbonate grains as seen by the petrographic and mineralogic (XRD) investigations in the form of calcite and/or dolomite in addition to calcium in the structure of both gypsum and anhydrites as well as sulfate which is represented by high values of SO_3 (**Table 1**). Trace element distribution of barium and strontium shows high values in mosaic, nodular, and nodular gypsum as compared to laminated and secondary selenite gypsum.

4.4 Efficiency as seal rocks

Porosity and permeability for selected intercalated evaporates and limestone samples from the Fat'ha Formation show that nodular gypsum lithofacies has higher capacity to lock hydrocarbons than the limestone due to very low porosity and permeability (see **Table 2**).

Size of gypsum/anhydrite nodules is an index to the porosity of their groundmass or matrix [30]. In the current work, it seems that chicken wire and

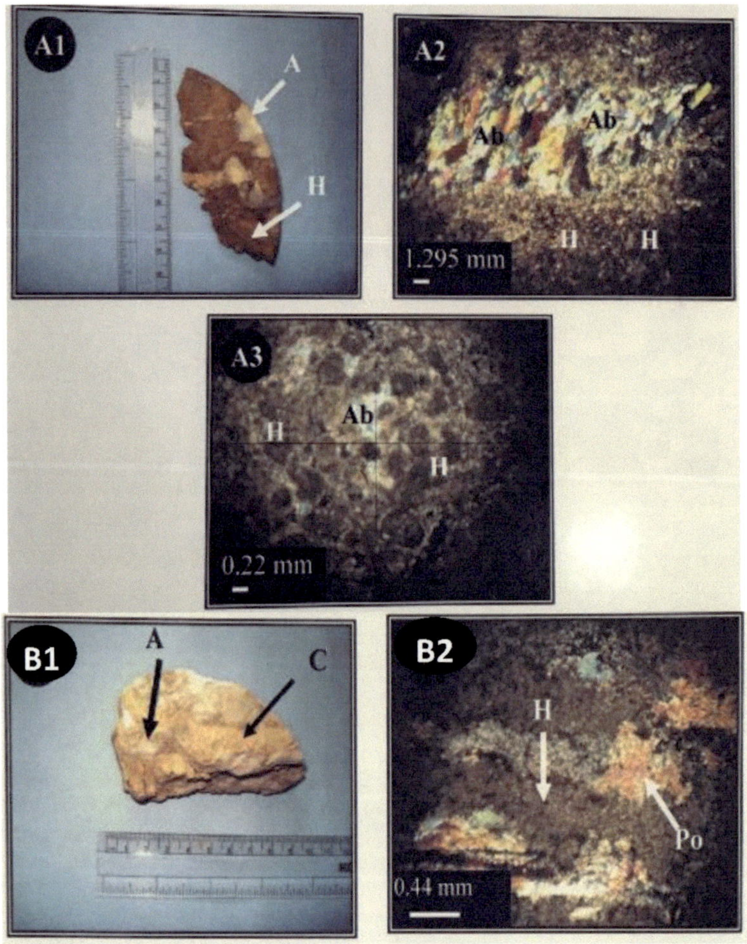

Figure 11.
(A1) Carbonate grain highly enriched in hydrocarbons (H) in which anhedral pure anhydrite nodules (A) are present, Kirkuk well; (A2–A3) thin section of the same sample in A1, showing basilar texture (Ab) and hydrocarbons (H); (B1) carbonate grain (C) with fine anhydrite, Kirkuk well; (B2) porphyroblastic texture of anhydrite embedded in hydrocarbon-rich materials.

enterolithic structures are common in the granular porous matrix; these structures required porous materials with solution movements to form [30]. The Fat'ha Formation evaporites are commonly of large-sized nodules embedded in granular matrix. This matrix could be principally porous that allow some hydrocarbons to disseminate. Consequently, when nodules grow and are compacted as a result of dehydration and compaction, the matrix porosity decreased, and the hydrocarbons were locked.

Petrographic study revealed that bituminous materials are locked in between anhydrite nodules within basilar (**Figure 11A1** and **A2**) and porphyoblastic (**Figure 11B1** and **B2**) textures that may refer to the important role of these anhydritic nodules in locking hydrocarbons.

However, gypsum nodules that formed by hydration of anhydrite, bituminous materials were found in the contact between alabastrine gypsum nodules (**Figure 6C1** and **C2**) that are represented by massive and wispy lithofacies, which may play a role in locking hydrocarbons.

Field study revealed that thick limestones (units A and C) enriched with bitumen in the lower member of the Fat'ha Formation are common below the mosaic

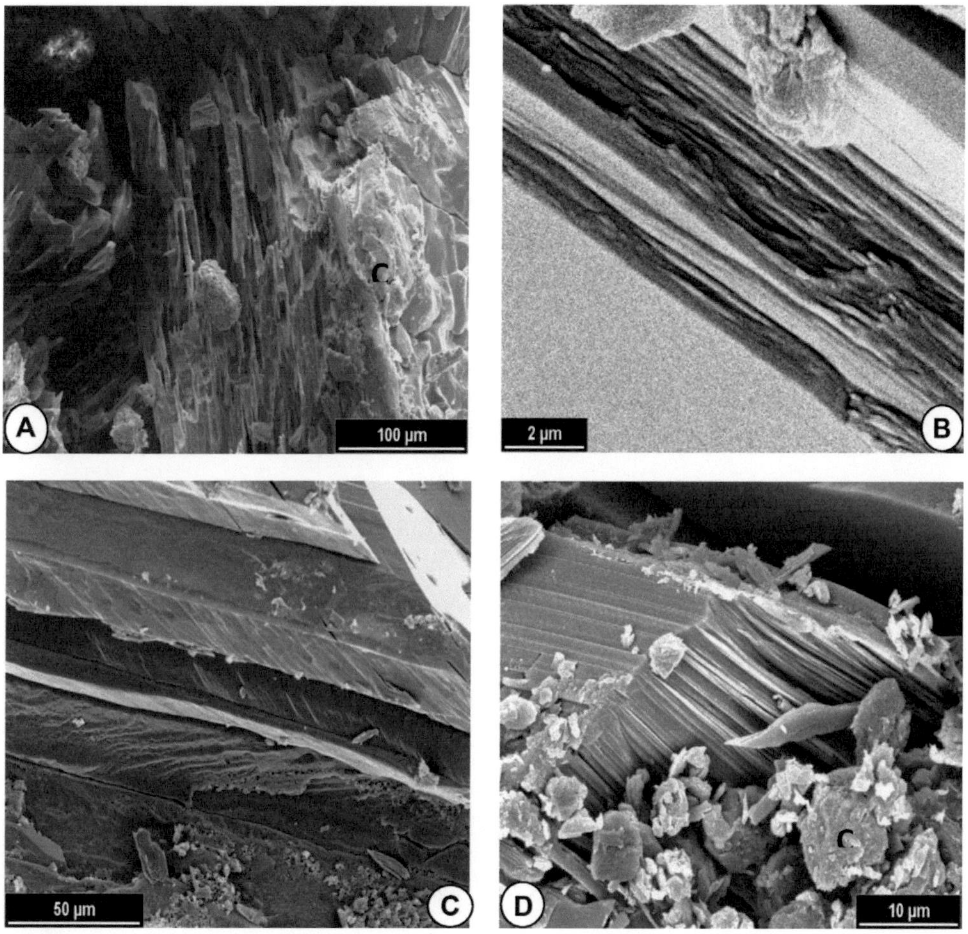

Figure 12.
SEM images show (A) coarse crystalline gypsum with scattered fine calcite (C) in a band that may be responsible for the gray color of the gypsum. (B) Alternating white and dark folia in selenitic gypsum. (C) Foliated nature of selenitic gypsum. (D) Broken folias of selenite with carbonate inclusions (C), Sheikh Ibrahim section.

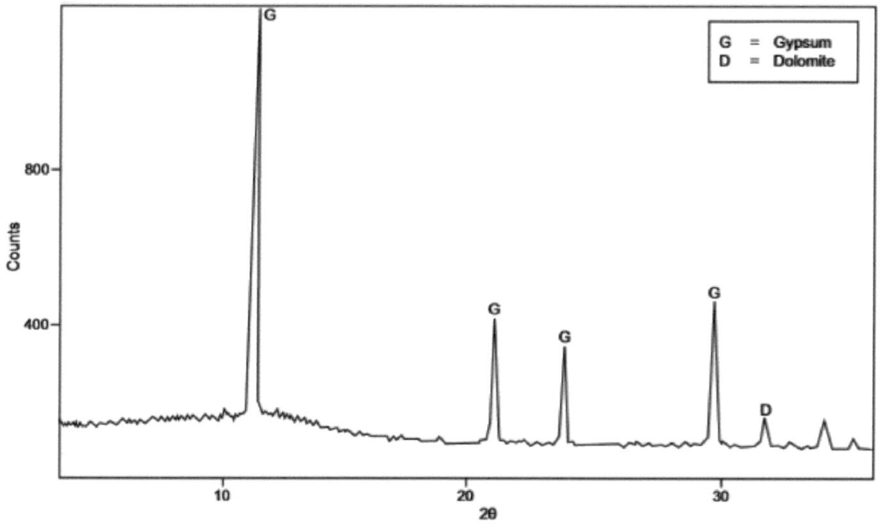

Figure 13.
XRD scan of nodular gypsum sample from the Fat'ha Formation in sheikh Ibrahim section illustrating the common presence of gypsum with rare dolomite.

Gypsum type	SiO_2 (%)	Al_2O (%)	TiO_2 (%)	MnO (%)	MgO (%)	CaO (%)	K_2O (%)	Na_2O (%)	Fe_2O_3 (%)	P_2O_5 (%)	SO_3 (%)	Ba ppm	Sr ppm
Nodular	1.4	0.4	0.02	0.01	0.2	18	0.01	0.1	0.02	0.03	48	2.8	246
Laminated	3.0	0.9	0.03	0.01	1.9	19.5	0.08	0.1	0.1	0.03	46	1.4	110
Massive	1.6	0.5	0.02	0.01	0.4	18.2	0.01	0.14	0.02	0.03	48	0.6	245
Gypsum-anhydrite mosaic	1.8	1.4	0.01	0.01	0.3	18.4	0.01	0.08	0.03	0.03	47	4.6	615
Brown massive	2.2	0.7	0.03	0.01	0.7	18.2	0.02	0.2	0.04	0.03	48	4.5	201
Wispy	2.9	0.9	0.03	0.01	1.8	19	0.07	0.1	0.11	0.03	46	1.2	113
Selenite	2.4	0.7	0.03	0.01	0.7	18.6	0.03	0.2	0.05	0.03	48	1.8	83

Table 1.
Geochemical data of selected gypsum samples of the Fat'ha Formation in Sheikh Ibrahim section.

Sample	Description	Porosity (%)
K1	Pure anhydrite	0.25
K2	Anhydrite with impurities	4.0
K3	Limestone	17
T1	Surface gypsum sample	0.9

Table 2.
Porosity data for selected samples.

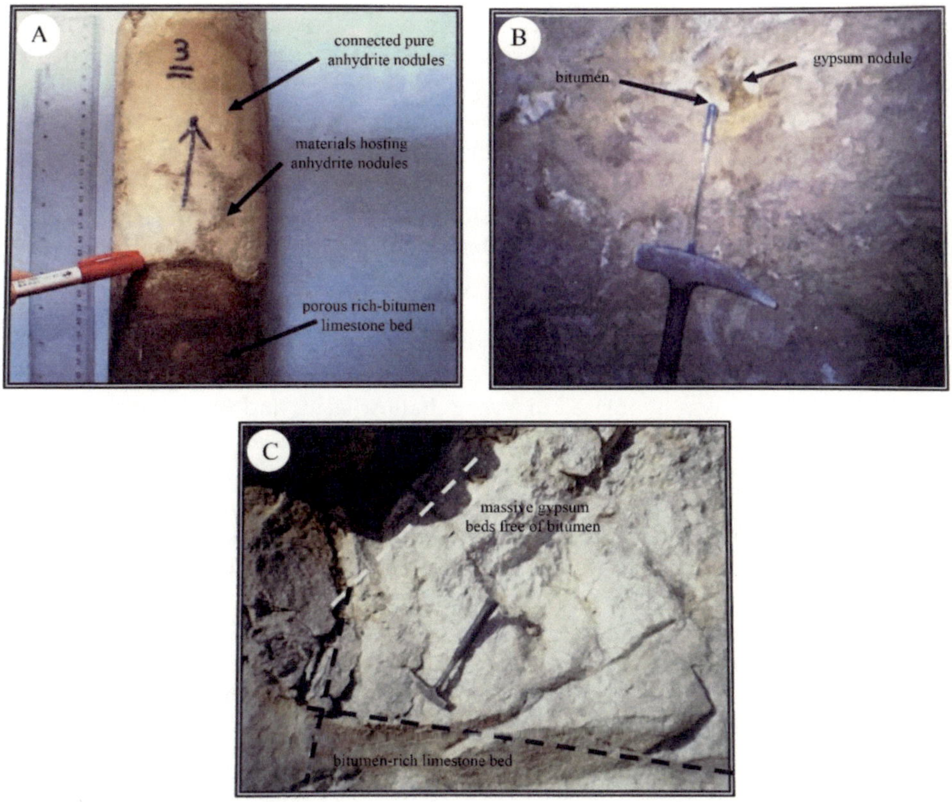

Figure 14.
Core and field images showing (A) core from Kirkuk well illustrates the nature of contact between the porous bitumen-rich limestone in the lower part and compact pore-free anhydrite bed; note some early disseminated hydrocarbons in the host materials (arrows). (B) Bed of nodular gypsum hosting brown hydrocarbon-rich matrix, Sheikh Ibrahim section. (C) Bitumen-rich limestone overlaid by hydrocarbon-free gypsum bed, Sheikh Ibrahim section.

gypsum beds. This gypsum beds may represent seal or cap rocks of the Fat'ha Formation (**Figure 14**). Permeability data show that it is low in the studied gypsum rocks.

5. Discussion and conclusions

Evaporites are indicative for arid continental environments [29], and their formation in sedimentary basins depends mostly on the connection of this basin with oceanic or sea water. Where this connection is periodically interrupted within arid settings, this may led to high evaporation of the basin and cyclic deposition of evaporitic successions in the sedimentary basins [31].

Lithofacies analysis of the studied evaporates revealed the presence of nodular and massive gypsum/anhydrite, laminated gypsum and secondary selenite, and satin spar lithofacies with several sublithofacies; these are representative of relict basin evaporate deposition based on their tectonic setting which they deposited during closure periods of the Neo-Tethys basin on the northern Arabian Plate passive margins [32].

Due to wide distribution of the Fat'ha Formation, several ideas have been proposed for the depositional cycles of gypsum formation. Semi-restricted lagoonal environments such as lakes which were connected to the open sea through narrow channels coincide with the brine-filled basin model suggested by [33, 34], while sabkha or supratidal flat depositional setting and coastal or inland sabkhas with semiarid shallow lagoon were favored by [18, 32], respectively. These models could be comparable with the Messinian basin evaporites of the Mediterranean [35] and Middle Miocene (Badenian) basin-marginal evaporites of the Carpathian Foredeep basin of western Ukraine [36].

Petrographic investigation of the gypsum and anhydritic rocks of the Middle Miocene Fat'ha Formation has revealed that nodular gypsum is the dominant type and is composed of granular integrated gypsum texture with evidence of recrystallization, whereas alabastrine texture is the common type in the laminated gypsum. Secondary gypsum of selenite and satin spar shows alabastrine, fine to coarse fibrous, and porphyroblastic textures with the alabastrine type being predominant.

Nodular gypsum was deposited in a very shallow, arid, and semi-restricted lagoonal environment which has undergone influx and reflux processes, while laminated gypsum may represent pulses of freshwater into the lagoonal basin of Fat'ha Formation.

The chemical composition of selected nodular, laminated, and secondary (selenite) and mosaic gypsum shows low values of strontium (Sr) in the secondary and laminated types due to their secondary origin by the hydration from the original anhydrite through which Sr. in the original anhydrite was expelled. The impoverishment in Sr. commonly occurs in secondary-type gypsum as compared with primary ones [37]. High values in some of gypsum types (see **Table 1**) may be attributed to diagenetic processes and the sea salinity.

Hydrocarbons present mainly in the limestone beds underlie gypsum beds and in materials hosting gypsum nodules. Porous granular texture of these materials allowed hydrocarbon inclusion, later on, during compaction and growth of nodular to compound mosaic due to recrystallization resulted in prevent hydrocarbon dissemination, then these materials were locked in these materials and partly in accompanied gypsum nodules. These results were revealed by low porosity and permeability of the studied gypsum nodules as compared to those of the limestone beds.

Author details

Ali I. Al-Juboury[1]*, Rana A. Mahmood[2] and Abulaziz M. Al-Hamdani[1]

1 Department of Geology, College of Sciences, University of Mosul, Mosul, Iraq

2 College of Electrical Engineering, Ninevah University, Mosul, Iraq

*Address all correspondence to: alialjubory@yahoo.com

References

[1] Yi-gang Z. Cool shallow origin of petroleum-microbial genesis and subsequent degradation. Journal of Petroleum Geology. 1981;**3**:427-444

[2] Warren JK. Evaporite Sedimentology: Importance in Hydrocarbon Accumulation. Englewood Cliffs: Prentice Hall; 2014. p. 285

[3] Al-Juboury AI, McCann T. The middle Miocene Fat'ha (lower Fars) Formation of Iraq. GeoArabia. 2008;**13**(3):141-174

[4] Dunnington HV. Generation, accumulation and dissipation of oil in northern Iraq. In: Weeks LG, editor. Habitat of Oil. USA: American Association of Petroleum Geologists; 1958. pp. 1194-1251

[5] Buday T. The regional of Iraq. In: Stratigraphy and Palaeogeography. Vol. 1. Baghdad, Iraq: State Organization for Minerals; 1980. p. 445

[6] Goff JC, Jones RW, Horbury AD. Cenozoic basin evolution of the northern part of the Arabian Plate and its control on hydrocarbon habitat. In: Al-Husseini MI, editor. Middle East Petroleum Geosciences Conference, GEO'94. Vol. 1. Bahrain: Gulf PetroLink; 1995. pp. 402-412

[7] Numan NMS. A plate tectonic scenario for the Phanerozoic succession in Iraq. Journal of the Geological Society of Iraq. 1997;**30**:85-110

[8] Al-Juboury AI, Al-Naqib SQ, Al-Juboury AMS. Sedimentology, mineralogy and depositional environments of the clastic units, Fat'ha Formation (middle Miocene), south of Mosul, Iraq. Dirasat, Pure Sciences, Jordan. 2001;**28**:80-105

[9] Van Bellen RC, Dunnington H, Wetzel R, Morton DM. Lexique Stratigraphique International. Paris: Centre National Recherché Scientifique, Fasc 10a, Iraq; 1959. p. 333

[10] Metwalli MH, Philip G, Moussly MM. Petroleum-bearing formations in northeastern Syria and northern Iraq. American Association of Petroleum Geologists Bulletin. 1974;**58**:1781-1796

[11] Beydoun ZR. Arabian plate hydrocarbon geology and potential: A plate tectonic approach. American Association of Petroleum Geologists, Studies in Geology. 1991;**33**:1-77

[12] Sharland PR, Archer R, Casey DM, Davies RB, Hall SH, Heward AP, et al. Arabian Plate Sequence Stratigraphy. Bahrain: GeoArabia Special Publication 2, Gulf PetroLink; 2001. p. 371

[13] Buday T, Jassim SZ. The Regional Geology of Iraq, Tectonism, Magmatism and Metamorphism. Baghdad, Iraq: Publication of the Geological Survey of Iraq; 1987. p. 352

[14] Al-Sharhan AS, Nairn AEM. Sedimentary Basins and Petroleum Geology of the Middle East. Amsterdam: Elsevier; 1997. p. 843

[15] Jordan TE. Thrust loads and foreland basin evolution, cretaceous, western United States. Bulletin of the American Association of Petroleum Geologists. 1981;**65**:2506-2520

[16] Allen PA, Homewood P, Williams GD. Foreland basins: An introduction. In: Allen PA, Homewood P, editors. Foreland Basins. Vol. 8. International Association of Sedimentologists, Special Publication. Berlin, New York: Springer; 1986. pp. 3-12

[17] Al-Sawaf FDS. Sulfate reduction and sulfur deposition in the lower Fars formation, northern Iraq. Economic Geology. 1977;**72**:608-618

[18] Shawkat MG, Tucker ME. 1978. Stromatolites and sabkha cycles from the lower Fars formation (Miocene)

of Iraq. Geologische Rundschau. 1978;**67**:1-14

[19] Holliday DW. Origin of lower Eocene gypsum-anhydrite rocks, south eastern St Andrew, Jamaica. Applied Earth Science. 1971;**80**:305-315

[20] Maiklem WR, Bebout DC, Glaister RP. Classification of anhydrite—A practical approach. Bulletin of Canadian Petroleum Geology. 1969;**17**:194-233

[21] Meyer FO. Anhydrite Classification According to Structure. 2005. Available from: http://www.crienterprises.com/Edu_Classif_Evahtml

[22] Sarg JF. Petrology of the carbonate-evaporite facies transition of the seven Rivers formation (Guadalupian, Permian), Southeast New Mexico. Journal of Sedimentary Petrology. 1981;**51**:73-93

[23] Paz JDS, Rossetti DF. Petrography of gypsum-bearing facies of the Codo formation (late Aptian), northern Brazil. Annals of the Brazilian Academy of Science. 2006;**78**(3):557-572

[24] Orti F, Rosell L. The Ninyerola gypsum unit: An example of cyclic, lacustrine sedimentation (middle Miocene, E Spain). Journal of Iberian Geology. 2007;**33**(2):249-260

[25] Bedelean H. Considerations on the parent material in the soil developed on the evaporate deposits from Stana. Geologica. 2003;**XL**(2):59-66

[26] Murray RC. Origin and diagenesis of gypsum and anhydrite. Journal of Sedimentray Research. 1964;**34**(3):512-523

[27] Warren JK. Evaporites: Sediments, Resources and Hydrocarbons. Berlin, New York: Springer; 2006. p. 1035

[28] Blatt H, Middeleton G, Murray R. Origin of Sedimentary Rocks. New Jersey: Prentice-Hall Inc.; 1972. p. 634

[29] Boggs SJ. Principles of Sedimentology and Stratigraphy. New Jersey: Pearson Prentice-Hall; 2006. p. 662

[30] Shearman DJ, Fuller JG. Anhydrite diagenesis, calcitization and organic laminates, Winnipegosis formation, middle Devonian, Saskatchewan. Bulletin the Canadian Petroleum Geology. 1969;**17**:496-525

[31] Nichols G. Sedimentology and Stratigraphy. Oxford: Blackwell Scientific Publications; 1999. p. 355

[32] Mustafa AAM. Sedimentological studies of the lower fars formation in the Sinjar Basin, Iraq [MSc thesis]. Iraq: Mosul University; 1980; p. 243 [Unpublished]

[33] Sulayman MD. Geochemistry, petrology, origin and diagenesis of gypsum rocks of lower fars formation at Butma West area, northern Iraq [MSc thesis]. Iraq: Mosul University; 1990; p. 175 [Unpublished]

[34] Jassim RZ. Gypsum deposits in Iraq: An overview. Iraqi Bulletin of Geology and Mining. 2019;**8**:241-261

[35] Rouchy JM, Taberner C, Blanc-Vallcron MM, Sprovieri R, Russell M, Pierre C, et al. Sedimentary and diagenetic markers of the restriction in a marine basin: The Lorca Basin (SE Spain) during the Messinian. Sedimentary Geology. 1998;**121**:23-55

[36] Peryt TM. Gypsum facies transitions in basin-marginal evaporites: Middle Miocene (Badenian) of West Ukraine. Sedimentology. 2001;**48**:1103-1119

[37] Playa E, Orti F, Rosell FL. Marine to non-marine sedimentation in the upper Miocene evaporites of the eastern Betics, SE Spain: Sedimentological and geochemical evidence. Sedimentary Geology. 2000;133(1-2):135-166

Chapter 14

Bioclastic Deposits in the NW Gulf of Naples (Southern Tyrrhenian Sea, Italy): A Focus on New Sedimentological and Stratigraphic Data around the Island of Ischia

Gemma Aiello

Abstract

Bioclastic deposits in the Gulf of Naples have been studied and compared based on new sedimentological and stratigraphic data, particularly referring to the rhodolith layers. They represent detrital facies deriving mainly from in situ rearrangement processes of organogenic material on rocky sea bottoms. These deposits are composed of medium-coarse-grained sands and bioclastic gravels in a scarce pelitic matrix and crop out at the sea bottom in a portion of the inner shelf located at water depths between −20 m and −50 m. Below water depths of −30 m the bioclastic deposits are rhodolith, characterized by gravels and lithoclastic sands. Rhodolith deposits are often found near the *Posidonia oceanica* meadows and/or in protected areas near the rocky outcrops. The Ischia Bank represents an excellent natural laboratory for studying the rhodolith layers. On the Ischia Bank, below the *Posidonia oceanica* meadow, both bioclastic sands immersed in a muddy matrix and volcaniclastic gravels were sampled. Both the Mollusk shells and the volcaniclastic fragments, where the contribution of the silty and sandy fractions is lower than 20%, were colonized by some species of red algae, while in the marine areas with a low gradient a maërl facies was deposited.

Keywords: bioclastic deposits, rhodolith layers, *Posidonia oceanica* meadow, sedimentological analyses, Gulf of Naples

1. Introduction

Rhodolith or maërl deposits consist of either alive or dead aggregations of coralline algae, which blanket wide coastal zones in the present-day oceans [1–3] and represent shared facies in carbonate platform settings. In some cases, the rhodolith layers indicate the transition from bioclastic-to-rocky sea bottoms, but they can form also on mobile sea bottoms [3–5]. The rhodoliths are the main components of the rhodalgal skeletal assemblage that characterizes the carbonate production in the oligophotic zone of Cenozoic and modern carbonate platforms [4, 6–10]. In contrast with the chloralgal and molechfor assemblages, respectively characterized by the lacking of hermatipic corals and by benthic foraminifers, mollusks, echinoids and bryozoans the rhodalgal assemblage is mainly composed of coralline algae [4]. The zonation of benthic assemblages of the Mediterranean sea performed by Peres and Picard [11] has improved the knowledge on lithology and facies interpretation of rhodolith layers (**Figure 1**). In the Mediterranean sea the bioclastic deposits occur at water depths ranging between – 40 m and – 100 m ("Détritique Cotier" of Peres and Picard; **Figure 1**) [11]. In particular, the rhodolith layers are concentrated in the marine sectors exposed to strong current regimes, such as the top of plateaus or banks. The main components of the "Détritique Cotier" are composed of the reworking and deposition of benthic communities on both mobile sea bottoms (biocoenosis of the "Détritique Cotier") and on hard sea bottoms (biocoenosis of the "Détritique Du Large"), more than the assemblages of *Posidonia oceanica* meadows and maërl deposits. As a consequence of the Holocene sea level rise the deep seafloor was covered by relict and drowned sediments ("Détritique Du Large"; **Figure 1**), characterized by low rates of sedimentation and by the occurrence of glauconite.

The global dominance of coralline algae forming the rhodalgal lithofacies from the Burdigalian to the Early Tortonian has been demonstrated based on stratigraphic data [12]. In particular, during this time interval the rhodalgal lithofacies has reached peak abundance, replacing the coral reef deposits. The prevalence of coralline algae over coral reefs was suggested as being controlled by the enhancement of the trophic resources and associated with an increase of biological productivity to a global scale [12]. This evidence was shown by geochemical data computed on carbon isotopes. During the Middle Miocene the rhodalgal lithofacies increased its extension, due to the upwelling triggered by the establishment of East Antarctic Ice Sheet led to enhanced. These stratigraphic studies performed to a global scale have confirmed the importance of rhodolith deposits as proxies of past oceanographic conditions [12].

This chapter provides new sedimentological and stratigraphic data on the bioclastic deposits and in particular on the rhodolith deposits, occurring in the offshore island of Ischia (Gulf of Naples, Italy); it is based on data coming from sea bottom samples collected during the CARG project aimed at the realization and informatization of the marine geological cartography of the geological sheet No. 464 [13]. The island of Ischia represents the emerged part of a large volcanic field, which extends from the island of Procida to the submerged volcanoes of the western offshore of Ischia [13–17].

The occurrence of rodolith deposits in the Gulf of Naples has been suggested by several studies [18–29]. The sedimentological analysis of the sea bottom samples,

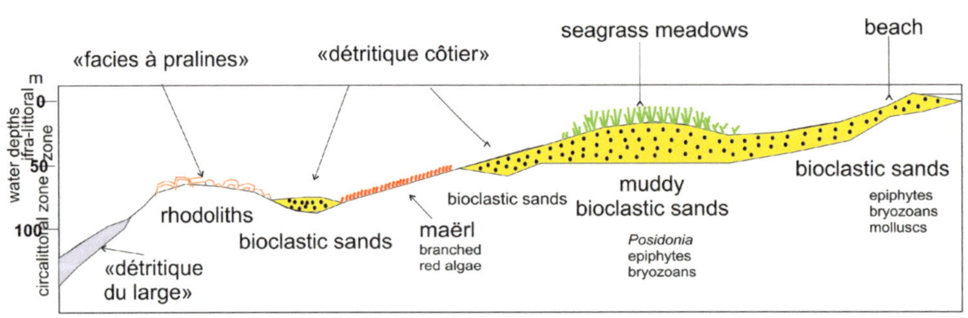

Figure 1.
Sketch diagram showing the main biocenosis of mobile sea bottoms (modified after Carannante et al. [4]).

together with the realization of geological cartography at the 1: 10.000 scale has allowed to study the bioclastic deposits and in particular, the rhodolith deposits occurring in the Ischia offshore. Moreover, the seismo-stratigraphic data have allowed for the calibration of the rhodalgal deposits on previously interpreted seismic Sparker profiles [15].

Previous studies focused on rhodolith deposits in the marine areas of the Gulf of Naples, in particular Nisida-Posillipo (Nisida Bank and La Cavallara saddle) and the Gulf of Pozzuoli (Miseno Bank), while only a single sample came from the Ischia Bank [29]. Otherwise the present work is based on a dense network of samples collected in the Ischia offshore. Moreover, the recognition of bioclastic deposits on some Sparker seismic profiles let us to perform a qualitative calibration of data coming from sampling.

Rhodolith deposits have previously been reported in the offshore of Ischia [29–31]. In particular, Toscano et al. [29] have shown the variability of the rhodalgal facies in the Gulfs of Naples and Pozzuoli, which is closely connected with the location of the platform, with the morpho-bathymetric structure, with the morphology of the sea bottom and with the hydrodynamic conditions on the submerged volcanic banks (Nisida, Miseno and Pentapalummo Banks; Ischia Bank). The top of these submerged volcanic banks is located at water depths of −28/−30 m and is overlain by the *Posidonia oceanica* meadow, growing up to water depths of −35/−40 m. In the marine areas where the *Posidonia meadow* is lacking, the action of the currents has controlled the formation of large fields of sandy ripples [32].

The rhodolith deposits form well-developed layers. They often lie on a sandy-gravelly sea bottom formed by both the shells of mollusks that live in the meadow and by the abundant pyroclastic granules, that derive from the erosion of sea bottoms on which the *Posidonia* meadow grows up [32–34]. Brandano and Civitelli [33] have analyzed the interactions between carbonate and siliciclastic sedimentation on the Pontinian shelf of the Tyrrhenian sea focusing on the relationships between the carbonate deposits and the *Posidonia* meadow. Six sedimentary facies and ten microfacies have been identified using the component analysis, the grain-size percentage, the sorting, the carbonate content and the rate of authigenic mineralization. The maërl deposits and the skeletal sands are located in the circalittoral zone (82 m to 112 m of water depth), also displaying relict facies. These authors have highlighted that the *Posidonia* meadows represent the main facies of the mobile infra-littoral substratum. In the area surrounding the Pontinian islands the *Posidonia* meadows, ranging at water depths between 30 m and 40 m water depth show a rich epiphytic flora and fauna, living on the seagrass leaves and rhizomes. The unattached coralline algal branches gravel facies indicates the environment representing the highest carbonate production rate. This facies consists of red algae (rhodalgal association) that are the main carbonate-producing biota in the Mediterranean. Brandano et al. [34] have highlighted that the seagrasses represent important carbonate factories, being characterized by important carbonate producing biota, as epiphytes on the leaves and infaunal forms. These authors have determined the skeletal assemblage of both modern (Maldivian and western Mediterranean) and fossil seagrass examples (Eocene; Apula and Oman carbonate platforms and Oligocene; Malta platform). In both Maldivian and western Mediterranean the bioclastic deposits are mainly composed of calcareous algae and foraminifera. As a difference, in the tropical setting they are represented by green algae (*Halimeda*), while in the Mediterranean they consist of red algae [34]. The performed stratigraphic study has shown that the green algae–foralgal assemblage is typical of tropical seagrass meadows. On the contrary, the red algae-foralgal assemblage is typical of tropical to subtropical seagrass meadows.

In the Gulf of Naples the red algae are widespread especially in the marine areas where the currents removing fine-grained fraction from soft sea bottom leave in place and the abundant coarse grains (lapilli in origin) become an important substratum for the growth of the coralline algae. The red algae are well developed along the western side of the Ischia bank, forming thick deposits near the tributary channel which joins the head of the Magnaghi canyon [29].

The rhodalgal deposits have their optimum bathymetric distribution between −30 m and −44 m of water depth, where a maximum biodiversity has been observed. Furthermore, Babbini et al. [30] have reported the occurrence of a maërl facies in the coastal area of Ischia, in particular in the north-western sector of Ischia, in the area between Punta Imperatore and the town of Forio and between Forio and Punta Caruso (S. Francesco). The studied sampling transects have been carried out at three different water depths (− 50 m, − 65 m, − 80 m). The taxonomic analysis of the macro-phyto-benthic component of the red coral algae has revealed well-pigmented thalli, with a various growth-form (crusty, lumpy, mammellate, arborescent). The free living branches of coralline algae have been attributed to the maërl facies [11], consisting of alive and dead thalli belonging to the species *Lithothamnion corallioides* and *Phymatolithon calcareum*.

Furthermore, the filming carried out with ROV has shown that there are areas of accumulation of calcareous algae in the ripples concavities [30]. In the same area, Gambi et al. [31] have evidenced the occurrence of rhodolith deposits between −50 m and −80 m of water depth, while the maërl facies was found in three samples, in a well-defined belt about one nautical mile long ("the pink mile") [31], located between −50 m and −65 of water depth. The identified zoo-benthic species are typical of the "détritique còtier" of Peres & Picard [11] and of muddy sea bottoms, and the rhodolith deposits of the offshore of Ischia show a very rich and diverse benthic flora and fauna, especially in the maërl facies [31].

2. Materials and methods

The geological and geophysical data were acquired in the framework of the realization of the geological map n. 464 "Ischia Island" at the 1:10.000 scale [13, 17, 35]. Detailed geological maps, showing the distribution of sea bottom sediments, were built on the basis of the previous geological survey. Furthermore, the new sedimentological analyses of sea bottom samples collected during oceanographic cruises in 2002 and in 2006 have allowed to reconstruct the facies distribution of the sea bottom and to compare the obtained sedimentological and geological results with the previous ones [29, 36]. The stratigraphic framework of the investigated area is based on both high resolution seismic profiles calibrated by cores and on high resolution sequence stratigraphy. Geological and geomorphological data collected at the 1: 10.000 and 1: 5000 scales have been reported on the 1:10.000 geological maps of Campania [35] in order to later produce national geological maps at the 1: 50.000 scale. The previously interpreted Sparker seismic profiles available around Ischia [15, 17, 35], were the subject of a new detailed interpretation focused on the Ischia Bank and Ischia Channel areas, aimed at the identification of bioclastic deposits and at the definition of their stratigraphic relationships with the volcanic and other sedimentary seismic units detected in the offshore of Ischia. The location of the samples analyzed to highlight the rhodolith deposits in the Ischia offshore was superimposed on the Ischia Digital Elevation Model (DEM; **Figure 2**).

During a first work phase, the Multibeam data processing let us to realize bathymetric maps with contour isobaths and shaded-relief maps for the geological

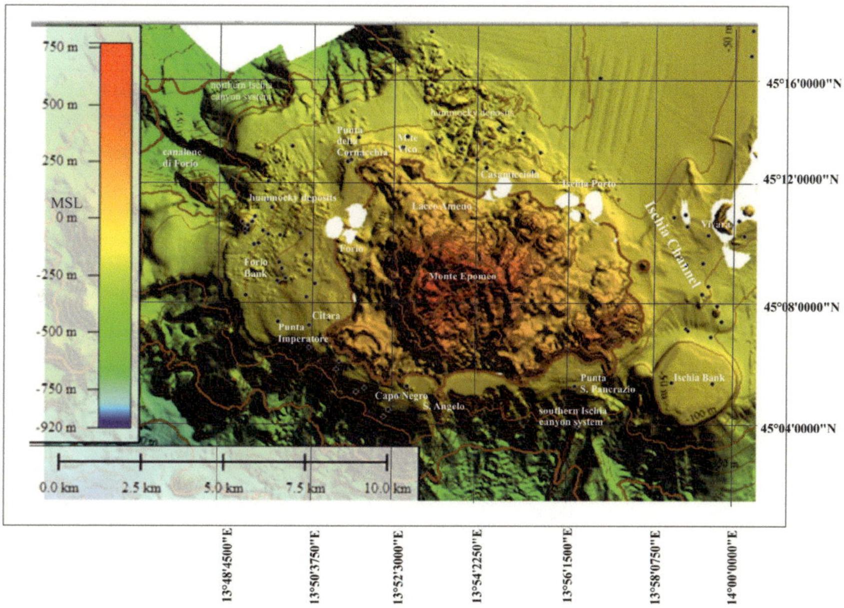

Figure 2.
Onshore-offshore DEM of Ischia reporting the sample location.

interpretation of the morpho-structural features. During a second phase of work, granulometric analyses were carried out on the sea bottom samples accordingly with the Folk classification. The geological interpretation was based on the identification of the acoustic facies, performed with the integrated interpretation of the multibeam and sidescan sonar data and through the calibration of the acoustic facies in lithological terms, using the results obtained from the granulometric analyses of the sea bottom samples (**Figure 2**; **Table 1**). The sedimentological analyses were performed at the sedimentology laboratory of the CNR-ISMAR in Naples, Italy, using a laser granulometer. The list of analyzed samples is reported in **Table 1**, including their description.

3. Results

3.1 Marine and coastal geology

The sedimentological data come from samples collected from sea bottoms located between −30 and – 200 m of water depth; they were placed in the framework of the marine geological survey at the 1: 10.000 scale (**Figure 3**). The morpho-bathymetric characteristics of the different sectors of Ischia, covered by geological survey, are highly variable (eastern, western, southern and northern offshore). The eastern offshore of Ischia, which includes the Ischia Channel, is characterized by sea bottoms with low gradients, locally interrupted by relict volcanic edifices ("I Ruommoli"; "La Catena"; "Vivara ants"). The southern offshore of Ischia is characterized by a narrow continental shelf, cut by submarine canyons and their tributary channels. Its physiography is strongly controlled by the onshore topography, characterized by alternating rocky promontories and inlets. The northern and western offshore of Ischia are characterized by a rough topography, genetically linked to the volcano-tectonic evolution of the island and to the emplacement of the debris avalanche deposits on land and in the sea [16].

Sample	Lithology	Location	Water depth	Description
B1082	Silty sands	Northern Ischia; M.te Vico promontory	−57 m	Middle-grained sand in a silty matrix. Color 5y4/2.*Posidonia oceanic* leaves, lamellibranch fragments and valves. Small pumices.
B1085	Sandy muds	Northern Ischia; Punta La Scrofa promontory	−65 m	Heterometric bioclastic sand in a silty matrix. Fragments of gastropods,bivalves,calcareous algae. Occurrence of superficial silt veil.
B1090	Sandy silts	North-western Ischia – Punta Caruso promontory	−149 m	Silty mud. Color 5y4/2. Occurrence of superficial silt veil, oxidized, thick about 1 cm.
B1094	Sands	South-western Ischia – Punta Imperatore promontory	−38 m	Medium-grained volcanic sand. Fragments of echinoids, whole irregular echinoids, lamellibranch shells. Lithic fragments.
B1096	Sands	South-western Ischia – Citara beach	−27 m	Coarse-grained volcanic sand. Scattered pebbles encrusted with worms. Fragments of calcareous algae. Small complete bivalves. Agglutinating worm tube. Pumice.
B1098	Sandy silts	Western Ischia – Punta del Soccorso promontory	−100 m	Fine-grained sand. *Posidonia oceanica* scales. Small bivalves, gastropods. Occurrence of superficial veil, thick 1 cm, composed of oxidized, fine-grained silt.
B1101	Silts	Northern Ischia – Monte Vico promontory	−95.6 m	Sandy silt with fragments of bioclasts. Color 5y 4/3. Lamellibranch valves, corals, bryozoans fragments. Occurrence of superficial veil composed of silty clay.
B1103	Silty sands	South-western Ischia – Punta dello Schiavo promontory	−88 m	Bioclastic sands in a fine-grained sandy matrix. Red algae, annelids, lamellibranch valves. *Posidonia oceanica* rhizomes. The superficial veil is covered by a net made of *Posidonia oceanica* flakes.
B1105	Silty sands	South-western Ischia – Capo Negro promontory	−74 m	Concretions of red algae, having dimensions in the order of 7–10 cm, overlying fine-grained sands. Tubes of worms, bryozoans, corals. Occurrence of a superficial veil, composed of fine-grained sand, *Posidonia oceanica* leaves, gastropods (*Turritella*), lamellibranchs, rounded pumice clasts and fragments of echinoids.
B1107	Sandy silts	South-western Ischia – Punta del Chiarito promontory	−168 m	Slightly sandy silt. Color 5y4/2. Reworked *Posidonia oceanica*.
B1111	Silty sands	Northern Ischia	−42 m	Bioclastic medium-grained sand with a silty matrix. Occurrence of bivalves, gastropods and bryozoan fragments.
B1794	Sands and gravelly sands	Ischia Bank	−27 m	Description lacking

Sample	Lithology	Location	Water depth	Description
B1796	Silty sands	Ischia Bank	−30 m	Coarse-grained sand with a silty matrix. Occurrence of small bivalves and live worms. Small lithic fragments.
B1798	Sandy silts	Ischia Bank	−74 m	Clay with a dark green sandy fraction. Occurrence of bivalves, rhizomes of *Posidonia oceanica*, gastropods and echinoids. Volcanic lithics, small pumice and remains of coal. Yellow superficial veil, slightly oxidized.
B1800	Sands and gravelly sands	Ischia Channel	−34 m	Sands with a silty matrix, overlying coarse-grained bioclastic sands. Fragments of bivalves, small lamellibranch, red algae, small gastropods, rhizomes of *Posidonia oceanica*, branched bryozoans.
B1802	Sands and gravelly sands	Ischia Bank	−24 m	Coarse-grained sand with a silty matrix. Occurrence of shell fragments, small regular echinoids and scarce lithics.
B1804	Sands and gravelly sands	Vivara (Procida)	−18 m	Bioclastic medium-coarse-grained sand in a scarce muddy matrix. Occurrence of *Posidonia oceanica* and lamellibranchs.
B1806	Silty sands	Ischia Channel	−63 m	Compact mud with medium-fine-grained sandy fraction, consisting of bioclasts, small pumiceous clasts and lithics. Very residual of *Posidonia oceanica* in the first subfloor with masses of rhizomes.
B1808	Silty sands	Ischia	−70 m	Gravelly mud ("Détritique Cotier"). Occurrence of lamellibranchs, red algae, single corals and pumiceous clasts.
B1810	Silty sands	Ischia harbor-Casamicciola	−99 m	Gravelly mud ("Détritique Cotier"). The gravelly fraction is composed of pumice and tuff fragments.
B1815	Sands	Forio-Casamicciola	−38 m	Sands with a silty-muddy matrix. The sandy fraction consists of lithoclasts and bioclasts. Living organisms (small crabs and lamellibranchs). Occurrence of tuff pebbles.
B1816bis	Sandy silts	Forio Bank	−159 m	Superficial veil saturated with water, towards the bottom mud. Occurrence of a bioclastic component composed of shells of marine organisms.
B1818	Sands and gravelly sands	Western Ischia offshore (Citara)	−36 m	Coarse-to-middle grained litho-bioclastic sand with a scarce muddy matrix.
B1820	Sandy silts	Maronti	−96 m	Mud with rare sedentary polychaetes.
B1084	Sands	Northern Ischia (Punta Cornacchia promontory)	−30 m	Bioclastic medium-grained sand with a silty matrix. Occurrence of bivalves and fragments of lamellibranch shells.
B1088	Silts	Northern Ischia	−73.2 m	Bioclastic coarse-grained sand on a silty matrix. Color 5y4/2. Occurrence of small bivalves, fragments of calcareous algae and branched bryozoans.

Sample	Lithology	Location	Water depth	Description
B1092	Silts	Northern Ischia – Punta Caruso promontory	–133 m	Superficial veil composed of oxidized silt of about 1 cm in thickness. Proceeding downwards plastic clay. Color 5y4/2.
B1095	Silty sands	South-western Ischia – Punta Imperatore Promontory	–71 m	Centimetric pebbles in a sandy matrix composed of fine-grained sand. Occurrence of lamellibranch and pectinidae valves.
B1097	Silty sands	South-western Ischia – Citara beach	–40 m	Volcanic rocky substratum with superficial algal incrustations colonized by red algae.
B1100	Sandy silts	South-western Ischia –Citara beach	–100 m	Fine-grained sand. Occurrence of fragments of calcareous algae and reworked *Posidonia oceanica* flakes. Occurrence of surface veil, about ½ cm thick, oxidized, silty.
B1102	Silts	Northern Ischia – Monte Vico promontory	–136 m	Towards the base clay. Color 5y4/1. Superficial veil composed of silty mud. Color 5y 4/3.
B1104	Sandy silts	South-western Ischia – Chianare di Spadera	–109 m	Fine-grained sand with a silty matrix. Occurrence of *Posidonia oceanica* flakes.
B1106	Sands and gravelly sands	South-western Ischia – Punta del Chiarito promontory	–40 m	Medium-coarse-grained volcanic sand. Occurrence of red algae, gastropods and lamellibranch fragments.
B1107	Sandy silts	South-western Ischia – Punta del Chiarito promontory	–168 m	Slightly sandy silt. Color 5y4/2. Reworked *Posidonia oceanica*.
B1796	Silty sands	Ischia Bank	Not recorded in the navigation data	Coarse-grained sand with a silty matrix. Occurrence of small bivalves and live worms. Small lithic fragments
B1798	Sandy silts	Ischia Bank	–74 m	Clay with a dark green sandy fraction. Occurrence of bivalves, rhizomes of *Posidonia oceanica*, gastropods and echinoids. Volcanic lithics, small pumice and remains of coal. Yellow superficial veil, slightly oxidized.
B1800	Sands and gravelly sands	Ischia Channel	–34 m	Sands with a silty matrix, overlying coarse-grained bioclastic sands. Fragments of bivalves, small lamellibranch, red algae, small gastropods, rhizomes of *Posidonia oceanica*, branched bryozoans.
B1802	Sands and gravelly sands	Ischia Bank	–24 m	Coarse-grained sand with a silty matrix. Occurrence of shell fragments, small regular echinoids and scarce lithics.

Sample	Lithology	Location	Water depth	Description
B1804	Sands and gravelly sands	Vivara (Procida)	−18 m	Bioclastic medium-coarse-grained sand in a scarce muddy matrix. Occurrence of *Posidonia oceanica* and lamellibranchs.
B1806	Silty sands	Ischia Channel	−63 m	Compact mud with medium-fine-grained sandy fraction, consisting of bioclasts, small pumiceous clasts and lithics. Very residual of *Posidonia oceanica* in the first subfloor with masses of rhizomes.
B1808	Silty sands	Ischia	−67 m	Gravelly mud ("Détritique Cotier"). Occurrence of lamellibranchs, red algae, single corals and pumiceous clasts.
B1810	Silty sands	Ischia harbor-Casamicciola	−99 m	Gravelly mud ("Détritique Cotier"). The gravelly fraction is composed of pumice and tuff fragments.
B1815	Sands	Forio-Casamicciola	−38 m	Sands with a silty-muddy matrix. The sandy fraction consists of lithoclasts and bioclasts. Living organisms (small crabs and lamellibranchs). Occurrence of tuff pebbles.
B1097	Silty sands	South-western Ischia – Citara beach	−40 m	Volcanic rocky substratum with superficial algal incrustations colonized by red algae.
B1110	Silty sands	North-western Ischia – Monte Vico promontory	−53 m	Coarse-grained bioclastic sands. Occurrence of bryozoans, bivalves and calcareous algae fragments. Occurrence of rocky fragments of centimeter dimensions.
B1112	Sands	Northern Ischia - Casamicciola	−32 m	Rocky substratum. Occurrence of volcanic blocks, immersed in a matrix composed of medium-coarse-grained bioclastic sands. Occurrence of branched bryozoans, lamellibranch valves and calcareous algae fragments.
B1795	Sands and gravelly sands	Ischia Bank	−30 m	Medium-coarse-grained sand with shell fragments overlain by *Posidonia oceanica* meadow. Occurrence of a poor fine-grained sandy matrix.
B1797	Sands and gravelly sands	Parasitic vent of the Ischia Bank	−37 m	Bioclastic sand in a silty matrix. Occurrence of glass fragments, branched bryozoans and red algae.
B1801	Sands	Ischia Channel	−39 m	Bioclastic sands, medium-coarse-grained. Occurrence of red algae and branched bryozoans.
B1803	Sands and gravelly sands	Ischia Channel	−16 m	*Posidonia oceanica* on a coarse-grained sandy bottom, with red algae. Occurrence of shell fragments in the sandy part.
B1805	Sands and gravelly sands	Ischia harbor	−24 m	Living *Posidonia oceanica* on a dark sandy ground with a muddy matrix. Occurrence of lamellibranch valves, fragments of lamellibranchs, live rhodoliths, bryozoans and echinoids.

Sample	Lithology	Location	Water depth	Description
B1807	Sandy silts	Ischia Channel	−53 m	Silt with fine-grained sands. Occurrence of reworked *Posidonia oceanica* and lamellibranchs. Oxidized surface veil.
B1809	Muddy sands	Ischia harbor	−51 m	Slightly oxidized superficial silty-sandy veil. Below bioclastic sands in abundant muddy matrix with living worms and fragments of mollusks.
B1813	Sands and gravelly sands	Casamicciola	−55 m	Slightly oxidized superficial silty-sandy veil. Below bioclastic sands in an abundant muddy matrix with living worms and fragments of mollusks.
B1816	Sandy silts	Forio Bank Base of the volcanic edifice	−191 m	Oxidized mud with gastropods.
B1817	Sandy silts	Western Ischia offshore (Citara)	−179 m	Compact oxidized surface mud with rare sub-rounded pumice.
B1819	Sandy silts	Cava Grado	−88 m	Mud with fine-grained sand. Downwards the sediment is full of valves of large lamellibranchs, colonized by single corals. Rare tufaceous pebbles, many polychaetes.
B1821	Sands	Cava Grado	−41 m	Medium-grained lithoclastic sand with a poor muddy matrix. Presence of glass, slag, tufaceous granules; worms and fragments of gastropods and lamellibranchs.

Table 1.
List of sea bottom samples and their lithology based on grain-size analysis. Location, water depths and description of sea bottom samples are indicated.

Marine geological survey on a scale of 1: 10.000 has allowed to map the Late Quaternary depositional sequence, which includes the TST-HST deposits (slope deposits) and the HST deposits (submerged beach deposits; inner and outer shelf deposits, either bioclastic or epiclastic in origin). This sequence covers significant accumulations of debris avalanche/debris flow deposits, which are located both in the northern offshore of Ischia between Lacco Ameno and Casamicciola and in the western offshore of Ischia, between the promontories of Punta del Soccorso and Punta Imperatore. These debris are composed of heterometric blocks and accumulations of blocks and lavas immersed in a coarse-to-fine grained matrix. Furthermore, significant outcrops of undifferentiated volcanic substratum have been recognized, whose precise attribution in the volcanic deposits of Ischia is problematic, due to the lacking of direct sampling.

The Section 40 at a 1:10.000 scale, located in the northern offshore of Ischia (promontory of Punta La Scrofa; **Figure 3**) has shown textures that include sands, pelitic sands and sandy pelites. The Late Quaternary marine deposits unconformably overlie the deposits of the northern debris avalanche and the undifferentiated volcanic substratum. Here the rhodolith deposits are represented by bioclastic detrital sands, the elements of which are composed of fragments of calcareous algae, bryozoans, mollusks and echinoids. These bioclastic sands, from a few decimeters to a few centimeters thick, cover pelitic drapes. The inner shelf deposits

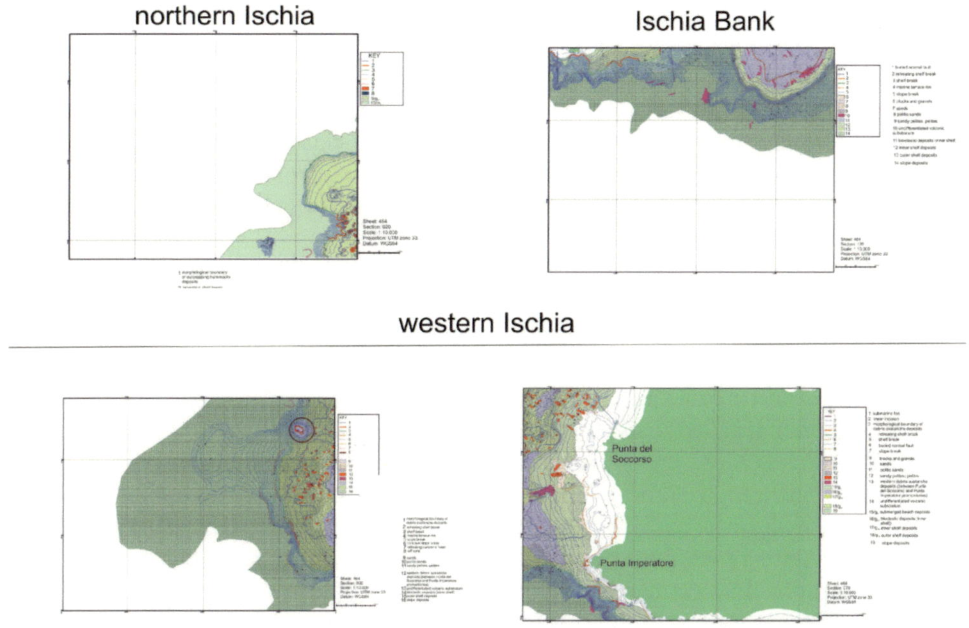

Figure 3.
Sketch frame of some cartographic sections of the Ischia marine geological survey (scale 1:10.000).

are composed of medium-coarse-grained sands and fine-grained pelitic sands. The outer shelf deposits are characterized by pelites with variable fractions of medium-fine-grained sands.

The Section 60 at the 1: 10.000 scale is located in the western offshore of Ischia, outside the promontory of Punta del Soccorso (**Figure 3**), and sands, pelitic sands, sandy and pelitic muds have been mapped. The Late Quaternary marine deposits unconformably overlie the deposits of the western debris avalanche and the undifferentiated volcanic substratum. The Forio Bank has been mapped as a tuff cone, the top of which is found at a water depth of −38 m. At the top of the tuff cone of the Forio Bank rhodolith deposits have been found, represented by bioclastic sands in a scarce pelitic matrix. Furthermore, the outer shelf deposits are composed of pelites with variable fractions of medium-fine-grained sands, with volcaniclasts, bioclasts and rhizomes of *Posidonia oceanica*.

The Section 70 at the 1:10.000 scale is located in the western offshore of Ischia between the promontories of Punta del Soccorso and Punta Imperatore (**Figure 3**). A large area of the sea bottom between the two promontories is characterized by the outcropping deposits of the western debris avalanche of Ischia. The submerged beach deposits are composed of well-sorted sands and pebbles, made up of volcanic lithic elements, from rounded to sub-rounded, with a scarce pelitic matrix and subordinately by bioclasts. The rhodolith deposits crop out on the Forio Bank and are composed of coarse-grained bioclastic sands and bioclastic gravels, similar to those ones found on the Ischia Bank. The inner shelf deposits are characterized by medium-coarse-grained sands and fine-grained pelitic sands. The outer shelf deposits are composed of pelites with variable fractions of medium-fine-grained sands, with volcaniclasts and bioclasts and subordinately, with rhizomes of *Posidonia oceanica*. Finally, the slope deposits are formed by pelites and sandy pelites.

In the Ischia offshore the inner shelf deposits, Holocene in age, are composed of four lithologic associations. The first lithologic association is characterized by heterometric blocks and pebbles, which come from the recent and current rearrangement of the adjacent deposits of debris avalanche/debris flow and crop out within depressed or protected areas.

The second lithologic association consists of medium-coarse-grained lithoclastic sands, sometimes pebbly, with scattered heterometric clasts and blocks, of pyroclastic composition (pumice, lithic and scoria) and lava, from rounded to subrounded, often immersed in a scarce sandy matrix, medium-fine-grained. Locally, a bioclastic component is present.

The third lithologic association is characterized by bio-lithoclastic sands, from medium-fine-grained to fine-grained, immersed in a scarce pelitic matrix, which include heterometric pebbles, from centimetric to pluri-centimetric, of lava and/or pyroclastic composition. The main components are represented by volcaniclasts (pumice, lithic and scoria) and bioclasts, mainly composed of fragments of mollusks. This association extensively crops out in large sectors of the proximal inner shelf and indicates a low-energy sedimentation on flat sea bottoms.

The fourth lithologic association, ranging in age between the Late Pleistocene and the Early Holocene, includes the palimpsest deposits, composed of gravels and sandy gravels with a prevalent pyroclastic composition, from rounded to subrounded, immersed in a scarce sandy matrix, from medium-fine-grained to fine-grained. In the offshore of Ischia this lithologic association is found in scattered outcrops in southern Ischia (La Guardiola), at water depths ranging between −12 m and −15 m and in western Ischia (Forio), at water depths of −30 m, where these deposits are associated with ancient shorelines.

3.2 Sedimentological data

Sedimentological analyses were performed with the aim of showing the main compositional and textural characters of sediments sampled at the sea bottom in Ischia. The sediment fractions recognized at the sea bottom based on particle size analyses include gravel sands, sands, silty sands, muddy sands, sandy silts and silts. Multibeam, Sidescan sonar and seismic data, together with samples, were acquired during the realization of sheet n. 464 "Ischia Island" of the new geological map of Italy [13].

Some textural classes have been identified on the basis of the integrated interpretation of geophysical and geological data. The recognized textures have provided additional information on the lithofacies associations, in order to identify the differences in the depositional elements that have been mapped. Moreover, ternary plots have been constructed for a better evaluation of different grain size (**Figure 4**), considering as variables respectively clay-sand-silt and gravel-sand-silt. The samples have been plotted in ternary diagrams subdividing them with respect to the oceanographic cruises GMS02_01 (diagrams in the upper part of **Figure 4**) and GMS06_03 (diagrams in the lower part of **Figure 4**).

In particular, the ternary diagram located in the upper left corner of **Figure 4** has considered as variables: clay, sand and silt. This plot has shown that the main lithologies are the clayey silts and the sandy silts (**Figure 4**). The ternary diagram located in the upper right corner of **Figure 4** has considered as variables: gravel, sand and silt. This plot has shown that the main lithologies are represented by silty sands and gravelly sands.

Moreover, the ternary diagram located in the lower left corner of **Figure 4** has considered as variables: clay, sand and silt. This plot has highlighted that the main lithologies are represented by sands and silty sands (**Figure 4**). The ternary diagram located in the lower right corner of **Figure 4** has considered as variables: gravel,

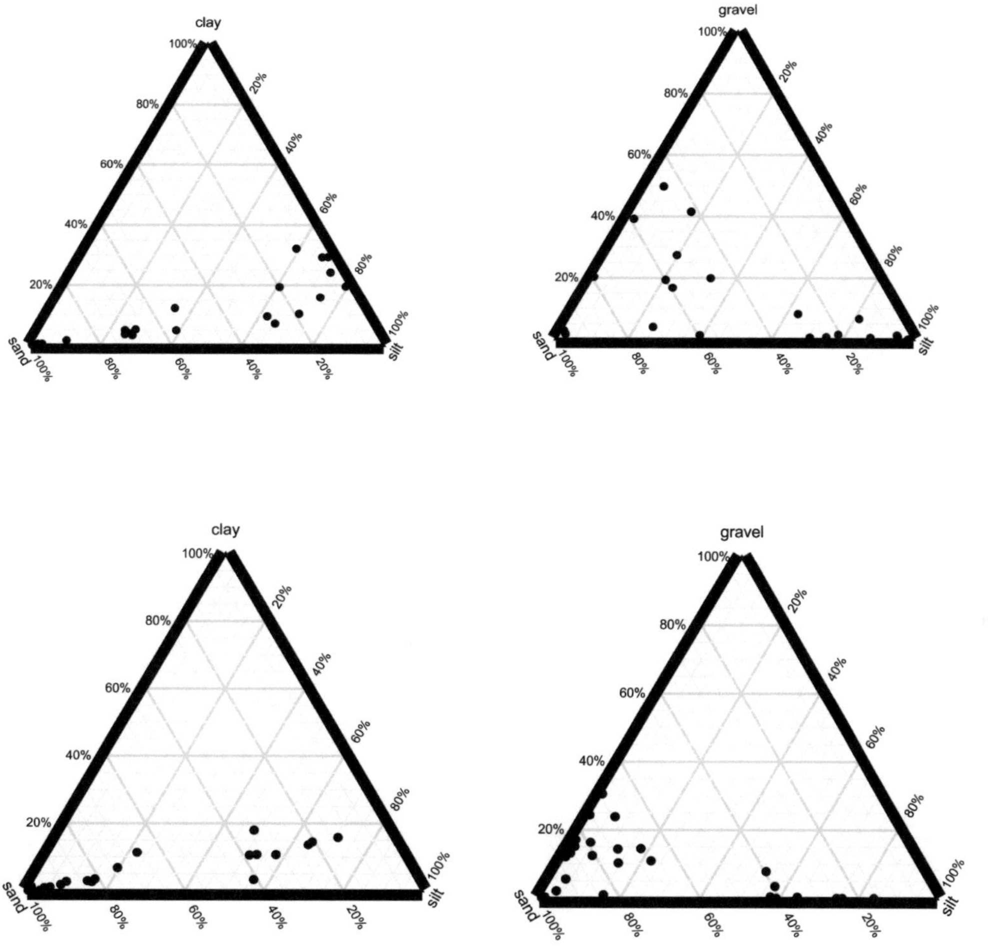

Figure 4.
Ternary plots of the processed sea-bottom samples.

sand and silt. This plot has highlighted that the main lithologies are represented by sands and gravelly sands and subordinately, by sandy silts (**Figure 4**).

3.3 Rhodolith deposits

On the Ischia Bank, a large volcanic edifice located in the south-eastern offshore of Ischia, an extensive meadow in *Posidonia oceanica* covers dark brown heterometric bioclastic sands (sample B1794; **Figure 5**) interpreted as rhodolith deposits due to the presence of living coralline algae. The sands cover a pebbly deposit consisting mainly of shells of mollusks, with fragments of echinoids, widespread on the *Posidonia oceanica* meadow. A low percentage of mud fraction occurs in these deposits.

The rhodolith deposits were also found on the parasitic vent, genetically connected to the main volcanic edifice of the Ischia Bank (sample B1797; **Figure 5**). Here bioclastic sands with glassy fragments, bryozoans and coralline algae have been found in a muddy lithoclastic matrix with a volcanic component. Rhodolith deposits were also found in sample B1799, located on the same adventitious cone, in which these deposits are associated with mollusk shells, small gastropods and scarce bryozoans (**Figure 5**).

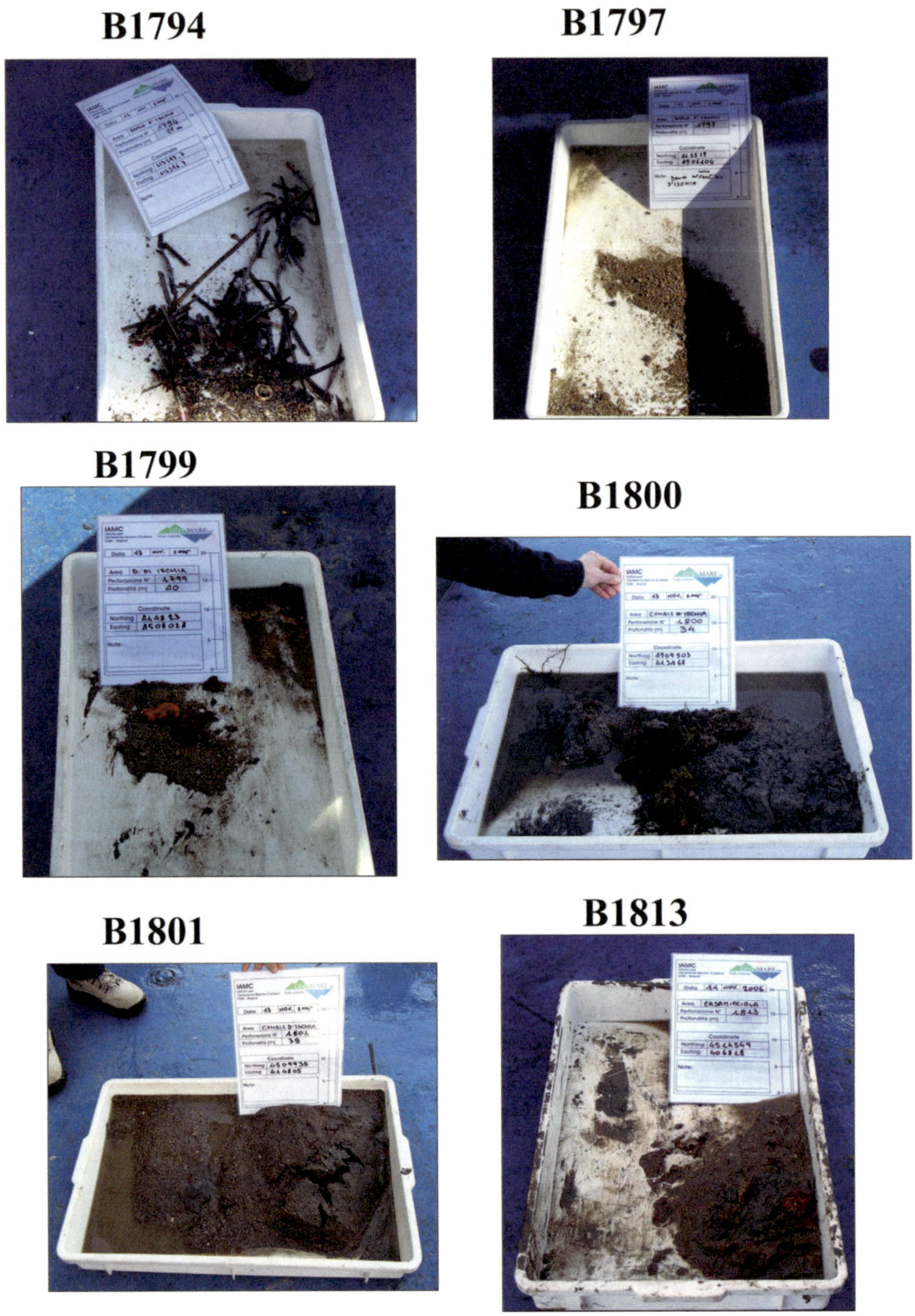

Figure 5.
Sea bottom samples showing the occurrence of bioclastic deposits in the Ischia offshore.

Rhodolith deposits were also found in the Ischia Channel, a morphological threshold located between the islands of Ischia and Procida, where they were sampled by sample B1800. Here the rhodolith deposits are covered by an extensive meadow in *Posidonia oceanica*. Coralline algae are associated with fragments of mollusks, small gastropods, rhizomes of *Posidonia oceanica* and branched bryozoans, as well as with small slags and volcanic glass (**Figure 5**).

Rhodolith deposits were also found in the offshore of Casamicciola (Ischia north; sample B1813; **Figure 5**). Here these deposits grade laterally to sandy and muddy sediments and to the debris avalanche deposits present in Casamicciola.

In the western offshore of Ischia, rhodolith deposits are present at the top of the volcanic edifice of the Forio Bank, where they are characterized by coarse-grained sands and bioclastic pebbles in a scarce pelitic matrix. These deposits are completely analogous to those found on the Ischia Bank, given that the fundamental genetic analogy between the two volcanic edifices.

3.4 Seismic stratigraphy

The sampling data on rhodolith deposits have allowed to review previous interpretations of seismic lines in the offshore of Ischia [15]. At Ischia, the occurrence of isolated volcanic bodies, which include intrusions, domes and volcanic chimneys, was particularly complex to apply seismic and sequence stratigraphy in the interpretation of seismic data [37–39]. Volcanic bodies, which include lava flows, volcanic domes and volcanic intrusions, cannot be investigated by reflection

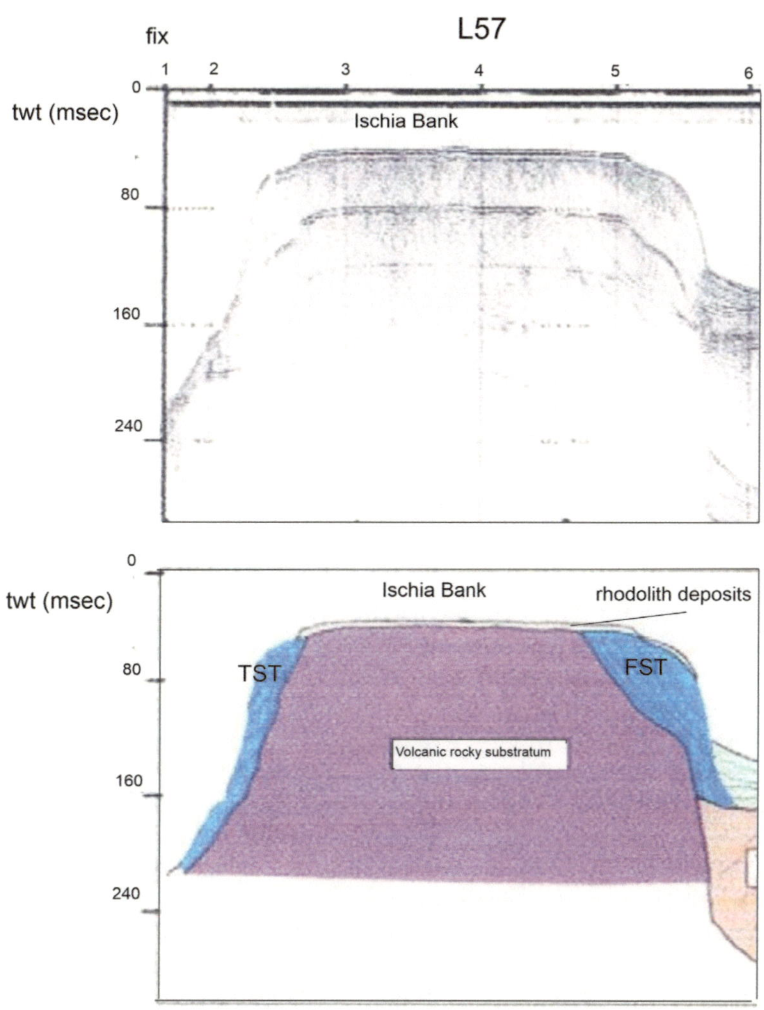

Figure 6.
Detail of the seismic profile L57 (Ischia Bank), showing the seismic unit cropping out at the sea bottom, including the rhodolith deposits occurring at the top of the volcanic edifice.

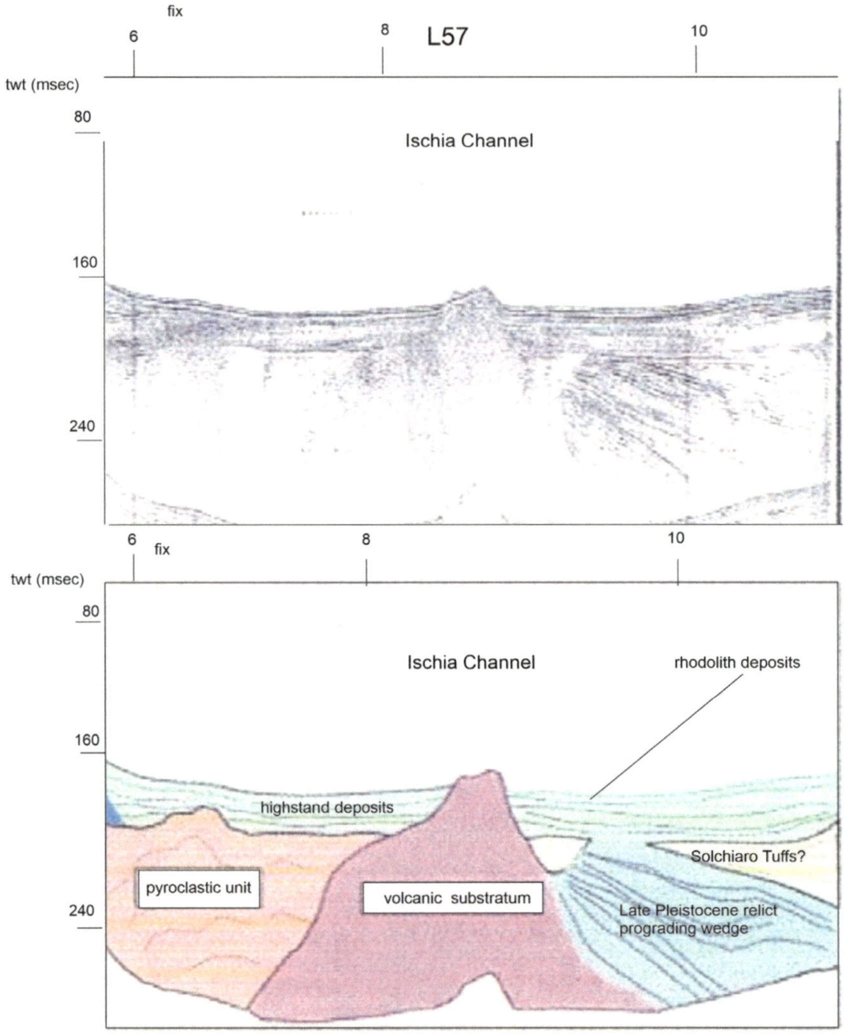

Figure 7.
Detail of the seismic profile L57 (Ischia Channel), showing the highstand deposits, Holocenic in age, within which the rhodolith deposits are inter-layered.

seismic data, except for their external geometry, since they are mainly acoustically-transparent. On the contrary, the seismic facies of pyroclastic edifices and buried pyroclastic deposits can be identified, thanks to the inner stratification of the pyroclastic deposits.

On the Ischia Bank, the rhodolith deposits are probably included within an extensive wedge-shaped unit located at the top of the volcanic edifice, which unconformably overlies the volcanic rocky substratum, which characterizes the bank (**Figure 6**). This unit crops out at the sea bottom and can be interpreted as a unit consisting of bioclastic and partially rhodolith deposits. The volcanic rocky substratum, which characterizes the main morpho-structure of the Ischia Bank, is characterized by an acoustically-transparent seismic facies, corresponding with lavas and pyroclastites (**Figure 6**). On the south-western and north-eastern slopes of the bank there are thick sedimentary wedges, which, accordingly to the geometries observed on the seismic line, are respectively interpreted as trangressive deposits (retrogradational) on the south-western slope and as forced regression

DOI: http://dx.doi.org/10.5772/intechopen.95083

wedges (progradational) on the northeastern escarpment (**Figure 6**). These deposits are not coeval, since the forced regression deposits are older than the transgressive deposits, which on the south-western slope of the bank have not been preserved.

In the Ischia Channel the rhodalgal facies are probably inter-stratified in the highstand deposits, which unconformably overlie the volcanic unit of the Ischia Channel. This unit was identified on the north-eastern section of the L57 seismic line, which runs from the Ischia Bank to the continental platform of Procida, crossing, in the Ischia Channel, the relict volcanic edifice of "Il Pertuso" (**Figure 7**). The volcanic unit of the Ischia Channel has been correlated with pyroclastites and lavas genetically connected with the relict volcanic edifices of the Ischia Channel.

In the Casamicciola offshore, the rhodolith deposits are represented by convex bodies, acoustically transparent, rooted in the top part of the seismic units interpreted as debris avalanches deposits (**Figure 8**). The acoustic facies and the rounded external appearance make them similar to algal bioconstructions.

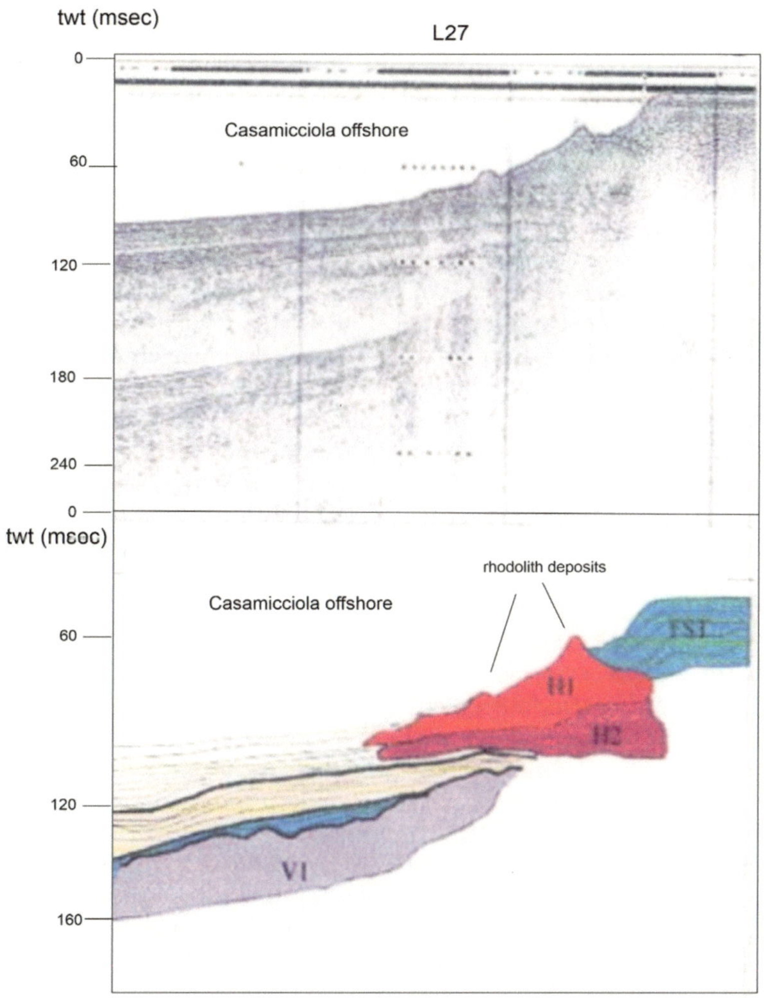

Figure 8.
Detail of the seismic profile L27 (Casamicciola offshore), showing the debris avalanche deposits (H1 and H2), within which the rhodolith deposits are inter-layered, appearing as mounds, representing algal bioconstructions.

4. Discussion

The bioclastic deposits of Ischia are here discussed and compared with similar deposits that are found in adjacent marine areas, with particular reference to the rhodolith layers. They represent detrital facies deriving mainly from in situ rearrangement processes of organogenic material on rocky sea bottoms. These deposits are composed of medium-coarse-grained sands and bioclastic gravels in a scarce pelitic matrix and crop out at the sea bottom in a portion of the inner shelf located at water depths between −20 m and −50 m, characterized by a prevalent carbonate sedimentation. Other significant outcrops are found on the morphological thresholds (Ischia Channel) and at the top of relict volcanic edifices, both in Ischia (Ischia and Forio Banks) and in Procida (La Catena, Il Pertuso and Vivara ants). Below water depths of −30 m the bioclastic deposits are rhodolith, characterized by gravels and lithoclastic sands, the biological component of which is made up of fragments of shells of mollusks (gastropods and lamellibranchs), echinoids and corals. Rhodolith deposits are often found near the *Posidonia oceanica* meadows and/or in protected areas near the rocky outcrops.

In particular, the Ischia Bank represents an excellent natural laboratory for the study of rhodolith deposits. On the Ischia Bank, below the *Posidonia oceanica* meadow, both bioclastic sands immersed in a muddy matrix and volcaniclastic gravels were sampled. The type of bioclastic sedimentation present in this area is characteristic, since the most of the carbonate shells, which come from the overlying meadow, settle at the net limit between the meadow and the sea bottom, where sands and bioclastic gravels crop out. Both the mollusk shells and the volcaniclastic fragments, where the contribution of the silty and sandy fractions is lower than 20%, were colonized by some species of red algae, while in the marine areas with a low gradient a maërl facies was deposited.

The sedimentological results obtained on the rhodolith layers are in agreement with the previous data on rhodolith deposits in the Mediterranean area [40], with particular reference to the southern Tyrrhenian Sea [5, 40], and to the Gulf of Naples [18, 19, 29–31]. In the Mediterranean area the rhodoliths were found in the eastern and western sub-basins at water depths of −30 m to −75 m, but also extend to water depths greater than −75 m [40, 41]. Rhodolith layers were reported in the most of the coastal sections of the Mediterranean, while they are missing along the coasts of the eastern Adriatic sea, Egypt, Syria, Lebanon and the Black Sea [40, 41]. Rindi et al. [41] that in the Mediterranean sea these deposits have shown a high spatial and bathymetric extension, also if the biocostructions of coralline algae virtually occur in all seas. Moreover, these authors have addressed specific research issues in future works, including more detailed paleontological studies, a more accurate taxonomic reassessment, and the extension of the studies on the effects of the climate change and acidification on a wider set of species.

Bracchi & Basso [5] have discussed the occurrence of calcareous algae on the Tyrrhenian continental shelf (Pontine Islands), finding two different carbonate facies, namely the coralline algae facies and the carbonate matrix facies. These authors have highlighted that the Pontine Islands represent a mobile sea bottom of the littoral zone, accordingly with the classification proposed by Peres & Picard [11]. The corresponding biocenosis has been called "Détritique Cotier" and is typically composed of a mixture of sands, gravels and muds. Furthermore, a moderate variability of sedimentary facies, in particular of sands dominated by biogenic carbonates, has been suggested in this area [33]. In this area of the Tyrrhenian Sea, coralline algae represent the most important control factor in the production of carbonate sediments and are typically found in a depth range between −40 m to −70 m. These water depths are consistent with the depths of the rhodolith deposits found in the Ischia offshore and discussed in this work.

In the marine area of the Nisida inlet and Posillipo offshore (Nisida Bank; Cavallara saddle), previously reported by Walther [18] as an area of massive discovery of rhodolith deposits on the rocky outcrops and on the surrounding sea bottom, pyroclastic gravels are frequently mixed with rhodolith deposits characterized by living red algae. The rhodalgal facies is mainly composed of dead, fallen or transported thalli from submerged rocky outcrops, which are colonized in a variable way. In addition, the biogenic fraction is composed of sandy skeletal assemblages, forming variable types of deposits, which have undergone an intense mechanical degradation.

Various types of rhodalgal facies have been found in the Miseno Bank area (Gulf of Pozzuoli). In correspondence with the rocky outcrops, live thalli were found, difficult to sample. On the surrounding sea bottom, characterized by gravels and bioclastic sands, palimpsest deposits were sampled, formed by bioclastic sands. In particular, the presence of a palimpsest drapes, consisting of fragments of algal thalli and invertebrate shells has been reported.

On the Ischia Bank, whose top is found at water depths between −28 m and −30 m, there is a living *Posidonia oceanica* meadow, extending down to −40 m of water depth. In correspondence with the channellised areas, where the *Posidonia oceanica* meadow is lacking, there are extensive fields of sandy ripples. Rhodolith deposits were found below the *Posidonia oceanica* meadow, characterized by living coralline algae (*Phymatolithon calcareum*, *Lithotamnion corallioides*, *Lithotamnion minervae*, *Lythophyllum racemes*) [29]. The rhodolith layers are well developed, covering a gravelly sea bottom, formed by mollusk shells.

The geological and sedimentological data have shown that around the Ischia island the rhodolith deposits have been controlled by different geomorphological and hydrological settings, which have influenced the variable structure of the coralline algae. Among these control factors, the most important one is represented by the topography of the seafloor, deeply influencing the stratigraphic architecture of the rhodolith deposits. Based on the studied data these deposits mainly occur next to the relict volcanic edifices, to the morphological thresholds and to the rough morphologies occurring at the sea bottom corresponding with the outcrops of debris avalanche deposits [42].

In the Ischia offshore the best developed rhodalgal carbonate factory has been found on the outermost bank (Ischia Bank), lacking of the fine-grained fraction and subjected to an intense action of the currents at the head of the adjacent tributary channel, where a part of the biogenic sands are locally transported towards the head of the Magnaghi canyon. Palimpsest deposits or partially remobilized deposits have been found on some other banks (Miseno and Nisida - La Cavallara), where the fine-grained fraction or the geomorphological characteristics have prevented the formation of an active carbonate production by red algae. The Forio Bank, located in the western offshore of Ischia, has shown rhodolith deposits similar to those ones found on the Ischia Bank, but to a smaller scale, having more limited dimensions with respect to the Ischia Bank, as suggested by Multibeam bathymetric data.

In the northern sector of Ischia (Casamicciola) the seismic data have suggested that the rhodolith deposits and, as a general rule, the bioclastic deposits have rounded-shaped morphologies, corresponding with algal bioconstructions and are rooted within the seismic units of debris avalanches. Further data on these deposits in the northern sector of Ischia have been highlighted by Babbini et al. [30] and Gambi et al. [31]. These authors have reported the occurrence of a maërl facies in the north-western sector of Ischia and in particular, in the offshore of Forio and in the San Francesco area on the basis of samples carried out at water depths between −50 m and −80 m. The microscopic characterization of these deposits has shown the occurrence of well-pigmented thalli and of variable morphologies (crusty, lumpy, mammellate, arborescent). The maërl facies looks like an accumulation of whole thalli of calcareous algae or fragments of calcareous algae, which often accumulate within the concavities of the rocky substratum. Furthermore, the ROV images collected in the Forio area (western offshore of Ischia) have shown the occurrence of important algal accumulations within the concavities of the ripple marks, occurring on the sandy bottoms [30]. Assemblages of very well diversified benthic organisms are associated with this facies [31], typical of the "Détritique Cotier" and "Détritique Du Large" biocenosis.

Moreover the obtained results have been compared with the distribution and characterization of rhodolith beds off the Campania region [43]. In the Gulf of Naples these authors have studied and described four selected sites, represented by Capri, Punta Campanella, Secchitiello, and Ischia. In particular, regarding Ischia, the authors have correlated their results with the data previously obtained by Babbini et al. [30] and Gambi et al. [31], which have singled out the occurrence of three morpho-types of rhodoliths, with a prevalence of unattached branches of *Phymatolithon calcareum* and *Lithothamnion corallioides*. In addition, Rendina et al. [43] have underlined a high percentage of dead thalli of red algae, accompanying the alive rhodoliths, suggesting that this percentage has been controlled by a high fraction of fine-grained sediments, triggering the burial of the rhodolith deposits.

The importance of the geomorphological and topographic control factors in controlling the stratigraphic architecture of the rhodolith deposits has been recently highlighted for extra-Mediterranean examples (Udo Island, Korea) [44], suggesting that the distribution of the rhodolith deposits is strongly affected by both the topography of the sea bottom and related physical energy and by different types of surface sediments. In particular, these authors have suggested that a bedrock exposed area is covered by alive rhodoliths (water depths up to – 10 m), where a rough topography of the sea bottom has prevented for a continuous growth of rhodoliths. An active growth area of alive rhodoliths and a sand dune area with dead rhodoliths have been suggested at water depths ranging between – 10 m and – 15 m. While the first area provides stable conditions for the growth of the rhodoliths, the second one represents an adverse environment for the development of rhodoliths. The seagrass covered area with alive rhodoliths develops at water depths greater than – 15 m, where various sizes of rhodoliths have been found.

5. Concluding remarks

The Ischia Bank is characterized by an active carbonate factory dominated by coralline algae, which have colonized an area where suitable environmental conditions have been established for the deposition of native and living rhodalgal deposits. The rhodalgal deposits are locally abundant and are mostly deposited in situ with a centimeter thickness.

Although in the Ischia offshore the investigated deposits were found in a similar bathymetric range, these deposits have shown how in different geomorphological and hydrological environment the coralline algae facies have different structures. Moreover, the topography of the sea bottom has controlled the stratigraphic architecture of these deposits (relict volcanic edifices, morphological tresholds, rough topographies controlled by the development of debris avalanche deposits).

Although qualitatively, the correlation between the sampling data and the interpretation of seismic profiles, previously interpreted and reviewed here, has suggested that the rhodolith deposits are inter-stratified within large seismic units, cropping out at the sea bottom or sub-surficial. In particular, on the Ischia Bank these deposits are inter-layered in a wedge-shaped unit located at the top of the volcanic bank. This unit unconformably overlies the acoustic rocky substratum, representing the main stratigraphic bulk of the bank. Although not documented by the seismic profiles, but only through sea bottom samples, the rhodolith deposits were also found at the top of the parasitic vent, genetically connected to the main building of the Ischia Bank volcanic edifice.

In the Ischia Channel the rhodolith deposits are presumably inter-layered with the highstand deposits, represented by a thick seismic unit partly cropping out at the sea bottom. This unit overlies volcanic seismic units of a probable pyroclastic nature, which are deposited within depressed palaeo-morphologies and which are probably correlated with the yellow tuffs of Solchiaro, cropping out onshore in Procida.

Author details

Gemma Aiello
Istituto di Scienze Marine (ISMAR), Consiglio Nazionale delle Ricerche (CNR), Calata Porta di Massa, Porto di Napoli, 80133, Napoli, Italy

*Address all correspondence to: gemma.aiello@cnr.it; gemmaiello@virgilio.it

References

[1] Kamenos N A, Burdett H L, Darrenougue N Coralline Algae as Recorders of Past Climatic and Environmental Conditions. In: Riosmena-Rodriguez R, Nelson W, Aguirre J, editors. Rhodolith/Maerl Beds: A Global Perspective. Coastal Research Library, 2017; 15, p. 27-54.

[2] Martin S, Hall-Spencer J M Effects of Ocean Warming and Acidification of Rhodolith/Maerl Beds. In: Riosmena-Rodriguez R, Nelson W, Aguirre J, editors. Rhodolith/Maerl Beds: A Global Perspective. Coastal Research Library, 2017; 15, p. 55-86.

[3] Riosmena-Rodriguez R, Nelson W, Aguirre J, editors. Rhodolith/Maerl Beds: A Global Perspective. Coastal Research Library, 2017; 15, p. 1-359.

[4] Carannante G, Esteban M, Milliman J D, Simone L: Carbonate lithofacies as a paleolatitude indicators: problems and limitations. Sedimentary Geology. 1988; 60:333-346.

[5] Bracchi V A, Basso D: The contribution of calcareous algae to the biogenic carbonates of the continental shelf: Pontine Islands, Tyrrhenian Sea, Italy. Geodiversitas, 2012; 34 (1): 61-76.

[6] Pomar L: Types of carbonate platforms: a genetic approach. Basin Research, 2001; 13 (3): 313-334.

[7] Basso D, Morbioli C, Corselli C Rhodolith facies evolution and burial as a response to Holocene transgression at the Pontiane Islands shelf break. In: Pedley H M., Carannante G, editors, Cool-water carbonates: depositional systems and palaeoenvironmental control. Geological Society of London Special Publication, 2006; 255, p. 23-34.

[8] Pomar L, Baceta J I, Hallock P, Mateu-Vicens G, Basso D: Reef building and carbonate production modes in the west-central Tethys during the Cenozoic. Marine and Petroleum Geology, 2017; 83: 261-304.

[9] Aguirre J, Braga J C, Bassi D Rhodolits and Rhodolith Beds in the Rock Record. In: Riosmena-Rodriguez R, Nelson W, Aguirre J, editors, Rhodolith/Maerl Beds: A Global Perspective. Coastal Research Library, 2017; 15, p. 105-138.

[10] Brandano M, Cornacchia I, Tomassetti L: Global versus regional influence on the carbonate factories of Oligo-Miocene carbonate platforms in the Mediterranean area. Marine and Petroleum Geology, 2017; 87: 188-202.

[11] Peres J M, Picard J Noveau manuel de bionomie marine benthique de la Mer Méditerranee. Recueil des Travaux de la Station Marine d'Endoume, 1964;

[12] Halfar J, Mutti M: Global dominance of coralline red-algal facies: A response to Miocene oceanographic events. Geology, 2005; 33 (6): 481-484; doi: 10.1130/G21462.1.

[13] Sbrana A, Toccaceli R M, Biagio G, Cubellis E, Faccenna C, Fedi M, Florio G, Fulignati P, Giordano F, Giudetti G, Grimaldi M, Italiano F, Luperini W, Marianelli P, Buia M C, Donadio C, Gambi M C, Putignano M L, Aiello G, Budillon F, Conforti A, D'Argenio B (2011) Geological sheet of Ischia, scale 1:10.000 – Geological sheet and explanatory notes. Region Campania, Sector of Soil Defense, Geothermics and Geotechnics, 2011; Naples, Italy.

[14] Bruno P P G, de Alteriis G, Florio G: The western undersea section of the Ischia volcanic complex (Italy, Tyrrhenian sea) inferred by marine geophysical data. Geophysical Research Letters, 2002; 29 (9): 1343, doi 10.1029/2001GL013904.

[15] Aiello G, Marsella E, Passaro S: Stratigraphic and structural setting of the Ischia volcanic complex (Naples Bay, southern Italy) revealed by submarine seismic reflection data. Rendiconti Lincei, 2012; 23 (4): 387-408.

[16] Sbrana A, Marianelli P, Pasquini G: Volcanology of Ischia (Italy). Journal of Maps, 2018; 14 (2): 494-503.

[17] Aiello G: New insights on the Late Quaternary geologic evolution of the Ischia Island coastal belt based on high-resolution seismic profiles. Italian Journal of Geosciences, 2018a; 137 (1): 87 106.

[18] Walther J: Le alghe calcarifere litoproduttrici del Golfo di Napoli e l'origine di certi calcarei compatti. Bollettino Regio Comitato Geologico d'Italia, 1885; 16: 360-369.

[19] Funk G: Die algenvegetation des Golf von Neapel. Pubblicazioni della Stazione Zoologica di Napoli, 1927; Suppl. 7: 1-507.

[20] Funk G Beitrage zur kenntnis der meeresalgen von Neapel. Pubblicazioni della Stazione Zoologica di Napoli, 1955; Suppl. 25: 1-178.

[21] Bacci G: Ricerche sulle zoocenosi bentoniche del Golfo di Napoli – La secca di Penta Palummo. Pubblicazioni della Stazione Zoologica di Napoli, 1947; 20: 158-178.

[22] Lewalle J: Determination macroscopique des algues rouges calcaires (Corallinaceae et Squamarciacae partim) du Golfe de Naples. Pubblicazioni della Stazione Zoologica di Napoli, 1961; 32: 241-271.

[23] Segre A G: Geological map of Italy, Sheet n. 183-184. Ischia, Naples. Geological Survey of Italy, 1967; Rome, Italy.

[24] Segre A G: The bathymetric map of the Gulf of Pozzuoli. Hydrographic Institute of the Italian Navy, Genova, Italy, 1972; F.C., 1053, 1-12.

[25] Buccheri G, De Stefano E: Contributi allo studio del Golfo di Pozzuoli. Pteropodi e nannoplancton calcareo contenuti in tre carote: considerazioni ambientali e biostratigrafiche. Memorie della Società Geologica Italiana, 1984; 27: 181-193.

[26] De Pippo T, De Cara A, Guida M, Pescatore T, Renda P: Contributi allo studio del Golfo di Pozzuoli: lineamenti di geomorfologia. Memorie della Società Geologica Italiana, 1984; 27:151-159.

[27] Pennetta M, Pescatore T, Vecchione C: Contributi allo studio del Golfo di Pozzuoli: caratteristiche tessiturali dei sedimenti superficiali. Memorie della Società Geologica Italiana, 1984; 27: 161-169.

[28] Pescatore T, Diplomatico G, Senatore M R, Tramutoli M, Mirabile L: Contributi allo studio del Golfo di Pozzuoli: aspetti stratigrafici e strutturali. Memorie della Società Geologica Italiana, 1984; 27: 133-149.

[29] Toscano F, Vigliotti M, Simone L Variety of coralline algal deposits (rhodalgal facies) from the Bays of Naples and Pozzuoli (northern Tyrrhenian sea, Italy). In: Pedley H M., Carannante G, editors, Cool-water carbonates: depositional systems and palaeoenvironmental control. Geological Society of London Special Publication, 2006; 255, p. 85-94.

[30] Babbini L, Bressan G, Massa-Gallucci A, Buia M C, Gambi M C: Segnalazione di una facies a maërl (*Rodophyta, Corallinales*) lungo le coste dell'isola d'Ischia. Biologia Marina Mediterranea, 2006; 13 (1): 548-552.

[31] Gambi M C, Buia M C, Massa-Gallucci A, Cigliano M, Lattanzi L, Patti F P The "pink mile": benthic assemblages of Rhodolith and

Mäerl beds (Corallinales) off the island of Ischia (Tyrrhenian Sea). UNEP-MAP-RAC/SPA Proceedings of the 1st Mediterranean Symposium on the Coralligenous and other calcareous bio-concretions of the Mediterranean Sea, Tabarka, 15-16 January 2009, CAR/ASP publication, Tunisi, p. 197-200.

[32] Colantoni P, Gallignani P, Fresi E, Cinelli F: Patterns of Posidonia oceanica (L.) DELILE beds around of the Island of Ischia (Gulf of Naples) and in adjacent waters. Marine Ecology, 1982; 3, (1): 53-74.

[33] Brandano M, Civitelli G: Non sea-grass meadow sedimentary facies on the Pontian Islands, Tyrrhenian sea: a modern example of mixed carbonate-siliciclastic sedimentation. Sedimentary Geology, 2007; 201: 286-310.

[34] Brandano M, Tomassetti L, Mateu-Vicens G, Gagliano G: The seagrass skeletal assemblage from modern to fossil and from tropical to temperate: Insight from Maldivian and Mediterranean examples. Sedimentology, 2019; 66 (6): 2268-2296.

[35] Aiello G: Marine geological maps of the Campania Region (Southern Tyrrhenian sea, Italy): considerations and contributions to a different scale of geological survey. Journal of Geography and Cartography, 2018b; 1 (3), doi:10.24294/jgc.v1i3.507.

[36] Ferraro L, Molisso F Sedimenti ed associazioni a foraminiferi bentonici di settori selezionati della piattaforma continentale dell'Isola d'Ischia (Tirreno meridionale). In: Gambi M C, De Lauro M, Iannuzzi F, editors, Ambiente Marino Costiero e Territorio delle Isole Flegree (Ischia Procida Vivara – Golfo di Napoli). Risultati di uno studio multidisciplinare. Memorie Accademia Scienze Fisiche e Matematiche in Napoli, 2003; 5, p. 67-82.

[37] Vail P R, Mitchum R M, Todd R G, Widmier J M, Thompson S, Sangree J B, Bubb J N Seismic stratigraphy and global change of sea level. In: Payton C E, editor, Seismic stratigraphy—applications to hydrocarbon exploration, AAPG Memory, 1977; 26, p. 49-116.

[38] Mitchum J R, Vail P R, Sangree J B (1977) Stratigraphic interpretation of seismic reflection pattern in depositional sequences. In: Payton C E, editor, Seismic stratigraphy—applications to hydrocarbon exploration, AAPG Memory, 1977; 26, p. 117-134.

[39] Catuneanu O, Galloway W E, Kendall C G, Miall A D, Posamentier H W, Strasser A, Tucker M E Sequence Stratigraphy: Methodology and Nomenclature. Newsletters on Stratigraphy, 2011; 44 (3): 173-245.

[40] Basso D, Babbini L, Ramos-Espla A A, Salomidi M Mediterranean Rhodolith Beds. In: Riosmena-Rodriguez R, Nelson W, Aguirre J, editors. Rhodolith/Maerl Beds: A Global Perspective. Coastal Research Library, 2017; 15, p. 281-298.

[41] Rindi F, Braga J C, Martin S, Pena V, Le Gall L, Caragnano A, Aguirre J Coralline Algae in a Changing Mediterranean Sea: How Can We Predict Their Future, if We Do Not Know Their Present? Frontiers in Marine Sciences, 29 November 2019, https://doi.org/10.3389/fmars.2019.00723.

[42] Aiello G New sedimentological and coastal and marine geological data on the Quaternary marine deposits of the Ischia Island (Gulf of Naples, Southern Tyrrhenian Sea, Italy). Geomarine Letters, 2020, 40, p. 593-618.

[43] Rendina F, Kaleb S, Caragnano A, Ferrigno F, Appolloni L, Donnarumma L, Russo G F, Sandulli R, Roviello V, Falace A Distribution and Characterization of Deep Rhodolith Beds of the Campania Coast (SW Italy,

Mediterranean Sea). Plants, 2020, 9, 985; doi:10.3390/plants9080985.

[44] Jeong J B, Kim S Y, Seo Y K, Shin J, Woo K S (2020) Influence of submarine topography and associated sedimentary processes on the distribution of live and dead rhodoliths near Udo Island, Korea. Geomarine Letters, 2020, 40, p. 35-51.